Understanding Periglacial Slope Deposits: Beyond the European Definition

Ambrose

Contents

1 Introduction

1.1 Research questions

Since using DZ from geomorphological layers for provenance is a pilot study, it was necessary in the previous investigations to find out what is possible. Thus, these were the objectives of this work:

I) Since a lot of Cenozoic volcanism can be found in the study area, we checked the reliable datability of DZ below the established age limit (>= 100 Ma). We tried U-Pb dating a tephra layer from the La Sal Mountains and two Jemez tephra layers. The latter were already dated with the common $^{40}Ar/^{39}Ar$ method. (Chapter 2)

II) Using this age handle (outlined in chapter 2), we next put the focus on the feasibility of age determinations of detrital zircons for decoding provenance and to test if we can differentiate cover beds (CB) regarding their zircon spectrum and find any regional matches. We wanted to know whether these age determinations allow for assigning minimum ages to the underlying CB. We also dated DZ in these older CB to evaluate whether they contain zircons of the same or even younger age than the tephra in between, i.e., whether there is evidence of reworking of the tephra material or whether the tephra layer may be regarded as being in situ. (Chapter 3)

III) The success of the approach applied in Chapter 3 raised the question whether we can transfer the finding to a much larger area to find sources of DZ and uncover regional similarities and differences. Despite there are plenty of DZ-based provenance studies from North American Archaean to Neogene rocks U-Pb-dating of DZ is not broadly established as a tool to reveal the provenance of Quaternary deposits, except some Loess studies mainly on the Chinese Loess Plateau and Quaternary fluvial deposits in South America, western USA and Germany (e.g., GÄRTNER 2011; KRIPPNER AND BAHLBURG 2013). To our knowledge, DZ studies have not yet been applied to slope deposits with admixed aeolian fines. In addition to dating with U-Pb, we use the morphology and the surface of the zircons, grain size analyses, density functions (PDF and KDE) as well as multidimensional scaling in order to be able to draw conclusions about the provenance. (Chapter 4)

IV) Since the results of the analyses were much more far-reaching than expected, many new questions have arisen that need to be discussed and investigated. (Chapter 5.2 and 5.3)

1.2 Cover beds

Periglacial slope deposits also known as cover beds (CB) are an accepted phenomenon in Europe. KLEBER AND TERHORST (2013b) defined:

"CBs are deposits that formed by unconcentrated dislocation processes chiefly from upslope materials, but that may contain admixed eolian matter. They cover undifferentiated slopes to a large extent or even completely rather than being restricted to drainage ways, linear discharges, or local failures. CBs typically consist of several layers separated by disconformities."

In Europe, CB consist of an upper layer ("Hauptlage", LH), an intermediate layer ("Mittellage", LM) and a basal layer ("Basislage", LB). According to KLEBER AND TERHORST (2013a) neither do all these types occur outside of Europe, nor are the CBs tied to the previous glaciation alone. In New Zealand the same term does not comply with the European definition but has a more generic meaning (RAESIDE; 1964) . Nonetheless, CBs are being studied in Europe, Asia and Australia, whereas in the USA the definition has not yet been established (HÜLLE ET AL. 2009; KLEBER 1997; WAROSZEWSKI ET AL. 2020).

1.3 Palaeosols

A paleosol is a former soil preserved by burial underneath sediments or one that has no relationship to the current factors of soil formation. When working with palaeosols, addressing the terrain and differentiating a sediment layer from a pedological horizon is extremely important. Here the terminologies must be kept apart and yet addressed in one profile for clarity. Only if there is evidence of in-situ soil formation processes, then it is a soil (LORZ ET AL. 2013). It is a palaeosol if there are evidences of different soil forming processes that cannot take place within a soil formation because they are physically or chemically mutually exclusive (e.g., clay illuviation), there are several palaeosols and thus cover layers.

1.4 Study area

Emergence

The North American continent is very old and consists of a wide variety of terranes and cratons. It has already arisen while the formation of other continents such as Europe and Asia had not even begun. The rocks show ages from the Hadean to the Quaternary. (BALLY et al. 1989; WHITMEYER AND KARLSTROM 2007; SIGLOCH AND MIHALYNUK 2013)

My study area is located in the south-western United States and includes mainly parts of Utah and Nevada (Fig. 1.1 and 4.1), between 42° to 37° North Latitude.

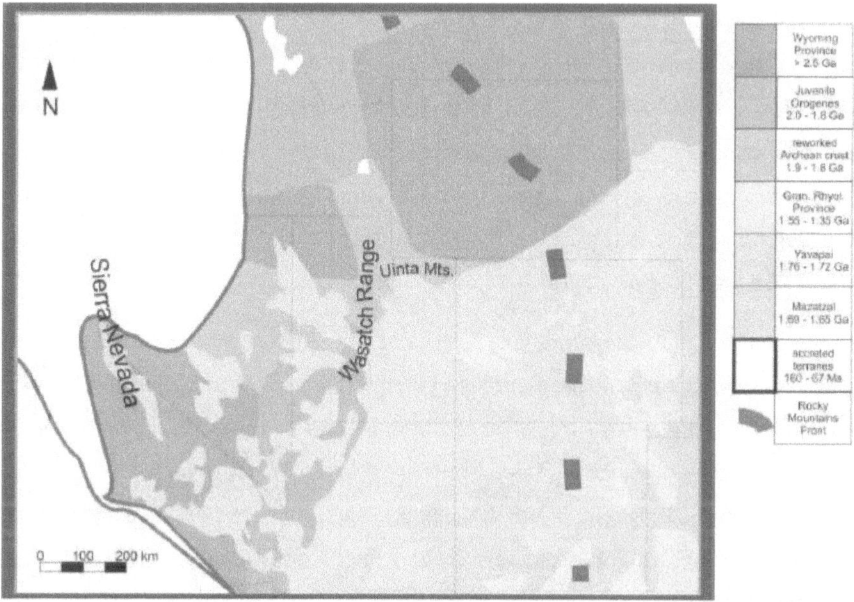

Figure 1.1 – Map of cratons, terranes and main mountain ranges of the study area (after WHITMEYER AND KARLSTROM 2007)

The oldest craton underlying the study area and, thus, the oldest supplier of autochthonous ages is the Wyoming province with over 2.5 Ga (Archaen). It underlies in the north and south-west. Juvenile orogenes with ages between 2.0 – 1.8 Ga (Orosirian) underly Idaho in the north (from there only aeolian input to the study area is possible) and the border region between Nevada, Utah, Arizona and California in the south. The

Trans-Hudson-Orogene in the north and north-east of our study area and other reworked Archean crust within Utah and Nevada give upper Orosirian ages (1.9 – 1.8 Ga), too. The Yavapai terranes dominate the south-east with ages between 1.76 – 1.72 Ga (Statherian), which is the same period as the Mazatzal terranes (1.69 – 1.65 Ga), outside our narrower study area. As well outside, only slightly touching it, is the Granite Rhyolite Province with ages from 1.55 – 1.35 Ga (Lower to Middle Mesoproterozoic, Calymmian to Ectasian). Then, the bedrock ages show a massive gap until the Cordilleran arc, which accreted from the west between 160 – 67 Ma (Upper Jurassic to Cretaceous). (Fig. 1.1)

Physiogeographically, nearly all of Nevada belongs to the Basin and Range Province. The ranges still show the folding of the terranes accreted through the subduction of the Farallon plate below the continental margin (SIGLOCH AND MIHALYNUK 2013; DECOURTEN AND BIGGAR 2017). Thus, most mountain ranges follow a north south strike. These mountains ranges show a high variety of Proterozoic to Cenozoic ages on the ridges but are all overlaid by Paleogene and Neogene sediments within the basins. Hydrogeographically, the region belongs to the Great Basin (GB), an endorheic basin. (DECOURTEN AND BIGGAR 2017)

Utah's west belongs to the same geographical units. The eastern and southern parts belong to the geologically and aesthetically highly impressive Colorado Plateau (CP), which owes its name to the main river. Most of the orogenies of the surrounding mountain ranges bypassed the CP since the Precambrian, so that the many canyons are carved into different (mainly) sandstone layers by uplifts. They show the diverse ages easily recognizable through different colours from dark purple, brown, red to beige and white. These layers testify to the existence of a shallow sea in the Cambrian, but also that afterwards only aeolian transported (and re-allocated) substrates allowed the plateaus to thicken. As far as we know today, the CP has not been completely covered by water since the Cambrian. Only during the Jurassic Period, a shallow sea covered western parts of the CP as well as within the Paleogene the Flagstaff Lake covered northern and western parts. However, the sandstones are mainly dune born. (WILLIAMS ET AL. 2014)

Important mountain ranges are the Sierra Nevada in the west, the Uinta Mountains in north-eastern Utah, the Wasatch Range, which delimits the GB from the CP with its north east-south west strike, in eastern Utah the north and south La Sal Mountains and the Henry Mountains as dominant elevations in the vast areas of the CP (for a detailed map view fig. 4.1). The elevations differ from 1200 m a.s.l. within the CP and the basins of the GB and ~ 4000 m a.s.l. in the GB (Wheeler Peak 3982 m a.s.l.) and ~3500 m a.s.l. in the CP (Abajo Peak 3465 m a.s.l., Mount Ellen 3512 m a.s.l.). The dominating ridge in the west, the Sierra Nevada, trumps this with the Mount Whitney (4421 m a.s.l.).

Due to the many rifts, fault zones, subductions and hot spots the North American continent has and had a high volcanic activity (JENSEN ET AL. 2020).

This is just an overview; the Geology of our study area is described more detailed within the manuscripts (2.4; 3.3.1, 4.3).

Climate and vegetation

The climate of the GB is mainly influenced by the foehn effect of the Sierra Nevada. The cold desert climate includes very different levels of precipitation, cold winters and hot summers (SNYDER ET AL. 2019). Although it belongs to the cold desert climates, too, climate and vegetation of the CP are influenced by El Niño and La Niña events due to the fact of the geographical location (USGS 2002).

The vegetation of the basins follows the semi-arid climate and is mainly cool desert shrubs (dominated by Great Basin Sagebrush, *Artemisia tridentata*, in the GB and Blackbush Scrub, *Coleogyne ramosissima*, in the CP) and grasses (including the neophyte Cheatgrass, *Bromus tectorum*) (BLIJ AND MULLER 2006). With elevation, forests start with the pinyon-pine formation (dominated by Utah Juniper, *Juniperus osteosperma*, and Pinyon Pine, *Pinus edulis*). Above, conifer (various species) forests accompanied by Quaking Aspen (*Populus tremuloides*) dominate. (FAGAN 2012) Our study sites were located in the shrub or pinyon-pine domains.

1.5 Zircons

Characteristics and scientific use

Zircon is an extraordinary mineral. Its chemical formula is $ZrSiO_4$. Usually, zircons derive from magmatic or subduction events. High pressure and temperature are needed to form a zircon and they differ in appearance. (CORFU ET AL. 2003)

Despite of its high closing temperature (> 800° C), it is extremely stable physically and chemically. The uranium trapped during the crystallization (instead of Zr) decays into its daughter isotopes, which, if one subtracts the common lead (Pb cannot be built-in) and other possible intruders, can calculate the time of closure (CORFU ET AL. 2003). These decay from U and Th to Pb is extremely long. The half life of ^{238}U is $4.5*10^9$ years (^{238}U > ^{234}Th > ^{234}Pb > ^{230}Th > ^{206}Pb (stable)). ^{235}U decays in $7.1*10^8$ years (^{235}U > ^{231}Th > ^{227}Th > ^{207}Pb) (Fig 1.2; DAVIS ET AL. 2003). Since these stable series of decays always take place at the same intervals, after the zircon has been ablated with the laser, the isotopes measured can be used to calculate when the zircon has closed and thus deduce the age.

Using these decay rates for the calculation of ages has got a very common tool in Geochronology. The calculation depends on the asumption that a decay chain is in secular equilibrium if the decay constant (λ) and the product of the abundance of an isotope are equal among all the intermediate daughter products and the parent isotope (N).

$$N_1\lambda_1 = N_2\lambda_2 = N_3\lambda_3 = \cdots$$

Following the decay law

$$N(t) = N_0 \cdot e^{-\lambda \cdot t}$$

shortend equation and and deductions of the ages for ^{235}U and ^{238}U:

$$^{238}U \longrightarrow {}^{206}Pb + 8\alpha + 6\beta^- + Q; \lambda_{238} = 1.55125e^{-10}\ year$$

$$^{235}U \longrightarrow {}^{207}Pb + 7\alpha + 4\beta^- + Q; \lambda_{235} = 9.8485e^{-10}\ year$$

Still, all the intermediate daughter isotopes are very important for material younger than 250 ka (cf. chapter 2.5). (SCHOENE 2014)

WETHERILL (1956) introduced a new complex form of representation of the Pb^{206}/Pb^{207} ages, which plot the $U^{238}/Pb206$ and U^{235}/Pb^{207} isotopes of the same analysis against each other into one diagram. All isotopes which, if one observes their half-lives, give the same age, result in the so-called *concordia curve*. Measurements that are individually reliable but not plot on or near the concordia are called discordant. These are therefore not necessarily wrong, but must be considered and discussed separately.

Depending on the results and research questions, one is analysing the zircons for, there are different methods to choose: SIMS, SHRIMP, LA ICP-MS. These are reliable methods for dating zircons older than 100 Ma (because of the half-life duration), younger zircons can be measured as well but because of the shorter half-life of ^{235}U the double-check is missing. Nevertheless, there are other analyses to testify the results.

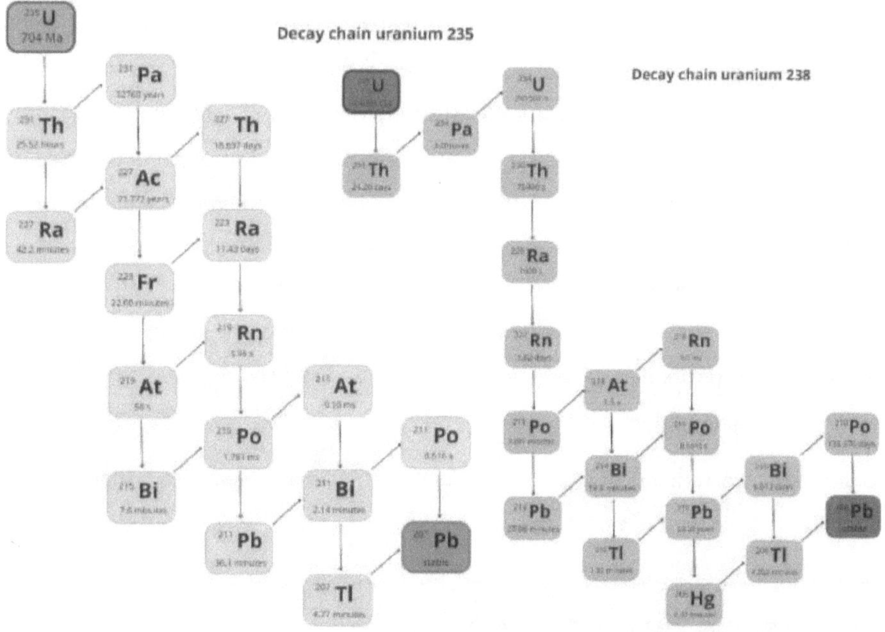

Figure 1.2 - Decay chains ^{235}U to ^{207}Pb (green) and ^{238}U to ^{206}Pb (red)

However, other built-in isotopes like hafnium (Hf) or thorium (Th) can be measured and analysed for dating and provenances as well. Further details about LA-ICP-MS U-Pb dating we used, may be found in chapter 4.4.2.

Provenance

For some time now, zircons have not only been used for dating in Geochronology, but also to conduct provenance studies, which, however, mainly focus on rocks and on ages as old as and older than Upper Cretaceous. Beside ages, the morphology of zircons is an interesting object for research due to the fact that the crystal lattice is very stable. Questions about what fractures, cracks, breaks and other features on the surface tell about the possible transportation of the single grains and how this can be a helpful tool to reconstruct them (GÄRTNER ET AL. 2013; NEIL ET AL. 2021).

1.6 book format

This cumulative book includes three published manuscripts (chapters 2, 3 and 4). All three journals (E&G Quaternary Science Journal, Quaternary Science Reviews, Earth Surface Processes and Landforms) guarantee an international peer review processing. Within this book, each manuscript provides an individual list of references. The three manuscripts are embedded into a general introduction and an extended summary (including a synthesis). Chapter 1 (Introduction), Chapter 5 (Extended Summary) and Chapter 6 (Summary information) together have a separate list of references (7 References). All manuscripts (including references, figures, tables and summary information) have been formatted to allow a uniform appearance within the book. The numbering and positioning of the figures and tables has been adjusted for the same reason and to allow easier reading.

Nevertheless, each manuscript stands for itself with its own research question, so that repetitions e.g., in methods or study area, cannot be avoided.

Last but not least, some terms have been adapted, as the book is written in British English (e.g., 'paleosol' converted to British English 'palaeosol').

2 Capability of U-Pb dating of zircons from Quaternary tephra: Jemez Mountains, NM, and La Sal Mountains, UT, USA

Chapter 2 is published in the peer-reviewed journal Eiszeitalter & Gegenwart, Quaternary Science Journal as:

Capability of U-Pb dating of zircons from Quaternary tephra: Jemez Mountains, NM, and La Sal Mountains, UT, USA

Authors: Jana Krautz[1], Mandy Hofmann[2], Andreas Gärtner[2], Ulf Linnemann[2], and Arno Kleber[1]

[1]Institute of Geography, Technische Universität Dresden, Helmholtzstr. 10, 01069 Dresden, Germany

[2]Senckenberg Naturhistorische Sammlungen Dresden, Museum für Mineralogie AND Geologie, Sektion Geochronologie, GeoPlasma Lab, Königsbrücker Landstraße 159, 01109 Dresden, Germany

Publication history: received 12 January 2018 / published 31 January 2018

Full reference:

Krautz, J., Hofmann, M., Gärtner, A., Linnemann, U., and Kleber, A.: Capability of U–Pb dating of zircons from Quaternary tephra: Jemez Mountains, NM, and La Sal Mountains, UT, USA, E&G Quaternary Sci. J., 67, 7–16, 2018.

Internet link: https://doi.org/10.5194/egqsj-67-7-2018

2.1 Kurzfassung

Zwei quartäre Tephren aus den Jemez Mountains, New Mexico, – Guaje- und Tsankawi-
Tephra – sind durch die ähnliche chemische Zusammensetzung ihrer Gläser nur schwer
zu unterscheiden. Dies gilt auch, bis auf geringfügige Unterschiede, für die Totalanalyse.
Wir haben die Möglichkeit untersucht, das Alter einer distalen Tephralage in den La Sal
Mountains, Utah, zu bestimmen und einer der Tephren aus den Jemez Mountains
zuzuordnen. Zur Vergleichbarkeit haben wir auch die Zirkone der Tephren aus den
Jemez Mountains U–Pb datiert. Obwohl die Tephren alle sehr jung sind, haben wir
reliable Alter durch das Cluster der jüngsten, im 2σ Fehler überlappenden Zirkone
erhalten. Demzufolge ist die Guaje-Tephra 1.513 ± 0.021 Ma[2] alt. Die Tephra aus den La
Sal

---

[1] Original publication: Myr; corrected because the context is wrong

[2] Original publication: Myr; corrected because the context is wrong

Mountains wurde mit der Tsankawi-Tephra korreliert: Drei Proben aus den Jemez Mountains (1×) und den La Sal Mountains (2×) ergaben eine Altersspanne von 1.31 – 1.40 Ma[2]. Alle Alter weichen etwas von bereits publizierten ^{40}Ar /^{39}Ar ab. Die Ergebnisse deuten darauf hin, dass die distalen Jemez-Tephren durch U–Pb Datierung unterschieden werden können. Wir wollen dazu ermutigen, diese Methode der Altersbestimmung auch für quartäres vulkanisches Material in Erwägung zu ziehen.

2.2 Introduction

Tephra is eruptive rock material deposited as airborne fallout often quite distant from its source volcano. Because of its chemical composition – usually obtained from glass shards – it often may be related to a particular volcanic eruption (WESTGATE ET AL. 1994). Therefore, tephrochronology has become an established method, using tephra intercalated between other deposits as a stratigraphic marker bed, provided the original eruption is well dated (LOWE 2011).

There are a lot of reliable methods for dating Quaternary tephra (DICKINSON AND GEHRELS 2009). Most commonly, the ^{40}Ar / ^{39}Ar method is applied using K-rich minerals (LOWE 2011). This utilizes the fact that embedded argon completely leaves the mineral lattice by disturbances such as a volcanic eruption. After this the enrichment by radioactive decay of K re-starts, and thenceforward the accrued isotopes may be measured. So ages can be calculated via the half-life of the isotopes (WORSLEY 1998).

A distal tephra layer discovered in the La Sal Mountains, Utah, was linked to the volcanic province of the Jemez Mountains, New Mexico, based on glass-shard chemistry. However, correlation with a particular eruption remained ambiguous (KLEBER 2013), because two tephras derived from there have closely similar chemical compositions (SLATE ET AL. 2007) – one of the major threads of tephrochronology (LOWE 2011). ZIMMERER ET AL. (2016) state that both tephras are difficult to date by Ar-Ar dating, asking for elaborate sample preparation and calculation of the results. Though still not done very often on such young zircons (LEE 2012), there have been a few successful applications of U-Pb dating of zircons to young material in recent years (e.g., ITO ET AL. 2016; SAKATA ET AL. 2017). Zircons have the advantage of being outstandingly chemically and physically robust. They are unsusceptible to alteration and weathering even under extreme conditions (WILSON ET AL. 2008). Thus, we tried dating the tephra layer using zircon dating.

Here we demonstrate reasonable age determinations of zircons from the La Sal Mountains tephra layer and of the two suspect tephras in the Jemez Mountains. Through this, the Jemez tephra layers may be discriminated with high certainty. Furthermore, we encourage giving the U-Pb method – which is available in a variety of labs worldwide – a try for dating volcanic material of undisclosed age even if the assumed age is as young as 1 Ma[3], after having tested the total uranium contents.

2.3 Geological setting

2.3.1 Jemez Mountains, New Mexico

The Jemez Mountains (Fig. 2.1) are calderas of various volcanic eruptions, among which the Valles, Antonio, and Toledo calderas are still recognizable as concentric mountain ranges. Their eruptive products, mainly basalt-andesite-dacite-rhyolite associations, range from about 15 Ma[3] (mid-Miocene) to < 2 Ma[3] (Pleistocene) (KUES ET AL. 2007). The Neogene and Quaternary formations are divided into three groups, named after Indian nations, from oldest to youngest: The Keres, the Polvadera, and the Tewa group (BAILEY ET AL. 1969). We took our samples from the Tewa group. This comprises the Bandelier Tuff, which is mainly the result of two large ignimbrite- and caldera-forming eruptions. The lower Otowi (including the Guaje tephra) and the upper Tshirege (including the Tsankawi tephra) sequences were deposited approximately 1.6 and 1.2 Ma, respectively (SELF ET AL. 1996; SLATE ET AL. 2007). Today large parts of these ignimbrite and tephra sequences belong to the Bandelier National Monument.

---

[3] Original publication: Myr; corrected because the context is wrong

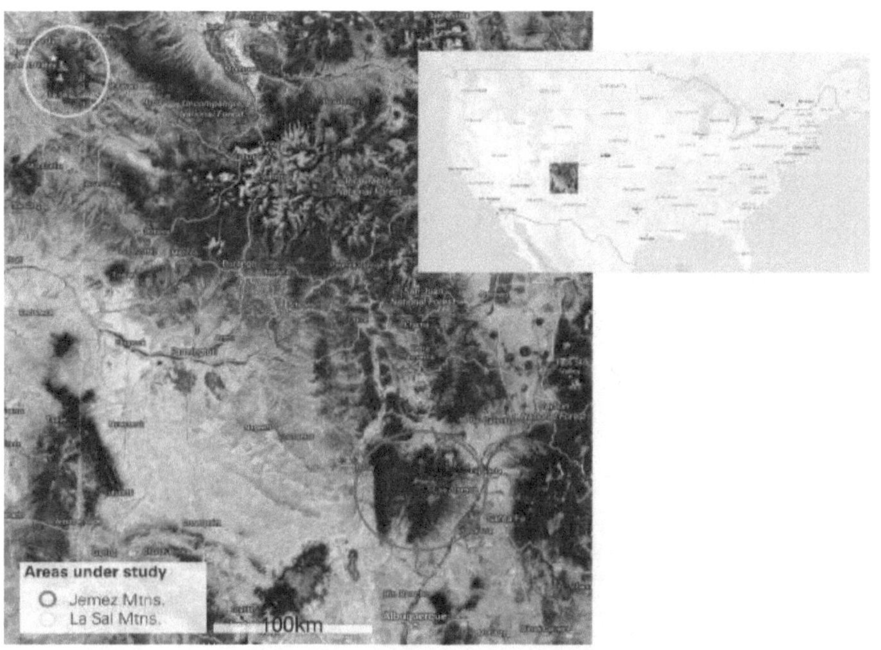

Figure 2.1 - Areas under study. Source of maps: Google Maps 2016 (http://maps.google.com).

The Jemez Mountains are known to be the source area of the La Sal Mountains tephra layer. We took samples approximately 8 km southeast of Los Alamos, New Mexico, from a slope along New Mexico State Road 502 (Guaje and Tsankawi tephras, located at 35°52′05″ N, 106°11′59″ W and at 35°52′05 N, 106°12′00″ W, respectively). The site is depicted in GOFF (2009) and in fig. 2.2a.

Figure 2.2 - (a) Sampling sites of tephras in the Jemez Mountains. All visible rocks are volcanic in origin. Photo: Jana Krautz (22 August 2014). (b) Sampling site in the La Sal Mountains. The whitish tephra intercalates between periglacial cover beds and is to the left of the picture cut by a gully fill. Photo: Arno Kleber (27 July 2009). The sampling spot visible in the La Sal Mountains tephra was for radiofluorescence dating, not for the present dating.

2.3.2 La Sal Mountains, Utah

The chain of the La Sal Mountains lies at the eastern border of Utah (Fig. 2.1). Like the Jemez Mountains, it is part of the Colorado Plateau Province and together with Mount Peale (3877 m a.s.l.) is the highest peak of the plateau (HENNING 1975; GRAHAME AND SISK 2002). The La Sal Mountains are remnants of laccoliths and mainly consist of granitoid rocks (HENNING 1975; ROSS 2006). The Precambrian basement is unconformably overlain by Paleozoic and Mesozoic sedimentary rocks, which were intruded by monzonite and diorite porphyry during the Paleogene (K-Ar ages are 25 – 28 Ma[3]; ROSS 2006). The laccolithic structures preserve Mesozoic rocks at the mountain flanks, mainly clays and sandstones (RICHMOND 1962; HENNING 1975). Within the adjacent Paradox Basin, the Mesozoic rock sequence is underlain by marine sediments, which include limestone, dolomite, slate, and a several-hundred-of-meters-thick diapiric layer of salt and gypsum (HENNING 1975). A distal tephra layer was found in the north-western La Sal Mountains, Utah, USA (located 38°34'33" N, 109°17'32" W), approximately 20 km linear distance from Moab, Utah, at 2130 m a.s.l., on a 22° steep slope, exposed by a road cut of the Manti-La Sal Circuit (KLEBER 2013 and fig. 2.2b). The tephra was identified by the US Geological Survey, Tephrochronology Laboratory, Menlo Park, CA, via the chemical composition of its glass shards. It was correlated with either the approximately 1.25 Ma old (PHILLIPS ET AL. 2007) Tsankawi tephra or – because of the Fe contents somewhat more likely – the approximately 1.65 Ma old (SPELL AND HARRISON 1993) Guaje tephra, both derived from the Jemez Mountains, New Mexico (KLEBER 2013).

2.4 Methods

We took two samples from the deposition area in the La Sal Mountains, UT, USA, and one from each original tephra layer, derived from the Toledo Caldera (Guaje tephra) and from the Valles Caldera (Tsankawi tephra). The latter two – taken from well-known tephra locations – were mainly measured to disclose whether the results of the U-Pb determinations are consistent with the aforementioned earlier $^{40}Ar/^{39}Ar$ datings and may, thus, yield reliable ages of distal tephra layers.

We performed sample preparation for cathodoluminescence (CL) images, LA-ICP-MS (laser ablation with inductively coupled plasma mass spectrometry) U-Pb analyses, and age calculations at the Geochronology Department of Senckenberg Naturhistorische Sammlungen Dresden, Germany. Circa 1 kg of material was collected for each sample. After crushing in a jaw crusher, the samples were sieved for the fraction 36 to 400 µm. Density separation of this fraction was accomplished with LST (solution of lithium heteropolytungstates in water). We used a Frantz isodynamic separator for the magnetic separation of the extracted heavy minerals. Single zircon grains of all grain sizes, colours, and morphological types were randomly picked under a binocular microscope and subsequently analysed regarding their morphology based on backscatter electron (BSE) images of the unmounted zircon grain surfaces using a Zeiss EVO50SEM at 20 kV and a spot size of 300 nm. Then the grains were mounted in resin blocks and polished to approximately half their thickness, in order to expose their internal structure. We obtained CL images using a Zeiss EVO50SEM coupled to a CL detector system at 20 kV and a spot size of 500 nm. Zircons were analysed for U, Th, and Pb isotopes by LA ICP-MS, utilizing a Thermo Scientific ELEMENT 2 XR sector field ICP-MS coupled to a New Wave UP-193 excimer laser system with laser spot sizes of 20 to 35 µm. Fifteen seconds of background acquisition was followed by 25 s of data acquisition during each analysis. The signal was tuned for a maximum sensitivity for Pb and U, whereas oxide production (^{235}UO vs. ^{238}U) was kept well below 1%. Raw data were corrected for background signal, common Pb, laser induced elemental fractionation, instrumental mass discrimination, and time- and depth-dependent elemental fractionation of Pb/Th and Pb/U using an Excel® macro developed by Axel Gerdes (Geosciences Inst., Goethe University Frankfurt, Germany). Reported uncertainties were propagated by quadratic addition of the external reproducibility obtained from the standard zircon GJ-1 (~0.6 and 0.5– 1% for $^{207}Pb/^{206}Pb$ and $^{206}Pb/^{238}U$, respectively) during individual analytical sessions and the within-run precision of each analysis. Concordia diagrams (2σ error ellipses) and concordia ages (95% confidence level) were created using Isoplot/Ex 2.49 (LUDWIG 2001). $^{207}Pb/^{206}Pb$ ages were used for concordant analyses of zircons above 1.0 Ga, and $^{206}Pb/^{238}U$ ages for younger ones. For ages younger than 10 Ma[4], we corrected for ^{230}Th disequilibrium using the formula of SIMON ET AL. (2008).

Geochemical analyses of bulk samples were performed at Activation Laboratories Ltd. (Ancaster, Ontario, Canada) using their standard protocols RX4 for sample preparation and 4LITHO-Quant Major Elements Fusion ICP (WRA)/Trace Elements Fusion ICP-MS (WRA4B2) for the analyses as described on their website (ACTLABS 2014). The samples from the La Sal Mountains were contaminated with pedogenic carbonates, whereas the samples from the Jemez Mountains were not, or at least not to the same degree. Therefore, the major elements (and the total percentages) were re-calculated on a carbonate-free basis, i.e., without considering MgO, CaO, and loss on ignition (LOI), though the original values of these three measurements are given so that one could re-assemble all original quantities. In addition to the aforementioned samples, we analysed a confirmed Guaje tephra sample provided by David B. Dethier (SLATE ET AL. 2007).

Microprobe analyses were conducted aided by a CAMECA SX51 electron microprobe with five wavelength dispersive spectrometers at the Earth Sciences Institute at Heidelberg University. The standard operating conditions were 15 kV accelerating voltage, 20 nA beam current, and a beam diameter of ca. 20 µm. Counting times during analyses were 10 s for Na and K; 20 s for Fe; 30s for Mn and P; and 50 s for Si, Ti, Al, Mg, and Ca. Detection limits were 0.02 wt% for Si, Al, and Ca, 0.001 wt% for Ti and Mn, 0.08 wt% for Fe, and 0.09 wt% for K and Na. Calibration was performed using natural and synthetic oxide and silicate standards. Values given are weight-percent oxide, re-calculated to be 100% fluid-free.

---

[4] Original publication: Myr; corrected because the context is wrong

2.4 **Results and discussion**

The microprobe analyses of glass shards corroborate the great similarity of the Guaje and the La Sal Mountains tephras (Table 2.1; cf. Supplement for raw data, chapter 6.1.1). Even the differences in Fe contents, typically acknowledged as the only clue to distinguish Guaje from Tsankawi tephras (Andrei M. Sarna-Wojcicki, personal communication, 1990), are within the standard deviations of the analyses.

Table 2.2 shows that the major and especially the trace element concentrations from bulk samples of the La Sal Mountains tephra are very close to the Tsankawi tephra from the Jemez Mountains but somewhat dissimilar to the Guaje tephra sample as well as to the Guaje sample DN-97-117 submitted by David B. Dethier. This holds especially true for the elements shaded in yellow in Table 2.2, with the most remarkable being Cr, Rb, Nb, and Th. The differences in the Sr and Ba contents between the La Sal Mountains and Jemez Mountains samples may be explained by eolian contamination, as both elements are frequent components of eolian deposits (JONES 1986). Similar differences in Tl contents may be due to different durations of sample materials being exposed to oxidation. These findings render the La Sal Mountains tephra correlative to the Tsankawi rather than the Guaje tephra.

In all samples, primary uranium contents in zircons were sufficiently high to allow reliable age determinations. Given the apparently young ages of the tephras, ^{207}Pb could not be accumulated in quantities remarkably above the detection limit of the instrument due to the extremely long half-life of ^{235}U and/or insufficiently high U contents to produce enough Pb in such short intervals of time (compare young grains in the Supplement). Thus, we could use only the ^{206}Pb/^{238}U for age estimations (cf. GEHRELS 2014). Therefore, ^{207}Pb/^{235}U and ^{207}Pb/^{206}Pb ratios for cross validation are not available; the degree of concordance cannot be calculated for these young zircon grains, and those data are left blank (Supplement, chapter 6.1). Accordingly, the ages we report are regarded as model ages.

Table 2.1 - Electron microprobe analyses of glass shards from tephra layers. ±: standard deviation. Values are weight-percent oxide, re-calculated to be 100 % fluid-free. Normalized data (raw data are available in Supplement, chapter 6.1.1).

source	phase	n	SiO_2	TiO_2	Al_2O_3	Fe_2O_3	MnO	MgO	CaO	Na_2O	K_2O	P_2O_5	Total
2014-NM-Gu	glass shards	31	77.12 ± 0.84	0.05 ± 0.02	12.29 ± 0.31	1.42 ± 0.07	0.09 ± 0.02	0.02 ± 0.04	0.26 ± 0.03	3.84 ± 0.28	4.90 ± 0.31	0.01 ± 0.01	100.00
2014-LSM-T	glass shards	22	77.44 ± 0.73	0.08 ± 0.02	12.24 ± 0.10	1.50 ± 0.14	0.07 ± 0.02	0.03 ± 0.01	0.29 ± 0.03	4.03 ± 0.18	4.30 ± 0.25	0.01 ± 0.01	99.99
2013-LSM-T	glass shards	44	77.41 ± 0.74	0.09 ± 0.03	12.21 ± 0.12	1.47 ± 0.16	0.07 ± 0.02	0.03 ± 0.01	0.28 ± 0.02	3.92 ± 0.24	4.51 ± 0.23	0.01 ± 0.01	100.00

Table 2.2 - Major and trace element concentrations of tephra samples. The table displays analyses from bulk samples. Major element percentages are calculated carbonate-free, i.e., without considering the colums in italic font. Trace elements which show remarkable differences between Guaje and Tsankawi tephras are identified in bold font. Source: Actlab report number A14-07544; report date 24th October 2014.

Analyte symbol	SiO_2	Al_2O_3	Fe_2O_3	MnO	MgO	CaO	Na_2O	K_2O	TiO_2	P_2O_5	LOI	Total	Sc	Be	V	Cr	Co	Ni	Cu
Unit symbol	%	%	%	%	%	%	%	%	%	%	%	%	ppm	ppm	ppm	ppm	ppm	ppm	ppm
Detection limit	0.01	0.01	0.01	0.001	0.01	0.01	0.01	0.01	0.001	0.01	–	–	1	1	5	20	1	20	10
Analysis method	ICP	ICP	ICP	ICP	ICP	ICP	ICP	ICP	ICP	ICP	ICP	–	ICP	ICP	ICP	MS	MS	MS	MS
2013 LSM-T	75.53	12.92	2.79	0.07	*0.22*	*1.06*	3.43	4.42	0.13	0.03	6.95	99.33	2	5	8	**30**	1	<20	<10
2014 LSM-T	74.67	13.81	2.56	0.08	*0.29*	*2.48*	3.46	4.22	0.14	0.04	8.07	98.99	2	7	12	**30**	1	<20	<10
2014 NM-GU	76.56	12.65	2.47	0.09	*0.39*	*0.48*	3.14	4.84	0.06	<0.01	6.84	99.82	1	13	7	**160**	1	<20	20
DN-97-117	74.52	13.78	3.09	0.10	*0.19*	*0.47*	2.70	5.39	0.11	0.02	*5.9*	99.70	2	11	13	**230**	2	<20	20
2014 NM-TS	76.12	11.77	2.54	0.07	*0.08*	*0.39*	4.00	4.26	0.09	<0.01	*0.78*	98.87	1	5	6	**30**	1	<20	<10

Analyte symbol	Zn	Ga	Ge	As	Rb	Sr	Y	Zr	Nb	Mo	Ag	In	Sn	Sb	Cs	Ba	Bi	La	Ce
Unit symbol	ppm	ppm	ppm	ppm	ppm	ppm	ppm	ppm	ppm	ppm	ppm	ppm	ppm	ppm	ppm	ppm	ppm	ppm	ppm
Detection limit	30	1	1	5	2	2	2	4	1	2	0.5	0.2	1	0.5	0.5	3	0.4	0.1	0.1
Analysis method	MS	MS	MS	MS	MS	ICP	ICP	ICP	MS	MS	MS	MS	MS	MS	MS	ICP	MS	MS	MS
2013 LSM-T	80	23	2	<5	**149**	90	55	244	**59**	6	0.9	<0.2	**5**	0.6	3.9	216	<0.4	**67.8**	133
2014 LSM-T	100	25	2	<5	**164**	151	76	272	**80**	6	1.1	<0.2	**6**	0.6	4.8	411	<0.4	**71.6**	135
2014 NM-GU	130	29	2	<5	**335**	17	93	245	**150**	8	0.8	<0.2	**11**	0.7	8.9	39	<0.4	**40.6**	93.3
DN-97-117	120	31	2	<5	**305**	29	92	243	**144**	8	0.8	<0.2	**10**	0.7	10.2	125	<0.4	**43.1**	105
2014 NM-TS	90	23	2	<5	**160**	23	59	200	**57**	2	0.7	<0.2	**3**	<0.5	2.9	78	<0.4	**58.6**	116

Analyte symbol	Pr	Nd	Sm	Eu	Gd	Tb	Dy	Ho	Er	Tm	Yb	Lu	Hf	Ta	W	Tl	Pb	Th	U
Unit symbol	ppm	ppm	ppm	ppm	ppm	ppm	ppm	ppm	ppm	ppm	ppm	ppm	ppm	ppm	ppm	ppm	ppm	ppm	ppm
Detection limit	0.05	0.1	0.1	0.05	0.1	0.1	0.1	0.1	0.1	0.05	0.1	0.04	0.2	0.1	1	0.1	5	0.1	0.1
Analysis method	MS	MS	MS	MS	MS	MS	MS	MS	MS	MS	MS	MS	MS	MS	MS	MS	MS	MS	MS

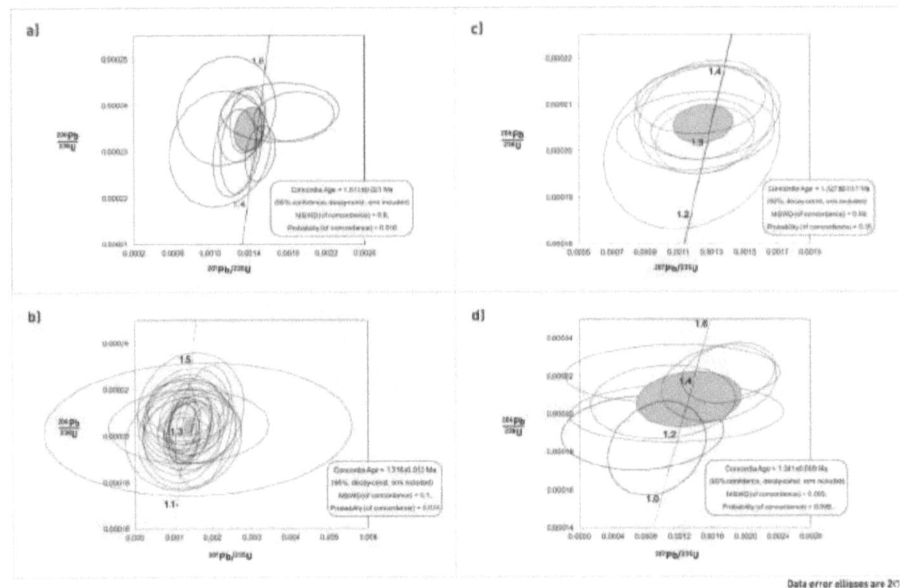

Figure 2.3 - Ages of tephra layers as derived from the youngest cluster of grain ages overlapping at the 2σ level. (a) Guaje tephra, (b) Tsankawi tephra, (c) La Sal Mountains tephra sampled in 2013, (d) same but sampled in 2014.

To establish the age of each tephra sample, we used the youngest cluster of zircon-derived U-Pb ages overlapping at 2σ. The mean age of the youngest cluster of grain ages that overlap in age at 2σ is regarded as the most conservative measure of age (DICKINSON AND GEHRELS 2009). These clusters may be seen as groups of analyses resulting in ages close together, thereby validating each other even without a reliable Pb-Pb age. Grains with younger $^{238}U/^{206}Pb$ ages than the ones used for the calculation of the concordia ages (cf. Supplement, chapter 6.1) are not part of such a cluster in the concordia plot and, thus, cannot be cross-validated. Accordingly, they were not considered sufficiently reliable.

The grains used for age determination are accentuated in tables in the Supplement (Chapter 6.2). The clusters are sufficiently large for the ages to be constrained to small confidence intervals (2σ); see also figs. 2.3 and 2.4: we assigned an age of 1.513 ± 0.021 Ma[5] to the Guaje

---

[5] Original publication: Myr; corrected because the context is wrong

tephra from the Jemez Mountains, which is somewhat younger than the published Ar-Ar-derived ages of 1.651 ± 0.011 Ma[5] (ZIMMERER ET AL, 2016) or 1.613 ± 0.011 Ma[5] (IZETT AND OBRADOVICH 1994). The other three samples yielded ages incompatible with the Guaje tephra: The Tsankawi tephra from the Jemez Mountains was determined to be as old as 1.316 ± 0.012 Ma[5]. The two samples from the La Sal Mountains yielded ages of 1.327 ± 0.017 Ma[5] (sample from the year 2013) and 1.341 ± 0.059 Ma[5] (2014 sample, which had the smallest number of zircon ages within the overlapping cluster). The confidence intervals of the latter three samples do all overlap within errors. Therefore, we correlate these tephra-layer samples with the same, the Tsankawi eruption. The common age range within 2σ of both samples is 1.31 – 1.40 Ma[5]. We assume this is the most likely age array. Zoning of zircons indicates steady growth. If there is a core depicted in the CL images, the measuring spot may not be located at a core's edge (Fig. 2.5).

The ages derived via Ar-Ar dating are 1.264 ± 0.010 Ma[6] (PHILLIPS ET AL. 2007; recalculated by ZIMMERER ET AL. 2016) and 1.223 ± 0.018 Ma[6] (IZETT AND OBRADOVICH 1994); i.e., they are slightly younger than ours. Though being very close to each other, the U-Pb ages are slightly older. The common notion is that Ar-Ar ages approximate the eruption ages and U-Pb ages indicate the (earlier) time of crystal closure (SIMON ET AL. 2008). However, this does not work for the Guaje tephra. ZIMMERER ET AL. (2016) observed similar differences between $^{40}Ar/^{39}Ar$ and uranium-series (U/Th) ages for other tephras of the Jemez Mountains. They explain their findings with a complicated crystallization history of the magma, leading to disequilibrium between the uranium isotopes in the melt. Another explanation could be that the zircon crystal lattices of the Guaje tephra were not completely closed during eruption, as our sample was taken close to an underlying mafic lava bed which still could have been hot enough to achieve this effect. Or there still are problems with the Ar–Ar dating of some Jemez tephras not yet understood. Older zircons (cf. Supplement for raw data, chapter 6.1.2) are assumed to be inherited from rocks melted during magma rise, with those zircons being their most temperature-resistant components.

---

[6] Original publication: Myr; corrected because the context is wrong

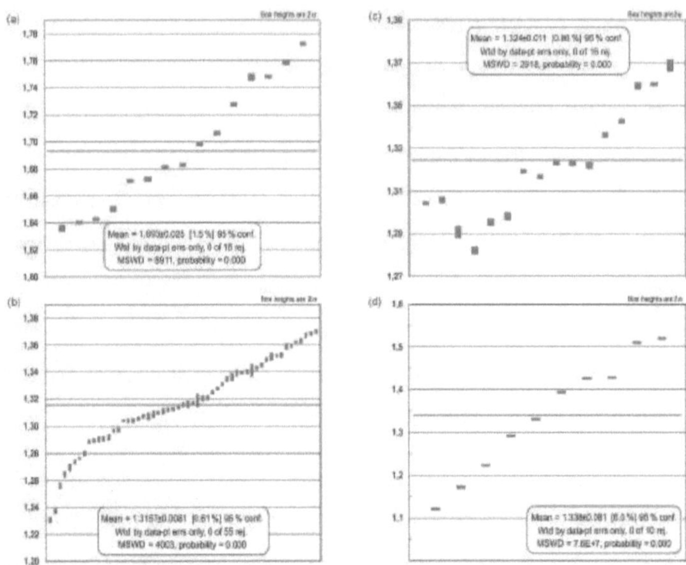

Figure 2.4 - Weighted average ages of tephra layers to compare with the age displays. (a) Guaje tephra, (b) Tsankawi tephra, (c) La Sal Mountains tephra sampled in 2013, (d) same but sampled in 2014.

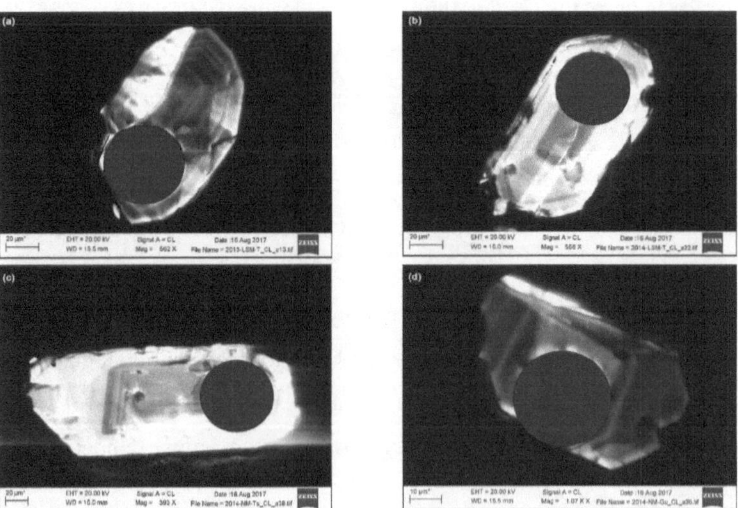

Figure 2.5 - CL images of selected zircons which have been included in the age displays (including laser ablation mark). (a) La Sal Mountains tephra sampled in 2013: c13; (b) same but sampled in 2014: a22; (c) Tsankawi tephra: a38; (d) Guaje tephra: a36.

2.5 Conclusions

Our findings demonstrate that U-Pb dating of zircons from Quaternary volcanic material may result in valuable age determination. U-Pb dating of zircons seems to allow – at least combined with bulk geochemical analyses – confident distinction between the two tephras derived from the Jemez Mountains, which are too similar to be clearly kept apart by glass-shard chemistry alone. This approach avoids the complications accompanying the Ar-Ar dating of Bandelier tephras (PHILLIPS ET AL. 2007; ZIMMERER ET AL. 2016).

We recommend considering U-Pb dating as a possible approach to identifying rather young tephras or to distinguish such tephras, as in our study. However, before application, we recommend measuring total uranium contents in zircon minerals, which might indicate whether this dating method will be applicable.

In Quaternary research, dating of zircons as young as 1 Ma[7] may well become a tool for better defining age models of sedimentary archives – such as loesses, cover beds, or paleosols – with interbedded or admixed tephra layers.

Data availability

All underlying data can be found in the Supplement[8].

The Supplement related to this article is available online at https://doi.org/10.5194/egqsj-67-7-2018-supplement.

Competing interests

The authors declare that they have no conflict of interest.

2.6 References

Actlabs: Geochemistry/Assay Overview, available at:

   http://www.actlabs.com/Page.Aspx?Menu=64&App= 210&Cat1=499&Tp=2&Lk=No, last

access: 22 December 2014.

Bailey, R. A., Smith, R. L., and Ross, C. S.: Stratigraphic nomenclature of volcanic rocks in the

Jemez

Mountains, New Mexico, Geological Survey Bulletin, 1274-P, U.S. Government Printing Office,
Washington, USA, 1969.

Dickinson, W. R. and Gehrels, G. E.: Use of U-Pb Ages of detrital zircons to infer maximum depositional ages of strata. A test against a Colorado Plateau Mesozoic database, Earth Planet. Sc.
Lett., 288, 115–125, https://doi.org/10.1016/J.Epsl.2009.09.013, 2009.

Gehrels, G. E.: Detrital zircon U-Pb geochronology applied to tectonics, Annu. Rev. Earth Pl. Sc.,
42, 127–149, https://doi.org/10.1146/Annurev-Earth-050212-124012, 2014.

Goff, F.: Valles Caldera. A geologic history, University of New Mexico Press, Albuquerque, USA,
2009.

Grahame, J. D. and Sisk, T. D.: La Sal Mountains, Utah, available at: https://web.archive.org/web/
20060208054415/http: //www.cpluhna.nau.edu/index.htm (last access: 10 December 2014),
2002.

Henning, I.: Die La Sal Mountains, Utah. Ein Beitrag zur Geoökologie der Colorado-Plateau-Pro-
vinz und zur vergleichenden Hochgebirgsgeographie, Akademie der Wissenschaften AND der
Literatur, Mainz, Germany, 88 pp., 1975.

Ito, H., Nanayama, F., and Nakazato, H.: Zircon U-Pb dating using LA-ICP-MS. Quaternary tephras
in Boso Peninsula, Japan, Quat. Geochronol., 40, 12–22,
https://doi.org/10.1016/j.quageo.2016.07.002, 2016.

Izett, G. A. and Obradovich, J. D.: $^{40}Ar/^{39}Ar$ age constraints for the Jaramillo Normal subchron and
the Matuyama-Brunhes geomagnetic boundary, J. Geophys. Res., 99, 2925–2934,
https://doi.org/10.1029/93JB03085, 1994.

Jones, R. L.: Barium in Illinois surface soils, Soil Sci. Soc. Am. J., 50, 1085–1087, 1986.

Kleber, A.: Heavy-mineral analysis as a tool in tephrochronology, with an example from the La Sal Mountains, Utah, USA, Geologos, 19, 87–94, https://doi.org/10.2478/logos-2013-0006, 2013.

Kues, B. S., Kelley, S. A., and Lueth, V. W. (Eds.): Geology of the Jemez region II. New Mexico Geological Society 58th Annual Field Conference, New Mexico Geological Society, 19– 22 September 2007, Socorro, NM, USA, 499 pp., 2007. Lee, M. S. (Ed.): Mass Spectrometry Handbook, John Wiley and Sons, Inc, Hoboken, NJ, USA, 2012.

Lowe, D. J.: Tephrochronology and its application: a review, Quat. Geochronol., 6, 107–153, https://doi.org/10.1016/J.Quageo.2010.08.003, 2011.

Ludwig, R. K.: User manual for Isoplot. Ex Rev. 2.49, Berkeley Geochronology Center Special Publication, 1a, 1–58, https://doi.org/10.1016/S1471-3918(01)80062-9, 2001.

Phillips, E. H., Goff, F., Kyle, P. R., Mcintosh, W. C., Dunbar, N. W., and Gardner, J. N.: The 40Ar/39Ar age constraints on the duration of resurgence at the Valles Caldera, New Mexico, J. Geophys. Res., 112, 2925–2934, https://doi.org/10.1029/2006JB004511, 2007.

Richmond, G. M.: Quaternary stratigraphy of the La Sal Mountains, Geological Survey, Professional Paper 450-D, United States Government Printing Office, Washington, USA, 1962.

Ross, M. L.: Preliminary geologic map of the Warner Lake Quadrangle, Grand County, Utah, Utah Department of Natural Resources, Salt Lake City, USA, 18 pp., 2006.

Sakata, S., Hirakawa, S., Iwano, H., Danhara, T., Guillong, M., and Hirata, T.: A new approach for constraining the magnitude of initial disequilibrium in Quaternary zircons by coupled uranium and thorium decay series dating, Quat. Geochronol., 38, 1–12, https://doi.org/10.1016/j.quageo.2016.11.002, 2017.

Self, S., Heiken, G., Sykes, M. L., Wohletz, K., Fisher, R. V., and Dethier, D. P. (Eds.): Field excursions to the Jemez Mountains, New Mexico, New Mexico Bureau of Mines and Mineral Resources, Socorro, NM, USA, 1996.

Simon, J. I., Renne, P. R., and Mundil, R.: Implications of preeruptive magmatic histories of zircons for U-Pb geochronology of silicic extrusions, Earth Planet. Sc. Lett., 266, 182–194, https://doi.org/10.1016/j.epsl.2007.11.014, 2008.

Slate, J. L., Sarna-Wojcicki, A. M., Wan, E., Dethier, D. P., Wahl, D. B., and Lavine, A.: A chronostratigraphic reference set of tephra layers from the Jemez Mountains volcanic source, New Mexico, in: Geology of the Jemez region II. New Mexico Geological Society 58th Annual Field Conference, New Mexico Geological Society, edited by: Kues, B. S., Kelley, S. A., and Lueth, V. W., 19– 22 September 2007, Socorro, NM, USA, 239–247, 2007.

Spell, T. L. and Harrison, T. M.: 40Ar/39Ar geochronology of post-Valles Caldera rhyolites, Jemez volcanic field, New Mexico, J. Geophys. Res., 98, 8031–8051, https://doi.org/10.1029/92JB01786, 1993.

Westgate, J. A., Perkins, W. T., Fuge, R., Pearce, N., and Wintle, A. G.: Trace-element analysis of volcanic glass shards by laser ablation inductively coupled plasma mass spectrometry. Application to tephrochronological studies, Appl. Geochem., 9, 323– 335, https://doi.org/10.1016/0883-2927(94)90042-6, 1994.

Wilson, C. J. N., Charlier, B. L. A., Fagan, C. J., Spinks, K. D., Gravley, D. M., Simmons, S. F., and Browne, P. R. L.: U–Pb dating of zircon in hydrothermally altered rocks as a correlation tool: application to the Mangakino geothermal field, New Zealand, J. Volcanol. Geoth. Res., 176, 191–198, 2008.

Worsley, P.: Altersbestimmung, in: Geomorphologie. Ein Methodenhandbuch für Studium und Praxis, edited by: Goudie, A., Springer, Berlin, Germany, 447–558, 1998.

Zimmerer, M. J., Lafferty, J., and Coble, M. A.: The eruptive and magmatic history of the youngest pulse of volcanism at the Valles Caldera. Implications for successfully dating Late Quaternary eruptions, J. Volcanol. Geoth. Res., 310, 50–57, https://doi.org/10.1016/J.Jvolgeores.2015.11.021, 20.

3 Cover beds older than the mid-Pleistocene revolution and the provenance of their eolian components, La Sal Mountains, Utah, USA

Chapter 3 is published in the peer-reviewed journal Quarternary Science Reviews as:

Cover beds older than the mid-pleistocene revolution and the provenance of their eolian components, La Sal Mountains, Utah, USA

Authors: Jana Krautz[1], Andreas Gärtner[2], Mandy Hofmann[2], Ulf Linnemann[2], and Arno Kleber[1]

[1]Institute of Geography, Technische Universität Dresden, Helmholtzstr. 10, 01069 Dresden, Germany

[2]Senckenberg Naturhistorische Sammlungen Dresden, Museum für Mineralogie AND Geologie, Sektion Geochronologie, GeoPlasma Lab, Königsbrücker Landstraße 159, 01109 Dresden, Germany

Publication history: received 1 September 2017 / received in revised form 8 January 2018 / accepted 21 January 2018 / published 1 April 2018

Full reference:

Krautz, J., Gärtner, A., Hofmann, M., Linnemann, U., and Kleber, A. Cover beds older than the mid-pleistocene revolution and the provenance of their eolian components, La Sal Mountains, Utah, USA. Quaternary Science Reviews 185. 1-8, 2018.

Internet link: https://doi.org/10.1016/j.quascirev.2018.01.012

3.1 Introduction

In mountainous environments of the mid-latitudes cover beds are common features and are supposed to be the most abundant surficial materials on slopes of low to intermediate gradient (KLEBER 1997; KLEBER ET AL. 2013). Hence, they are a decisive component of 'Earth's critical zone' (KLEBER AND TERHORST, 2013). Cover beds are defined as deposits formed by processes of unconcentrated dislocation chiefly from upslope materials, which may be mixed with eolian matter. They cover slopes to a large extent, rather than being restricted to drainage ways or local failures. Cover beds typically consist of layers, which are separated by disconformities (KLEBER AND TERHORST 2013). In the vicinity of the northern Great Basin, western USA, cover beds typically contain eolian particles (KLEBER 1994).

Cover beds are rarely well dated, but by far the most instances appear to relate to the last glaciation (HÜLLE AND KLEBER 2013). Moreover, in the western USA there is also strong evidence of cover beds that have formed during the termination of the penultimate glaciation (KLEBER 1994). There, discriminating layers of cover beds is made possible by examining soil properties, especially by the overprint of argillic features by later carbonate

43

enrichment (KLEBER 2000). Cover beds may be considered as archives of past environments. As with their age, the duration of the formation of cover beds is an open question, though this would be an important aspect to judge what they may tell about the paleoenvironmental conditions under which they have formed. In this context, it would be helpful to know a minimum age of the cover-bed formation, especially, whether their formation was possible during the much shorter-lasting glaciations before the "mid-Pleistocene revolution" almost 1 Ma ago. Before this revolution each glacial cycle lasted c. 40 ka on average, whereas cycles persisted for c. 100 ka afterwards (PAILLARD 2001). However, as yet no evidence of cover beds that old has been reported.

In the La Sal Mountains, Colorado Plateau, southeastern Utah, a cover bed has incorporated a layer of tephra derived from the Jemez Mountains, New Mexico, USA (KLEBER 2013). KLEBER (2013) speculated that the tephra layer may have been reworked much later than its primary deposition, so that the encompassing cover bed as well as underlying, older cover beds might be of much younger age than the original tephra. Here we present ages of this tephra using U-Pb series age determinations of zircons. To test whether these determinations allow for assigning minimum ages to the underlying cover beds, we also dated detrital zircons (DZ) in these older cover beds to evaluate whether they contain zircons of the same or even younger age than the tephra, i.e., whether there is evidence of reworking of the tephra material or whether the tephra layer may be regarded as being in situ.

3.2 Material and methods

3.2.1 The La Sal Mountains tephra layer

Referred to hereafter as the La Sal Mountains (LSM) tephra layer, the tephra was found in the north-western LSM, Utah, U.S.A. (located 38°34'33" N, 109°17'32" W), at an elevation of 2130 m a.s.l. on a 22° steep slope, exposed by a road cut of the Manti-La Sal circuit (cf. fig. 3.1). The tephra was identified by the US Geological Survey, Tephrochronology Laboratory, Menlo Park, CA, via the major-element composition of its glass shards as either the approx. 1.65 Ma old (SPELL ET AL, 1990) Guaje Tephra or the approx. 1.25 Ma old (PHILLIPS ET AL, 2007) Tsankawi Tephra, both derived from the Jemez Mountains, NM (the geochemical data of the glass-shards are published by KLEBER 2013 and KRAUTZ ET AL. 2018). In the field and under the microscope, the tephra layer shows only slight indication of weathering; there is no such indication at all in its core. Such preservation is unlikely, if the tephra layer had been as close to the surface, i.e. 90 cm, as it is now since more than 1 Ma. Thus, the question arises whether material originally overlying the LSM tephra layer has been eroded or whether the tephra layer, despite its pure appearance, was reworked considerably after it had been originally deposited and is in a secondary position in this profile.

3.2.2 Cover beds and palaeosols

If the LSM tephra layer, dated at either 1.25 or 1.65 Ma on the basis of its correlation with a Jemez Mountains eruptive, is largely in its primary position of deposition, i.e., in situ, the underlying deposits would predate the mid-Pleistocene revolution. Beneath the LSM tephra layer, the exposure consists of several soil horizons that were formed mainly from loess- and gravel-rich cover beds during various soil-forming episodes (Fig. 3.1). Various paleosols may be distinguished by means of their compound clay- and carbonate-enriched (argillic and calcic, respectively) soil horizons (KLEBER 2013): the carbonate in these compound horizons is supposed to have accumulated after the argillic properties had been formed, because simultaneous carbonate enrichment and clay illuviation are mutually exclusive in the same horizon. Hence, clay translocation was only possible during or after the carbonate had been depleted (cf. KLEBER, 2000 for detailed reasoning). Furthermore, a much warmer soil-temperature regime than the area is experiencing at present would have been needed to form calcic horizons reaching as deep as in this profile, provided the distance of the horizon from the surface was the same as it is at present (cf. MCFADDEN AND TINSLEY 1985), let alone materials that might have been eroded off the top of the profile. Both lines of reasoning lead to the interpretation that the parent material of each soil was deposited after the argillic properties in the respective underlying soil had already been formed, which implies that the latter is part of an even older paleosol (cf. KLEBER 2000). Accordingly, the soils in this profile most likely formed during a considerable span of time before the deposit containing the LSM tephra layer arrived.

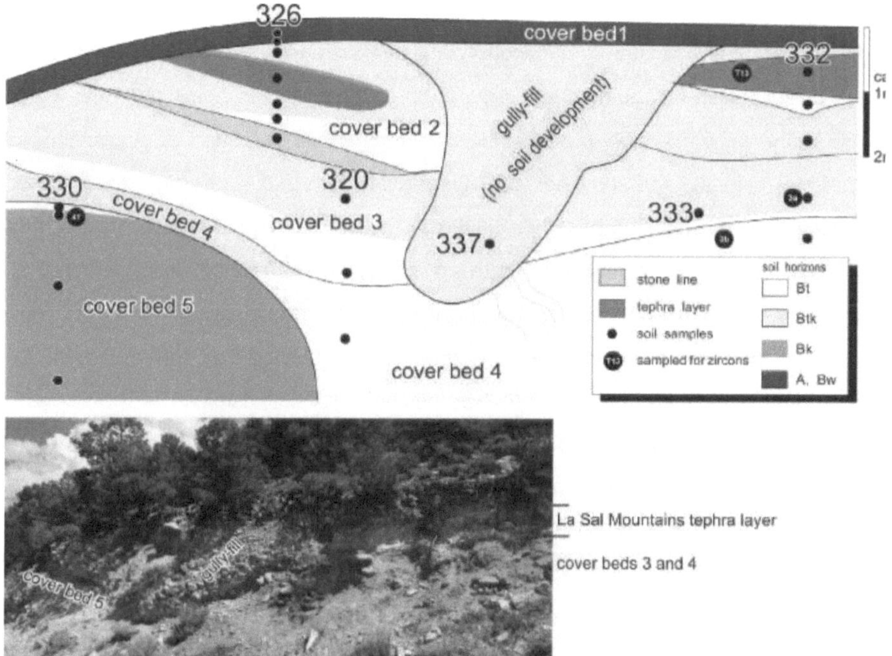

Figure 3.1 - Top: Sketch of soils and deposits of the exposure under study. The scale is approximate. There is no visible soil development in the gully fill and in the tephra layer in the right part of the exposure, whereas there is some carbonate and clay enrichment in the left occurrence of the tephra layer. The samples taken for U-Pb dating of zircons are 2013 LSM-T (T13) and 2014 LSM-3a, -3b, and -4T within cover beds 3, 4, and 5, respectively. The numbers at soil samples refer to the start of the particular columns in Table 3. Bottom: The exposure under study. Photograph by Florian Schneider (on student field trip, Aug. 18, 2015).

3.2.3 Samples and analyses

We analyzed zircons from two samples of the tephra layer and DZ from three samples from three different underlying cover beds (Fig. 3.1) and determined their U-Pb ratios.

Sample preparation, backscattered electron (BSE), and cathodoluminescence (CL) images of external and internal zircon textures using scanning electron microscope (SEM) techniques, Laser Ablation with Inductively Coupled Plasma Mass Spectrometry (LA-ICP-MS) U-Pb analyses, and age calculations were performed at the Geochronology Dept. of Senckenberg Naturhistorische Sammlungen Dresden (Germany). Around 1 kg of material was collected for each sample. The samples were crushed in a jaw crusher and sieved for the fraction 36 mm to 400 mm. We accomplished the density separation of this fraction with LST (solution of lithium heteropolytungstates in water). A Frantz isodynamic separator was used for the magnetic separation of the extracted heavy minerals. Single zircon grains were manually picked under a binocular microscope. Zircon grains of all grain sizes, colours, and morphological types were selected randomly and analysed regarding their morphology based on BSE imaging using a Zeiss EVO50 SEM. After this, the zircon grains were mounted in resin blocks (i.e., mounts), and polished to approximately half their thickness to expose their internal structure. CL images were obtained using the aforementioned SEM coupled to a CL detector system (HONOLD). Zircons were analysed for U, Th, and Pb isotopes by LA ICP-MS, using a Thermo-Scientific Element 2 XR sector field ICP-MS coupled to a New Wave UP-193 Excimer Laser System with laser spot sizes of 20 to 35 µm. During each analysis, 15 s of background acquisition were followed by 25 s of data acquisition with an energy density of 2e3 J/ cm^2. The signal was tuned for a maximum sensitivity for Pb and U, while keeping oxide production (^{254}UO/^{238}U) well below 1%. Raw data were corrected for background signal, common-Pb (^{204}Pb), laser induced elemental fractionation, instrumental mass discrimination, time- and depth-dependent elemental fractionation of Pb/ Th and Pb/U, using an Excel® macro developed by AXEL GERDES (Geosciences Inst. University Frankfurt a. M. Germany). For the common lead (^{204}Pb) correction, we calculated the concentration of the isotope ^{204}Hg, which shares the mass of common lead, from measurements of ^{202}Hg using the chart of Rosman and Taylor (1999). After this we monitored the amount of ^{204}Pb during background and during laser ablation. If the amount during laser ablation is increased compared to the background, a common lead correction was conducted, using the given abundances of the naturally occurring isotopes of ^{204}Pb, ^{206}Pb and ^{207}Pb, using the aforementioned chart.

For ages younger than 10 Ma, we corrected for ^{230}Th disequilibrium according to SIMON ET AL. (2008). Reported uncertainties were propagated by quadratic addition of the external reproducibility obtained from the standard zircon GJ-1 (~0.6% and 0.5e1% for ^{207}Pb/^{206}Pb and ^{206}Pb/^{238}U, respectively) during individual analytical sessions and the within-run precision of each analysis. Concordia diagrams (2σ-error ellipses) were created and concordia ages (95% confidence level) were calculated with Isoplot/Ex 3.75 (LUDWIG 2012) and probability-density plots with AgeDisplay (SIRCOMBE 2004). The $^{207}Pb/^{206}Pb$ ages were used for zircons above 1.0 Ga, $^{206}Pb/^{238}U$ ages for younger ones. We used the mean age of the youngest grains overlapping in age at 2 σ to infer the age of the tephra layer. For this layer and the underlying soil parent materials we also determined the age distribution of all grains that are concordant within ±10%. Given the relatively young age of this tephra, only the $^{206}Pb/^{238}U$ ratios provided reliable ages. Radiogenic $^{207}Pb/^{206}Pb$ is almost constant over the age range 1-10 Ma, leading to enormous error magnification when determining an age using $^{207}Pb/^{206}Pb$ for such young grains. For the young zircons of our samples, there was no need to cope with this, because the detection of ^{207}Pb failed for most analyses due to the small percentage of ^{235}U within the grains. Due to this, there was insufficient time to generate a measurable amount of ^{207}Pb. Therefore, the $^{207}Pb/^{206}Pb$ ages and the degree of concordance could not be calculated for the young zircon grains and were left clear in Tab. 3.3 (chapter 6.2 Supplemental Material).

Geochemical analyses of bulk samples were performed at Activation Laboratories Ltd. (Ancaster, Ontario, Canada) using their standard protocols RX4 for sample preparation and 4LITHO-Quant Major Elements Fusion ICP (WRA)/ Trace Elements Fusion ICP/MS (WRA4B2) for the analyses as described on their website (www.actlabs.com).

Soil data were obtained in the Geomorphological Laboratory of the University of Bayreuth. Particle sizes below 2 mm were determined - after removal of organic material and carbonate - by wet sieving and the pipet method (dispersant $Na_4P_2O_7$, KLUTE 1986: 393, 399-404, but with the sand/silt boundary at 0.064 mm), larger ones were determined by volume in the field (BIRKELAND 1984). Soil organic C (C_{org}) was measured by rapid dichromate oxidation with $K_2Cr_2O_7$ (colorimetrical method, PAGE ET AL.1982: 565-570), and pH was measured in $CaCl_2$ (PAGE ET AL. 1982: 206-207). Carbonate contents were determined using a gas-volumetric Scheibler apparatus (PAGE ET AL. 1982: 183-187).

Horizon designations and other soil terminology follow SOIL SURVEY STAFF (2014), except for that the suffix "b" for buried horizons is omitted as it would apply to all horizons. The designation "Bt" is based on evidence of clay translocation, because eluvial horizons are not preserved. Such evidence was provided by clay cutans identified by hand lens on ped faces. The carbonate morphology is described according to the classification of BACHMAN AND MACHETTE (1977).

3.3 Results and discussion

The tephra layer is intercalated into the second cover bed and soil of the exposed profile (Fig. 3.1). Due to their high primary uranium contents, its zircons yield reliable ages. The U-Pb age of zircons from the LSM tephra layer (1.327 ± 0.017 Ma, fig. 3.2) renders it correlative to the Tsankawi Tephra (KRAUTZ ET AL. 2018).

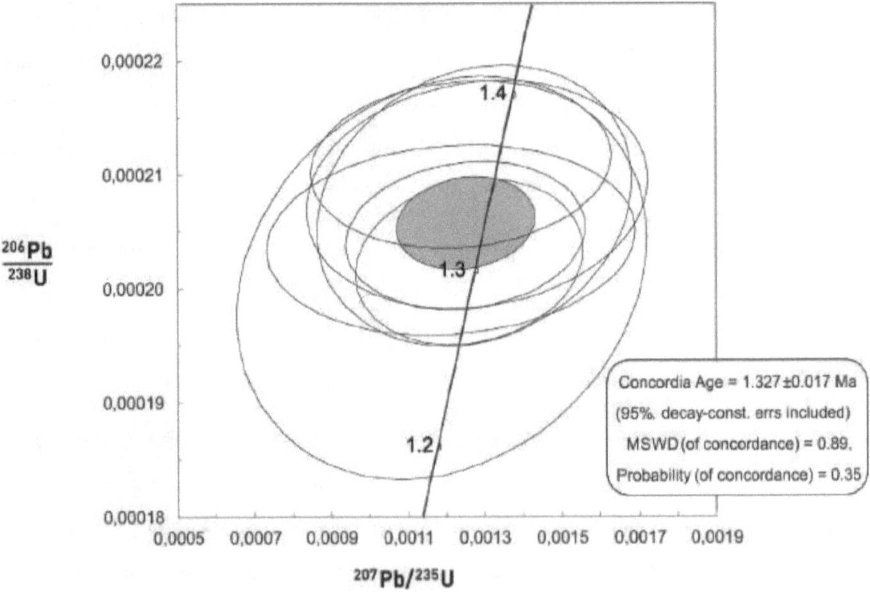

Figure 3.2 - Concordia age of sample T13 based on seven single measurements. Data-point error ellipses are 2 s confidence level. Raw data and ratios for this Concordia Age plot are given in the Supplementary Material 6.2.

Two more cover beds occur below the layer enclosing the tephra. These cover beds have sediment and soil properties (cf. Table .1 for standard analyses of cover beds and their soils) very similar to the surficial substrates typical for much of the northern Great Basin and its rim (KLEBER1994). Most of these beds have fine earth dominated by sand, assumed to be derived from the local rhyolite. There is also a prominent coarse fraction of locally-sourced rock fragments indicating dislocation by slope processes.

However, silt is another important component of the fine matter, which is assumed to be mainly eolian in origin. This assumption is supported by heavy-mineral analyses, which show allochthonous components (i.e., minerals different from the source rocks, see KLEBER 1990, 1994; cf. DAHMS 1993). Silt contents typically do not decrease with depth within a particular cover bed. This renders post-depositional enrichment with eolian matter unlikely (contrary to, e.g., SHROBA AND BIRKELAND 1983, APPLEGARTH AND DAHMS 2004), so that the silt most likely was admixed with the local materials before or during their final dislocation. In contrast, the stages of carbonate enrichment as visible in the field (BACHMAN AND MACHETTE 1977) are most prominent in the upper part of each particular palaeosol, indicating that the carbonate was leached down to where it has precipitated during pedogenesis, thereby overprinting older clay-illuviated horizons (cf. KLEBER 2000).

Below these, further cover beds are identified (fig. 3.1). The uppermost one hosts a thin horizon with a pinkish cast atypical for the surficial soils of the area, characterized by illuviated clay and some dispersed carbonate. The clay contents of the underlying horizons are remarkably higher than in all overlying soil horizons (Tab. 3.1, next pageTable). These underlying cover beds are oversaturated by carbonate, a stage of enrichment (stage IV after BACHMAN AND MACHETTE 1977) that is much more advanced than in other soils observed during several weeks of field work in the La Sal Mountains.

DZ in these underlying cover beds yield a wide range of ages from 30 Ma to 3.2 Ga (Fig. 3.3). None of these ages comes close to the age of the LSM tephra layer. The bulk chemical compositions of the cover beds underlying the tephra layer show some characteristics remarkably dissimilar to the tephra layer. This difference holds true especially for Ti and various trace elements such as V, Cr, Zn, and Rb (Tab. 3.2). These findings render reworking of the LSM tephra layer during the time of the deposition of its underlying cover beds unlikely. Rather, those cover beds definitely pre-date the primary deposition of the tephra layer.

Table 3.1   Grain size, colour and chemical properties of soils in the exposure under study.

sample	horiz	depth	>20mm	>2-20	>0.6-2	>0.2-0.6	>0.06-0.2	>0.02-0.06	>0.006-0.02	0.002-0.006	<0.002	color dry	color wet	pH CaCl	CaCO3	Corg
			% of total						*% of fine earth*							
			coarse clasts	fine	coarse	medium sand	fine	coarse	medium silt	fine	clay					
326	A	-10	10	9.1	15.5	11.2	19.6	26.6	8.9	3.7	14.4	7.5YR4/3	7.5YR3/2	7.7	3.0	1.8
315	Bw	-40	15	6.6	16.4	14.4	23.5	16.7	8.9	3.3	14.9	7.5YR6/4	7.5YR4/4	7.6	15.0	1.4
316	2Dbk1	-63	2	0.0	6.4	7.3	23.1	22.2	6.6	5.4	27.0	7.5YR6/4	7.5YR5/5	7.7	17.1	0.5
317	3Bk	-100			6.0	7.6	24.7	26.4	8.0	5.9	21.4	7.5YR7/4	7.5YR5/5	8.0	19.1	0.2
327	4Bk2	-105	2	1.7	4.2	6.3	26.1	25.6	9.4	4.7	23.7	7.5YR7/4	7.5YR6/4	8.1	17.7	0.2
318	4Bk3	-245	2		4.1	7.1	26.8	29.2	7.1	5.2	20.6	7.5YR6/4	7.5YR5/5	8.0	14.4	0.2
319	5Bk	-275	25	3.2	11.7	10.3	27.9	21.8	7.4	2.5	16.5	5YR5/6	5YR4/6	8.0	6.9	0.1
320	6Bt1	s.fig.	10	4.6	13.0	12.7	21.2	19.9	5.4	5.9	16.9	7.5YR7/4	7.5YR6/5	8.1	14.5	0.2
329	6Bt2	s.fig.	1	0.3	7.6	12.9	30.6	18.2	7.6	3.9	19.3	5YR6/5	5YR5/5	8.1	1.1	0.2
321	7Bt	s.fig.	15	4.3	13.1	11.0	21.0	25.1	10.6	1.6	17.5	7.5YR6/6	5YR4/6	8.1	2.6	0.1
330	4Bt	s.fig.	5	6.5	15.3	10.9	22.0	23.3	6.5	3.3	18.6	5YR4/6	5YR4/6	8.1	6.9	0.2
328	5Bk	s.fig.	10	0.4	10.4	15.6	15.1	8.8	11.3	8.2	30.5	7.5YR8/4	7.5YR7/4	8.5	63.6	0.4
323	5Bk	s.fig.	10	0.7	10.5	13.5	17.6	12.4	7.4	5.9	31.7	5YR7/3	7.5YR6/5	8.1	44.3	0.2
324	6Bk	s.fig.	15	1.2	23.1	20.8	13.0	7.0	10.0	7.4	16.7	10YR8/3	10YR7/4	8.2	69.0	0.4
332	3C	40-115			0.4	13.8	54.7	14.3	8.3	2.3	6.2	10YR8/1	10YR6/2	8.0	1.9	0.2
334	4Bw	-140	1	5.9	8.1	9.8	36.0	11.4	10.7	4.0	19.8	7.5YR6/3	7.5YR4/4	8.0	7.3	0.6
335	4Bk	-150	1	3.9	13.9	14.0	34.9	7.7	8.1	3.5	17.0	7.5YR7/3	7.5YR5/4	7.9	10.7	0.5
336	5Bk	-170	5		7.3	9.6	27.5	18.9	8.3	4.8	23.6	5YR6/5	5YR6/5	7.9	6.3	0.4
338	7Bt	-260	2		1.4	3.8	22.2	22.3	10.2	6.0	32.1	5YR6/5	5YR4/6	7.9	0.2	0.1
333	3Bt	s.fig.	5	0.3	4.5	6.1	28.8	29.0	10.3	4.6	16.7	5YR6/5	5YR5/6	8.0	2.1	0.2
337	2C	s.fig.	40	4.7	14.8	21.7	20.8	9.1	8.6	6.6	19.4	7.5YR7/4	7.5YR6/4	7.9	8.9	0.4

Depth: measured from top of profile where applicable, otherwise see figure 3.1. Horizon: designation according to Soil Survey Staff (2014). Grain size classes in mm. Coarse clasts = vol% of whole soil (field estimate); fine clasts = wt.% of whole soil after removal of cGr; other particle classes = wt.% of fine earth <2 mm. Color: Munsell scale. CaCO3 = total carbonate. Corg.: organic carbon.

Table 3.2   Concentrations of selected elements in tephra layer and soils.

Analyte Symbol	TiO2	Be	V	Cr	Co	Zn	Ga	Rb	Sr	Y	Zr	Nb	Sn	La	Ce	Pr
Unit Symbol	%	ppm	ppm	ppm	ppm	ppm	ppm	ppm	ppm	ppm	ppm	ppm	ppm	ppm	ppm	ppm
Detection Limit	0,001	1	5	20	1	30	1	2	2	2	4	1	1	0,1	0,1	0,05
Analysis Method	FUS-ICP	FUS-ICP	FUS-ICP	FUS-MS	FUS-MS	FUS-MS	FUS-MS	FUS-MS	FUS-ICP	FUS-ICP	FUS-ICP	FUS-MS	FUS-MS	FUS-MS	FUS-MS	FUS-M
2013 LSM-T	0.117	5	8	30	1	80	23	149	90	55	244	59	5	67.8	133.0	14.00
2014 LSM-T	0.123	7	12	30	1	100	25	164	151	76	272	80	6	71.6	135.0	15.20
2014 LSM-2	0.490	1	63	110	7	40	12	85	179	24	358	7	2	28.3	53.4	6.36
2014 LSM-3a	0.454	1	62	60	7	40	13	77	202	18	370	6	2	24.9	55.2	5.60
2014 LSM-3b	0.519	2	66	160	9	60	14	92	160	26	306	8	3	34.3	63.9	7.65
2014 LSM-4	0.418	2	67	180	5	40	12	65	278	15	304	9	2	23.5	50.8	5.65
2014 LSM-4T	0.234	<1	37	40	3	<30	7	36	257	11	187	2	<1	15.2	29.1	3.43

Analyte Symbol	Nd	Sm	Eu	Gd	Dy	Ho	Er	Tm	Yb	Lu	Hf	Ta	Pb	Th	U
Unit Symbol	ppm	ppm	ppm	ppm	ppm	ppm	ppm	ppm	ppm	ppm	ppm	ppm	ppm	ppm	ppm
Detection Limit	0.1	0.1	0.05	0.1	0.1	0.1	0.1	0.05	0.1	0.04	0.2	0.1	5	0.1	0.1
Analysis Method	FUS-MS	FUS-MS	FUS-MS	FUS-MS	FUS-MS	FUS-MS	FUS-MS	FUS-MS	FUS-MS	FUS-MS	FUS-MS	FUS-MS	FUS-MS	FUS-MS	FUS-M
2013 LSM-T	48.8	9.9	0.24	8.3	9.0	1.8	5.4	0.87	5.2	0.72	7.3	4.7	28	19.9	5.6
2014 LSM-T	56.0	12.0	0.55	10.5	12	2.4	7.1	1.20	7.5	1.12	9.3	6.5	37	24.9	6.8
2014 LSM-2	23.8	4.8	0.96	3.7	3.6	0.8	2.4	0.39	2.8	0.39	8.2	0.7	14	7.8	2.4
2014 LSM-3a	21.1	4.3	0.91	3.0	3.0	0.6	1.8	0.30	1.8	0.27	8.7	0.7	12	7.8	2.4
2014 LSM-3b	31.0	6.0	1.28	4.4	4.1	0.8	2.2	0.38	2.5	0.36	7.1	1.0	18	9.2	2.6
2014 LSM-4	21.1	3.9	0.86	3	2.7	0.5	1.7	0.27	1.7	0.26	7.1	0.7	9	6.6	2.3
2014 LSM-4T	12.9	2.6	0.61	1.9	1.8	0.4	1.1	0.17	1.1	0.16	4.3	0.3	7	4.0	1.5

The table displays selected elements, which show remarkable differences between the La Sal Mountains tephra layer (samples 2013/2014 LSM-T) and enclosing (2014 LSM-2) and underlying cover beds (2014 LSM-3a, 2014 LSM-3b, 2014 LSM-4, 2014 LSM-4T ¼ cover beds 3, 4, Bt horizon in cover bed 5, Bk horizon in cover bed 5, respectively; the layers are labeled in figure 3.1).

To obtain information on provenances of the eolian matter, i.e., to reconstruct wind directions, we determined DZ age spectra of the cover beds and we compared them to published age spectra of the surrounding rocks. Each cover bed contains a few DZ in the Archean age range of the Kenoran Orogeny (ASPLER ET AL. 2001), probably derived from the Canadian Shield. The youngest concordant DZ ages from the third and the fourth cover bed are 218 ± 4 Ma and 366 ± 9 Ma, respectively, whereas the fifth cover bed shows a remarkably different age distribution (fig. 3.3c). Cover bed 3 (fig. 3.3a) is dominated by DZ at around and after 1.4 Ga, when the southwestern Laurentian continent was affected by strong plutonism (CROWLEY ET AL. 2006), whereas in cover bed 4, DZ in the age range of the amalgamation of the Mogollon and the Yavapai terranes between 1.8 and 1.6 Ga (CROWLEY ET AL. 2006) are more abundant (fig. 3.3b). These DZ must stem from sedimentary reworking of older material, because only small patches of Paleoproterozoic rocks are close to the surface in south-western Laurentia (CROWLEY ET AL. 2006). DZ correlative with the orogenic phases from about 570 Ma (WALSH AND ALEINIKOFF 1999) to 260 Ma (HATCHER JR. 2002) in the Appalachian Mountains and the accreted terranes are also common but do not dominate the composition. We assume that most of the eolian matter in this area has been blown off river deposits. Due to this, the source material is expected to be a multiply recycled mixture from a whole catchment. Despite of this complication, the age spectra of our analyses depict remarkable differences and some evidences of their derivation, possibly because upstream spectra tend to be diluted downstream. Furthermore, all rivers have flown to the Colorado River essentially having its modern course since at least 6 Ma (YOUNG 2008), i.e., away from our study site, so that the bedrock sources discussed in detail below represent the minimum distance of the eolian matter in our samples. The absence of younger grains (except for a single grain in cover bed 3) indicates that rocks with zircons younger than Permian are not common in the area that contributed to the eolian matter mixed into cover beds 3 and 4. The age distribution resembles that of the Permian rocks in the Grand Canyon area (cf. GEHRELS ET AL. 2011), with the distribution in cover beds 3 and 4 being more similar to that of the Coconino Formation and to that of the Tonoweap Formation (cf. GEHRELS ET AL. 2011), respectively (Fig. 3.3d). Large areas with outcropping Permian rocks are distributed to the southwest of the La Sal Mountains, whereas none are mapped farther north (USGS, 2017;

fig. 3.4). However, both cover beds depict DZ-age peaks which probably were not derived from the Permian rocks: cover bed 3 shows a remarkable maximum at 1.21 Ga (Fig. 3.3a). There is a prominent gap between the zircon ages derived from the Grenville orogen and those from synorogenic plutons just around 1.2 Ga (DICKINSON AND GEHRELS 2009). Accordingly, zircons of this age are rare in North America (CONDIE ET AL. 2009). They are known from the Appalachian forelands (PARK ET AL. 2010), and a matching peak dominates in Jurassic deposits north of the Grand Canyon (DICKINSON AND GEHRELS 2008; B in fig. 3.4). Other sources from the Colorado Plateau either do not contain zircons

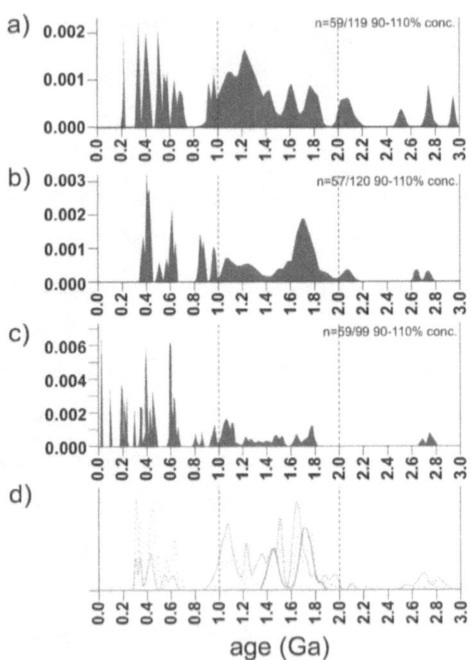

Figure 3.3 - Probability-density plots of U-Pb ages from zircons plotted using AgeDisplay (Sircombe, 2004). a) 2014 LSM-3a, b) 2014 LSM-3b, and c) 2014 LSM-4T, i.e. cover beds 3, 4, and 5 in figure 3.1, respectively. Raw data are given in the Supplemental Material (Table 3.3, chapter 6.2). d) Plots from the Grand Canyon area (dashed: Upper Permian; solid: Lower Permian; shaded: Cambrian) redrawn from Gehrels et al. (2011), not to scale.

of 1.2 Ga, or these are accompanied by lots of DZ of Mesozoic age (IEDA Data Facility, 2017), which are essentially absent in cover bed 3. This is regarded as further evidence for the southwestern provenance of cover bed 3.

Cover bed 4 has a prominent DZ age peak at 1.7 Ga (Fig. 3.3b). An equivalent peak is found in Grand Canyon deposits (GEHRELS ET AL. 2011; fig. 3.3d), but this is by far not as dominating as in cover bed 4. This age is especially prominent in Cretaceous deposits east of the Grand Canyon area (DICKINSON AND GEHRELS 2008, Supplemental Age Plots; C in fig. 3.4). Those DZ are probably derived from Paleoproterozoic rhyolites (COX ET AL., 2002). These rocks are absent farther north (JONES III ET AL. 2008), supporting a southern provenance of cover bed 4. The 1.7 Ga age peak has also been found farther north in Paleogene river deposits, but these deposits are rich in Mesozoic DZ (DAVIS ET AL. 2010, Data Repository), which are absent in cover bed 4.

Therefore, the cover beds 3 and 4 have their source area in the south-west and south of the profile under study, respectively.

In contrast, cover bed 5 (Fig. 3.3c) has a stronger share of DZ ages correlative to the Appalachian orogenies, i.e., from the Neoproterozoic until the Carboniferous. Furthermore, it contains a few DZ with Oligocene age, the youngest concordant grain at 30.52 ± 0.6 Ma, and from the Cretaceous. The latter age around 100 Ma is interpreted to have its ultimate source in the Cretaceous Cordilleran magmatic arcs (DICKINSON ET AL. 2012) to the west of our study site. This DZ age peak appears in rocks of the Colorado Plateau of Upper Cretaceous age (DICKINSON AND GEHRELS 2009; fig. 3.4) with occurrences in the vicinity of the La Sal Mountains. The Oligocene age is correlative to a major volcanic phase of the southwestern USA (ARMSTRONG 1969). This grain might stem from the La Sal Mountains laccolith which has a similar zircon-derived U-Pb age of 29.1 ± 0.3 Ma (RØNNEVIK ET AL. 2017). The latter rock forms the core of the modern La Sal Mountains and is the bedrock underlying the site under study up to the local crestline (ROSS 2006). The DZ grain may also be derived from one of the two super volcanoes, which erupted at about 30 Ma to the west and the east of our study site (BEST ET AL. 2013, WOTZLAW ET AL. 2013; D in fig. 3.4).

However, if the transport of the eolian component of cover bed 5 was directly from the west or the east, one would expect a much higher contribution from these massive eruptions, which renders these directions unlikely. A transport direction from the north is improbable as well, as this should have conveyed a greater proportion of Archean and Mesoproterozoic DZ from the Uinta Mountains (cf. DICKINSON ET AL. 2012; KINGSBURY-STEWART ET AL. 2013). Accordingly, the source areas of these redistributed DZ grains have younger rocks at the surface than those of the overlying - i.e., stratigraphically younger - cover beds. Thus, the question arises whether the zircons in cover bed 5 may stem from right around the site or were transported over some greater distance. However, a local derivation is unlikely because only one single grain (see above) represents the age of the local rhyolite and, thus, may stem from weathering of the local bedrock. This rock provides all clasts in the profile, and in the upslope vicinity of the site under study there is no other bedrock mapped until the local crest line (ROSS 2006). If the local rock had contributed remarkably to the matrix of the cover bed, grains of the age, size, and idiomorphic form of that single zircon (Fig. 3.5) should have a much greater concentration.

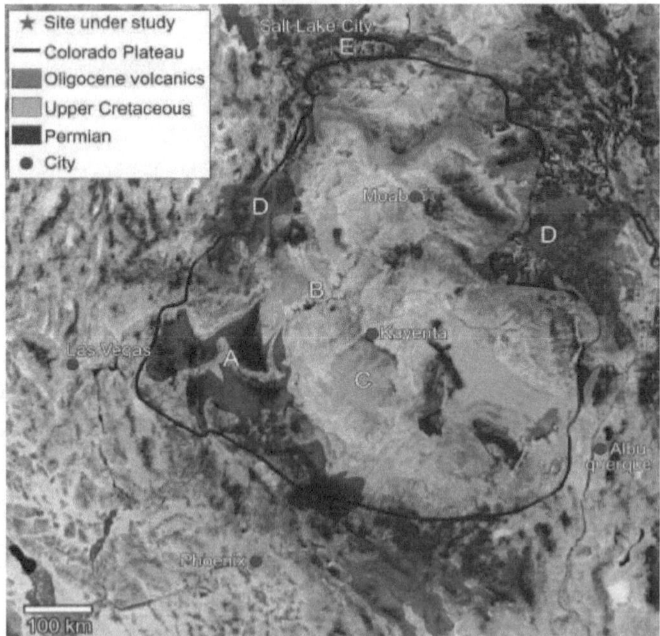

Figure 3.4 - Map of the Colorado Plateau with the site under study and selected geologic units. Letters A-E depict areas discussed in the text (map sources: Google Maps, USGS, 2017).

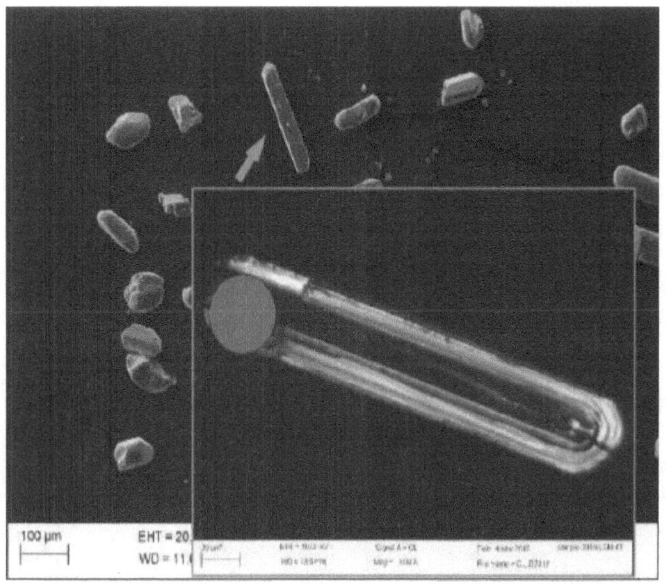

Figure 3.5 - BSE and CL (insert, incl. laser ablation mark) images of the youngest concordant zircon from cover bed 5 (grain 2014-LSM-4T-b23). The CL image was made after mounting and polishing of the zircon so that the view is mirror inverted.

3.4 Conclusions

The cover beds underlying the LSM tephra layer do not contain zircons of, or close to, the age of the LSM tephra layer itself. Similarly, the chemical signatures of the cover beds also do not indicate reworking of the tephra material. Both findings reveal that the tephra layer was not redistributed after its primary deposition, contrary to the conclusion reached by KLEBER (2013).

The zircon ages are consistent with the field-derived evidence that the cover beds underlying the LSM tephra layer comprise several cover beds of different ages, as the distributions of zircon U-Pb ages vary remarkably. Their DZ U-Pb-derived ages are due to the provenance of the admixed matter rather than to the ages of the particular cover beds as indicated by the fact that the age spectra are not in order, i.e., that the stratigraphically deepest sampled cover bed yielded younger zircons than overlying ones.

Cover beds have rarely been dated, because of their polygenetic evolution (HÜLLE AND KLEBER 2013). The exposure in the La Sal Mountains provides the first clue of cover-bed ages older than the mid-Pleistocene revolution. Obviously, the environmental conditions at that time allowed the formation of slope deposits very comparable to those that formed during later Pleistocene times. Pedogenesis was also remarkably similar, indicating that the available amount of time for the formation of soils was sufficient to reach approximately the same results, despite of the much shorter glacial-interglacial cycles. This finding does not hold true for the oldest exposed soil, which is much more mature than the younger soils. Our findings support MACHETTE (1985) who stated that stage four or higher carbonate-enriched horizons have formed since 500 ka or more. However, multiple cover beds older than the overlying, ca. 1.3 Ma old tephra have formed on top of the calcic horizon, each with its own (less mature) horizon of carbonate enrichment. Obviously, the formation of this calcic horizon does not cover the entire span of time since then. Rather, it may be assumed that its formation took place under environmental conditions different from those that formed the overlying layers and soils, possibly before the Quaternary. Such an intensity of pedogenesis is possible only during a long time of low geomorphic activity, especially because the preservation of calcic horizons indicates that precipitation was not exceptionally high during soil formation. The latter would have interrupted soil formation and would have induced its start-over in a new deposit. This long duration also points to a period preceding the frequent changes of conditions typical for the Quaternary.

The most surprising result of our study is the accuracy with which the provenance of the eolian components of some of the cover beds appears to be discernible. Especially both cover beds directly underlying the tephra may be traced back to a source area, i.e., the wider Grand Canyon area, and they are essentially not contaminated by grains from younger rocks. This is understandable by the fact that particles have lengths of trajectories in the air depending on their size and specific gravity - here the eolian matter is medium to short-travelled with diameters of 40-100 µm. A very constant direction of the geomorphologically active winds is another necessary condition for such a narrow definition of the source area, allowing rather precise reconstruction of winds and, alongside, of regional aspects of the atmospheric circulation. This in principle holds true for the underlying cover bed, which allows identifying a different contributing area and, thus, changes in wind direction or strength.

Summary information A. Supplementary data

Supplementary data[9] related to this article can be found at

https://doi.org/10.1016/j.quascirev.2018.01.012.

3.5 References

Applegarth, M.T., Dahms, D.E., 2004. Aeolian modification of moraine soils, Whiskey Basin, Wyoming, USA. Earth Surf. Proces. Landf. 29, 579-585. https://doi.org/10.1002/esp.1052.

Armstrong, R.L., 1969. K-Ar dating of laccolithic centers of the Colorado Plateau and vicinity. Geol. Soc. Am. Bull. 80, 2081-2086. https://doi.org/10.1130/00167606(1969)80[2081:KDOLCO]2.0. CO;2.

Aspler, Lawrence B., Wisotzek, Ira E., Chiarenzelli, Jeffrey R., Losonczy, Miklos F., Cousens, Brian L.,

McNicoll, Vicki J., Davis, William J., 2001. Paleoproterozoic intracratonic basin processes, from breakup of Kenorland to assembly of Laurentia. Hurwitz Basin, Nunavut, Canada. Sediment. Geol. 141-142. https:// doi.org/10.1016/S0037-0738(01)00080-X.

Bachman, G.O., Machette, M.N., 1977. Calcic Soils and Calcretes in the Southwestern United States. U.S, pp. 77-794. Geological Survey Report - Open file series.

Best, M.G., Christiansen, E.H., Gromme, S., 2013. Introduction. The 36-18 Ma southern Great Basin, USA, ignimbrite province and flareup: swarms of subduction-related supervolcanoes. Geosphere 9, 260-274. https://doi.org/ 10.1130/GES00870.1.

Birkeland, P.W., 1984. Soils and Geomorphology. Oxford University Press, New York.

---

[9] cf. chapter 6.2

Condie, K.C., Belousova, E., Griffin, W.L., Sircombe, K.N., 2009. Granitoid events in space and time: constraints from igneous and detrital zircon age spectra. Gondwana Res. 15, 228-242. https://doi.org/10.1016/j.gr.2008.06.001.

Cox, R., Martin, M.W., Comstock, J.C., Dickerson, L.S., Ekstrom, I.L., Sammons, J.H., 2002. Sedimentology, stratigraphy, and geochronology of the proterozoic mazatzal group, central Arizona. Geol. Soc. Am. Bull. 114, 1535-1549. https:// doi.org/10.1130/0016-7606(2002)114<1535:SSA-GOT>2.0.CO;2.

Crowley, J.L., Schmitz, M.D., Bowring, S.A., Williams, M.L., Karlstrom, K.E., 2006. U-Pb and Hf isotopic analysis of zircon in lower crustal xenoliths from the Navajo volcanic field: 1.4 Ga mafic magmatism and metamorphism beneath the Colorado Plateau. Contrib. Mineral. Petrol. 151, 313-330. https://doi.org/10.1007/ s00410-006-0061-z.

Dahms, D.E., 1993. Mineralogical evidence for eolian contribution to soils of late Quaternary moraines, Wind River Mountains, Wyoming, USA. Geoderma vol. 59, 175-196. https://doi.org/ 10.1016/0016-7061(93)90068-V.

Davis, S.J., Dickinson, W.R., Gehrels, G.E., Spencer, J.E., Lawton, T.F., Carroll, A.R., 2010. The Paleogene California River. Evidence of Mojave-Uinta paleodrainage from U-Pb ages of detrital zircons. Geology 38, 931-934. https://doi.org/ 10.1130/G31250.1.

Dickinson, W.R., Gehrels, G.E., 2008. Sediment delivery to the Cordilleran foreland basin: insights from U-Pb ages of detrital zircons in upper Jurassic and Cretaceous strata of the Colorado Plateau. Am. J. Sci. 308, 1041-1082.

Dickinson, W.R., Gehrels, G.E., 2009. U-Pb ages of detrital zircons in Jurassic eolian and associated sandstones of the Colorado Plateau: evidence for transcontinental dispersal and intraregional recycling of sediment. Geol. Soc. Am. Bull. 121, 408e433. https://doi.org/10.1130/B26406.1.

Dickinson, W.R., Lawton, T.F., Pecha, M., Davis, S.J., Gehrels, G.E., Young, R.A., 2012. Provenance of the Paleogene Colton Formation (Uinta Basin) and Cretaceous-Paleogene provenance evolution in the Utah foreland: evidence from U-Pb ages of detrital zircons, paleocurrent trends, and sandstone petrofacies. Geosphere 8, 854-880. https://doi.org/10.1130/GES00763.1.

Gehrels, G.E., Blakey, R., Karlstrom, K.E., Timmons, J.M., Dickinson, B., Pecha, M., 2011. Detrital zircon U-Pb geochronology of paleozoic strata in the Grand Canyon, Arizona. Lithosphere 3, 183-200. https://doi.org/10.1130/L121.1.

Hatcher Jr., R.D., 2002. Alleghanian (Appalachian) orogeny, a product of zipper tectonics: rotational transpressive continent-continent collision and closing of ancient oceans along irregular margins. Geol. Soc. Am. Spec. Pap 364, 199-208.

Hülle, D., Kleber, A., 2013. Chronology of periglacial cover beds. In: Kleber, A., Terhorst, B. (Eds.), Mid-latitude Slope Deposits (Cover Beds). Elsevier, Amsterdam, pp. 58-71 (Developments in Sedimentology 66).

IEDA Data Facility, 2017. EarthChem: geochron database (last retrieved 08-25-2017). http://www.geochron.org/detritalsearch.php.

Jones III, J.V., Connelly, J.N., Karlstrom, K.E., Williams, M.L., Doe, M.F., 2008. Age, provenance, and tectonic setting of Paleoproterozoic quartzite successions in the southwestern United States. Geol. Soc. Am. Bull. 121, 247-264. https:// doi.org/10.1130/B26351.1.

Kingsbury-Stewart, E.M., Osterhout, S.L., Link, P.K., Dehler, C.M., 2013. Sequence stratigraphy and formalization of the middle Uinta mountain group (neoproterozoic), central Uinta Mountains, Utah: a closer look at the western laurentian seaway at ca. 750Ma. Precambrian Res. 236, 65-84. https://doi.org/ 10.1016/j.precamres.2013.06.015.

Kleber, A., 1990. Upper Quaternary sediments and soils in the Great Salt Lake area, USA. Z. Geomorphol. N.F 34, 271-281.

Kleber, A., 1994. On the paleoenvironment of the northern Great Basin and adjacent Rocky Mountains. Z. Geomorphol. N.F 38, 421-434.

Kleber, A., 1997. Cover-beds as soil parent materials in midlatitude regions. Catena 30, 197-213. https://doi.org/10.1016/S0341-8162(97)00018-0.

Kleber, A., 2000. Compound soil horizons with mixed calcic and argillic properties examples from the northern Great Basin, USA. Catena 41, 111-131. https:// doi.org/10.1016/S0341-8162(00)00111-9.

Kleber, A., 2013. Heavy-mineral analysis as a tool in tephrochronology, with an example from the La Sal Mountains, Utah, U.S.A. Geologos 19 (6), 87e94. https://doi.org/10.2478/logos-2013-0006.

Kleber, A., Terhorst, B. (Eds.), 2013. Mid-latitude Slope Deposits (Cover Beds). Elsevier, Amsterdam (Developments in Sedimentology, 66).

Kleber, A., Leopold, M., Vonlanthen, C., Völkel, J., 2013. Transferring the concept of cover beds. In: Kleber, A., Terhorst, B. (Eds.), Mid-latitude Slope Deposits (Cover Beds). Elsevier, Amsterdam, pp. 171e228 (Developments in Sedimentology 66).

Klute, A. (Ed.), 1986, Methods of Soil Analysis. Part 1. Physical and Mineralogical Properties, second ed., vol. 9. Wisconsin: American Society of Agronomy Monograph, Madison.

Krautz, J., Hofmann, M., Gartner, A., Linnemann, U., Kleber, A., 2018. Capability of U-Pb dating of zircons from Quaternary tephra: Jemez Mountains, NM, and La Sal Mountains, UT, USA. E&G Quaternary Science Journal 67, 7-16. https://doi.org/ 10.5194/egqsj-67-7-2018.

Ludwig, K.R., 2012. User's Manual for Isoplot. Berkeley Geochronology Center Special Publication 5, 1-75.

Machette, M.N., 1985. Calcic soils of the southwestern United States. Geol. Soc. Am. Spec. Pap 203, 1-22. https://doi.org/10.1130/SPE203-p1.

McFadden, L.D., Tinsley, J.C., 1985. Rate and depth of pedogenic-carbonate accumulation in soils: Formulation and testing of a compartment model. Geol. Soc. Am. Spec. Pap 203, 23-42. https://doi.org/10.1130/SPE203-p23.

Page, A., Miller, R., Keeney, D. (Eds.), 1982. Methods of Soil Analysis. Part 2. Chemical and Microbiological Properties, second ed. Wisconsin: American Society of Agronomy and Soil, Madison.

Paillard, D., 2001. Glacial cycles: Toward a new paradigm. Rev. Geophys. 39, 325-346. https://doi.org/10.1029/2000RG000091.

Park, H., Barbeau Jr., D.L., Rickenbaker, A., Bachmann-Krug, D., Gehrels, G., 2010. Application of Foreland Basin Detrital-Zircon Geochronology to the Reconstruction of the Southern and Central Appalachian Orogen. J. Geol. 118, 23-44. https://doi.org/10.1086/648400.

Phillips, E.H., Goff, F., Kyle, P.R., McIntosh, W.C., Dunbar, N.W., Gardner, J.N., 2007. The 40Ar/39Ar age constraints on the duration of resurgence at the Valles caldera, New Mexico. J. Geophys. Res. 112. https://doi.org/10.1029/ 2006JB004511. B08201.

Rønnevik, C., Ksienzyk, A.K., Fossen, H., Jacobs, J., 2017. Thermal evolution and exhumation history of the Uncompahgre Plateau (northeastern Colorado Plateau), based on apatite fission track and (U-Th)-He thermochronology and zircon U-Pb dating. Geosphere 13, 518-537. https://doi.org/10.1130/GES01415.1.

Rosman, K.J.R., Taylor, P.D.P., 1999. 1997 report of the IUPAC Subcommittee for Isotopic Abundance Measurements. Pure Appl. Chem. 1999 (71), 1593-1607.

Ross, M.L., 2006. Preliminary Geologic Map of the Warner Lake Quadrangle. Salt Lake City: Utah Department of Natural Resources, Grand County, Utah.

Shroba, R.R., Birkeland, P.W., 1983. Trends in late-Quaternary soil development in the Rocky Mountains and Sierra Nevada of the western United States. In: Porter, S.C. (Ed.), Late Quaternary Environments of the United States, 1: the Late Pleistocene. Longman, London, pp. 145-156.

Simon, J.I., Renne, P.R., Mundil, R., 2008. Implications of pre-eruptive magmatic histories of zircons for UePb geochronology of silicic extrusions. Earth Planet Sci. Lett. 266, 182-194. https://doi.org/10.1016/j.epsl.2007.11.014.

Sircombe, K.N., 2004. AgeDisplay: an EXCEL workbook to evaluate and display univariate geochronological data using binned frequency histograms and probability density distributions. Comput. Geosci. 30, 21-31. https://doi.org/ 10.1016/j.cageo.2003.09.006.

Soil Survey Staff, 2014. Keys to Soil Taxonomy, twelfth ed. USDA-Natural Res. Conserv. Serv, Washington, DC.

Spell, T.L., Mark, H.T., Wolff, J.A., 1990. 40Ar/39Ar dating of the Bandelier Tuff and San Diego Canyon ignimbrites, Jemez Mountains, New Mexico: Temporal constraints on magmatic evolution. J. Volcanol. Geothermal Res 43, 175-193. https://doi.org/10.1016/0377-0273(90)90051-G.

USGS, 2017. Geologic Map of North America - South. https://ngmdb.usgs.gov/gmna/ gmna_images/gmna_South.jpg (last retrieved 08-27-2017).

Walsh, G.J., Aleinikoff, J.N., 1999. U-Pb zircon age of metafelsite from the Pinney Hollow Formation: Implications for the development of the Vermont Appalachians. Am. J. Sci. 299, 157-170.

Wotzlaw, J.-F., Schaltegger, U., Frick, D.A., Dungan, M.A., Gerdes, A., Gunther, D., 2013. Tracking the evolution of large-volume silicic magma reservoirs from assembly to supereruption. Geology 41, 867-870. https://doi.org/10.1130/ G34366.1.

Young, R.A., 2008. Pre-Colorado River drainage in western Grand Canyon: potential influence on Miocene stratigraphy in Grand Wash trough. Geol. Soc. Am. Spec. Pap 439, 319-334.

4 Zircon provenance of Quaternary cover beds using U-Pb dating: regional differences in the south-western USA

Chapter 4 is accepted in the peer-reviewed journal Earth Surface Processes and Landforms as:

Zircon provenance of Quaternary cover beds using U-Pb dating: regional differences in the south-western USA

Authors: Jana Richter-Krautz[1*], Mandy Hofmann[2], Johannes Zieger[3], Ulf Linnemann[2], and Arno Kleber[1]

[1] Geographisches Institut, Technische Universität Dresden, Helmholtzstr. 10, 01069 Dresden, Germany

[2] Senckenberg Naturhistorische Sammlungen Dresden, Museum für Mineralogie & Geologie, Sektion Geochronologie, Königsbrücker Landstraße 159, 01109 Dresden, Germany

[3] Senckenberg Museum für Naturkunde Görlitz, Am Museum 1, 02826 Görlitz, Germany

Publication history: received 1 June 2020 / revised 28 December 2020 / accepted 3 January 2021 / published 14 January 202

Full reference:

Richter-Krautz, J., Hofmann, M., Zieger, J., Linnemann, U., Kleber, A. Zircon provenance of Quaternary cover beds using U–Pb dating: Regional differences in the Southwestern USA. Earth Surf. Process.Landforms. 46: 968–989, 2021.

Internet link: https://doi.org/10.1002/esp.5073

4.1 Introduction

U-Pb dating of zircons is an established tool for decoding stages of regional geologic evolution because zircons form during magmatic events and may be overprinted by high-grade metamorphism but are resistant against anatexis. They are also durable if affected by chemical or physical treatment. Due to these properties, zircons redeposited by geomorphic processes, so-called detrital zircons (DZ), are also frequently used for provenance analyses of their host sedimentary rocks, utilising the "zircon facies" (LAMASKIN 2012) of their original source areas (LINNEMANN ET AL. 2011; DICKINSON ET AL. 2014; GÄRTNER ET AL. 2014; FENN ET AL., 2018; MCRIVETTE ET AL. 2019) and transportation regimes (NIEMI 2013).

There are plenty of DZ-based provenance studies from North American Archaean to Neogene rocks (e.g., CROWLEY ET AL. 2006; GEHRELS ET AL. 2011; LINK ET AL. 2014; KARLSTROM ET AL., 2018). Despite this, U-Pb-dating of DZ is not broadly established as a tool to reveal the provenance of Quaternary deposits. Quaternary fluvial deposits have been studied in the western USA (LINK ET AL., 2005; BERANEK ET AL. 2006; DICKINSON ET AL. 2014; KIMBROUGH ET AL. 2015). However, loess studies mostly focus on the Chinese Loess Plateau (STEVENS ET AL. 2010; VERMEESCH 2013; BIRD ET AL. 2015; NIE ET AL. 2015; LICHT ET AL. 2016; NIE ET AL. 2018; SHANG ET AL. 2018; SUN ET AL. 2018), whereas only a very few studies emphasize on European (ÚJVÁRI ET AL. 2012; 2013; ÚJVÁRI AND KLÖTZLI 2015) or North American Quaternary loess (GESLIN ET AL. 1999; LINK ET AL. 2005; ALEINIKOFF ET AL. 2008; MUHS ET AL. 2008; MUHS ET AL. 2013; KATSIAFICAS 2014; CONROY ET AL. 2016; AYERS ET AL. 2017). None of these are located in the arid to semi-arid south-western USA, except for KRAUTZ ET AL. (2018b) focusing on Early Pleistocene sediments.

Cover beds are common in mountainous environments of the mid-latitudes. They are assumed to be the most abundant surficial materials on slopes of low to intermediate gradient (KLEBER 1997; RAAB ET AL. 2007; KLEBER ET AL. 2013). Hence, they are a decisive component of "Earth's critical zone" with impact on pedogenesis or plant growth (KLEBER

AND TERHORST 2013). Cover beds are defined as sediments formed by unconcentrated dislocation processes chiefly from upslope materials, which may contain admixed eolian matter. They cover slopes largely, rather than being restricted to drainage ways or local failures. They typically consist of layers, separated by disconformities (KLEBER AND TERHORST 2013). Surficial substrates in and in the vicinity of the northern Great Basin and the Colorado Plateau typically contain eolian particles (CHADWICK AND DAVIS 1990; KLEBER 1993; PAVICH AND CHADWICK 2003; REYNOLDS ET AL. 2006; JACOBS AND MASON 2007).

Single-grain provenance tracing has significant advances over geochemical analyses of bulk samples as it differentiates sedimentary components and is largely unaffected by weathering. This approach is increasingly applied in loess research. To our knowledge, DZ studies have not yet been applied to slope deposits with admixed aeolian fines. Here we present a pilot study evaluating the feasibility of provenance analysis to the aeolian component of cover beds.

4.2 **Materials**

4.2.1 **Study areas**

The northern Great Basin Desert comprises a great number of endorheic basins in the southwestern United States within the Basin and Range Province. It encompasses broad desert-basins separated by horsts and half-horsts, most of which form elongate, south-north trending mountain ranges. Those ranges mainly consist of allochthonous terranes accreted mainly during the Paleozoic, and their derivative deposits. The basins usually are not connected but divided by passes; accordingly, there are only a few rivers of noteworthy length (Fig. 4.1). During the Quaternary, the highest mountains fed mountain glaciers. The foot zones were affected by pedimentation and fan deposition. Most basins were covered by paleolakes during times of global glaciations (Fig. 4.1). Those episodes of greater moisture also induced the formation of soils, the most mature of which in the mountain foot zones and basins would now be classified as Calciargids (soil terminology in the present text follows Soil Survey Staff, 2014).

The Colorado Plateau is an elevated plateau with isolated ranges of Paleogene volcanic intrusions mainly as laccoliths. Its basement was chiefly formed in the Paleoproterozoic as a series of collisions of island-arc terranes. This basement is covered by a thick sedimentary sequence dominated by sandstones of Paleozoic to Mesozoic age. The plateau was relatively stable during the tectonic events deforming all its surrounding provinces so that a wide cuesta landscape could develop (cf. BAARS 2000; FILLMORE 2011). The usually deeply incised Colorado River and its tributaries dominate the area. Main aspects of Quaternary research are river terraces, landslides, and pediments. The study of soils of the Colorado Plateau had a focus on dust incorporation (e.g., REYNOLDS ET AL. 2006), but in our study we found that the most mature soils are Calciargids as in the Great Basin.

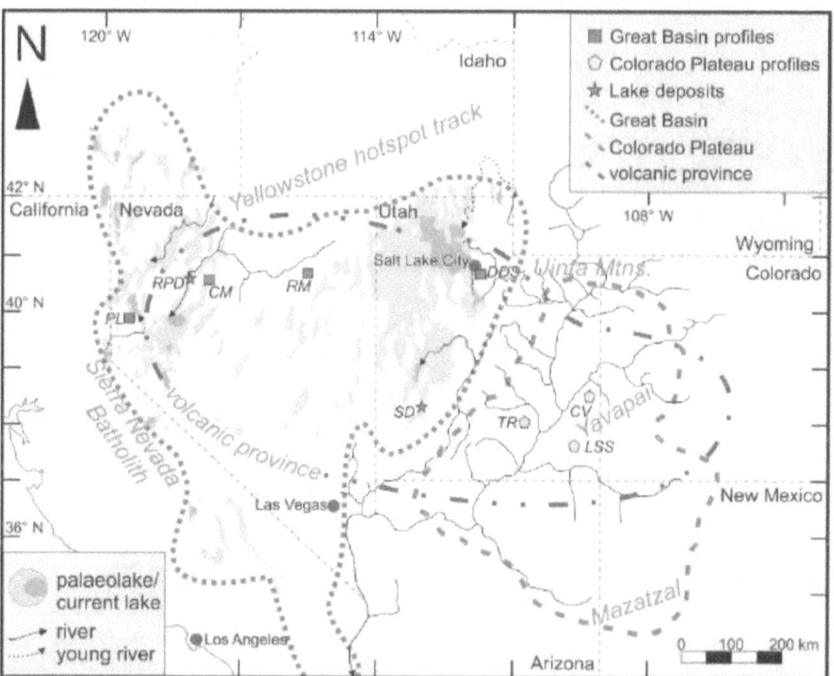

Fig. 4.1 - Studied sites and extent of Great Basin and Colorado Plateau in the southwestern USA. Approximate positions of geologic provinces mentioned in the text are printed in grey. Palaeolakes in the Great Basin catchment and modern (including ephemeral) lakes are redrawn from Oviatt (2016). Young rivers took their present course only after the deposition of both studied cover-bed layers (Truckee River draining into Pyramid Lake) or after the deposition of the LWL (Bear River draining north of Salt Lake City into the Great Salt Lake).

4.2.2 Stratigraphy and sampling sites

This study utilizes the recurring record of three major phases when slope deposits were laid down: KLEBER (1993) describes three layers, which are each separated from what is

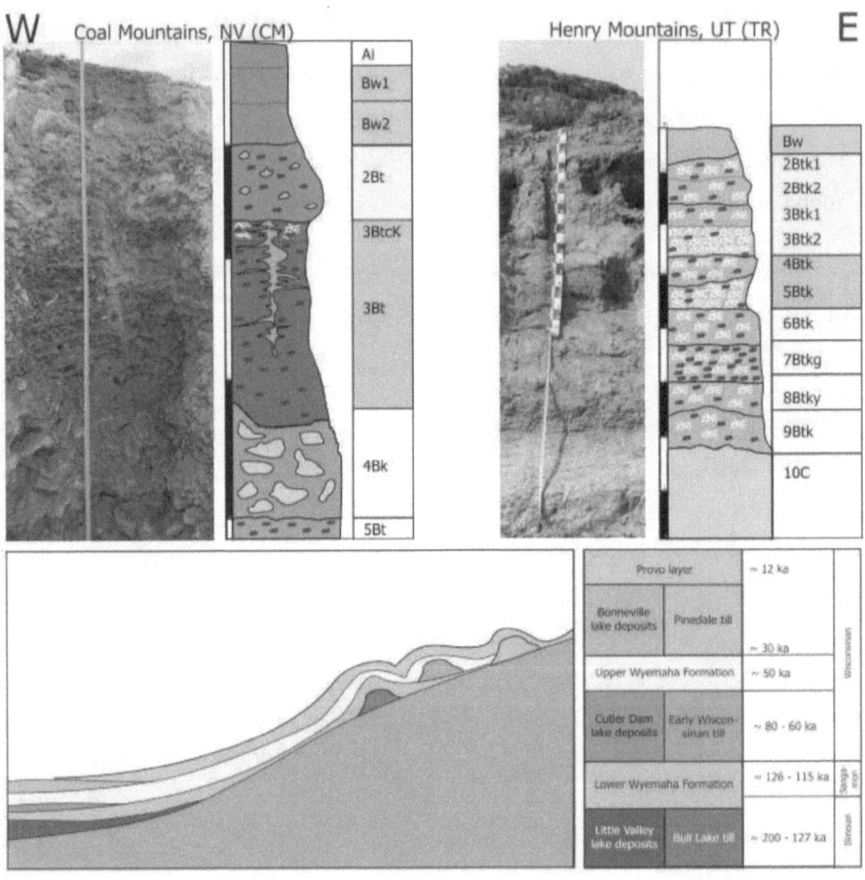

Fig. 4.2 - Background of the sequence stratigraphy of cover beds (bottom sketch) and two exemplary profiles (top sketches). Sketches of all profiles of this study are found in the Supplementary Material (SI2). The stratigraphy is based on the interfingering of cover beds with lake deposits (standard nomenclature for the Bonneville Basin; Sack, 2020) and tills (standard nomenclature for the Rocky Mountains; Pierce, 2003). The profile sketches show the profiles CM (Great Basin) and TR (Colorado Plateau). Please note the different scales of the profile sketches. The layers denoted by yellow and green colours in all parts of the figure are in the scope of our study. The profile sketches reflect relative hardness under field conditions and the colours are an abstraction of the natural colour.

underlying by a disconformity. The term "cover bed" has been proposed for such deposits (KLEBER 1990; KLEBER AND TERHORST 2013). Each cover bed of this area contains aeolian matter (evidenced by coarse silt and by allochthonous heavy minerals, KLEBER 1993).

In the south-western USA palaeolakes (Fig. 4.1) have left sediments. Relations to dated lake deposits and shorelines of these palaeolakes as well as to Rocky Mountain moraines allow to constrain the ages of sediments and soils (Fig. 4.2): the youngest layer (Provo layer) is developed on certain shorelines and on moraines of the last glacial maximum and, thus, assumed to be 14,000-13,000 years old (KLEBER ET AL. 2013). It bears a soil of intermediate maturity (Provo soil) formed after its deposition (KLEBER 1993). Two other layers (upper and lower Wyemaha layers, UWL and LWL) were deposited between lake deposits (Fig. 4.2) and, thus, have formed after the penultimate glaciation but prior to the last glacial maximum. They are separated by a disconformity or by intercalated other sediments. The maximum age of the UWL is limited by (1) traces of a Mt. St. Helens tephra dated at ca. 50 ka (BERGER 1991) dispersed in the layer in many cases (KLEBER ET AL. 2013), (2) a palaeolake deposit of assumed Early Wisconsinan age occasionally intercalated between LWL and UWL (MORRISON 1964; KLEBER ET AL. 2013), and (3) by underlying deposits of the Bear River, which has been diverted into the Great Basin at around 50 ka (BOUCHARD ET AL. 1998). Furthermore, to the south-east of our study area there is a palaeosol numerically dated at 50 – 18 ka by limiting deposits (HALL AND GOBLE 2012).

Contrary to most occurrences of cover beds in Central Europe, KLEBER (1997) shows that a chrono-stratigraphy of cover beds may be contained, if they intercalate with datable deposits or landforms, or if they contain discernible palaeosols. Since cover beds intercalated between lake deposits or overlying moraines fulfil the first, and all observed instances the second condition, a sequence-stratigraphic approach (CATUNEANU 2019) allows us tracing the cycles of deposition and soil formation to areas outside the former lakes and moraines (KLEBER 1993; KLEBER ET AL. 2013). The principle of sequence stratigraphy is derived from geological basin research and states that the succession of events throughout basins or even larger areas is mainly driven by environmental changes.

Basically, to establish an area-wide stratigraphy of the cover beds, we assume that an equivalent sequence of layers and soils between basins and mountains is a response to climate changes that were approximately contemporary throughout the entire study area. Utilising palaeosols as stratigraphic markers, such approaches have also been successfully applied to other cases (e.g., AMOROSI ET AL. 2017). Since our approach has chronological anchors at the western and eastern borders of the Great Basin as well as in the Rocky Mountains, our approach basically is an interpolation, suggesting validity of the age suggestions over a large area.

The main indicators of recurrent unconformities are calcic horizons engulfing argillic ones, typically welding the upper to the lower soil. The carbonate must have been engulfed after the argillic properties had been formed, because carbonate enrichment and clay illuviation cannot occur simultaneously in the same horizon, and clay translocation is only possible during or after the carbonate has been depleted (KLEBER 2000, 2013; KRAUTZ ET AL. 2018b). Composite horizons with pedogenic carbonate are separators of welded, multistoried paleosols, indicated by, e.g., radiocarbon datings (DEUTZ ET AL. 2001), and they are often found as uppermost horizons of argillic palaeosols worldwide (BRONGER 2003; HUANG ET AL. 2003; BELLOSI AND GONZÁLEZ 2010; PRESLEY ET AL. 2010; BAYAT ET AL. 2017). In this study, composite horizons were used to differentiate between both the Wyemaha layers (cf. KLEBER 2000; fig. 4.2). In some cases, particular layers consisted of two or three layers with abrupt textural contacts separating them, but without similarly abrupt soil horizon changes, indicating that these layers had been affected by the same pedogenic phase and do not establish a different stratigraphic cycle.

For this study, we sampled three profiles in the Colorado Plateau, four in the Great Basin, and we took three samples from lake deposits from two sites (Tab. 4.1). The site LSS (La Sal Mountains South) is located on the north-eastern footslope of the Abajo Mountains, UT, upon Late Cretaceous Dakota sandstone. TR (Trachyte Ranch) is on a pediment surface at the eastern flank of the Henry Mountains, UT, underlain by mudstones of the mid-Jurassic Summerville Formation. CV (Castle Valley) in located on alluvium relative-dated to the penultimate glacial epoch (KLEBER 1999) with mixed clasts from Mesozoic

Tab. 4.1 - Sample sites of cover beds and lake deposits.

Site	state	acronym	coordinates North	West	elevation
Colorado Plateau					
Abajo Mountains	UT	**LSS**	37°55'59.7"	109°27'15.1"	2391 m
Henry Mountains	UT	**TR**	37°58'02.6"	110°37'33.5"	1537 m
Castle Valley	UT	**CV**	38°35'42.5"	109°17'58.8"	1881 m
Great Basin					
Dimple Dell Creek	UT	**DDS**	40°33'29.7"	111°49'17.1"	1450 m
Ruby Mountains	NV	**RM**	40°48'24.4"	115°22'28.5"	1723 m
Coal Mountains	NV	**CM**	40°10'22.8"	118°13'29.2"	1342 m
Pyramid Lake	NV	**PL**	39°50'37.0"	119°27'14.6"	1182 m
Lake deposits					
Rye Patch Dam	NV	**RPD**	40°28'06.3"	118°18'12.3"	1250-1270 m
Sevier Desert	UT	**SD**	38°46'25.7"	112°55'51.8"	1204 m

sandstones and Paleogene rhyolite. DDS (Dimple Dell Soil) is located in Dimple Dell Creek Canyon close to the eastern edge of the Great Basin and is intercalated between lake deposits of the last two major lake highstands of Lake Bonneville (MORRISON 1964; MCCOY 1987; KLEBER 2000). The site RM (Ruby Mountains) is on the western footslope of the Ruby Mountains, NV, on Paleogene felsic igneous tuff. CM (Coal Mountains) is positioned on a pediment surface within the Coal Mountains, NV, and is underlain by gravel derived from Upper Triassic to Lower Jurassic sandstone.

4.2.3 Palaeolake deposits

Lake Lahontan, a nested system of lakes in the northwestern Great Basin, and Lake Bonneville, a large homogenous lake in the northeast, both aging 12 – >155 ka, with various highstands, were the most prominent palaeolakes in this area (Fig. 4.1). They may serve as potential zircon interim reservoirs. Therefore, we also sampled palaeolake deposits. These are the sample abbreviated SD (Sevier Desert) located in the Black Rock Desert, UT, from the Late Wisconsinan Bonneville lake highstand (OVIATT 1988). Sample RPD2 (Rye Patch Dam) is from an unnamed lake deposit between two colluvial layers (MORRISON 1964), probably equivalent to the Early Wisconsinan Cutler Dam lake advance of Lake Bonneville (KLEBER 2000); and RPD3 is from so-called Eetza deposits equivalent to the Illinoian Little Valley highstand (MORRISON 1964; MCCOY 1987; KLEBER 2000).

4.2.4 Potential sources of detrital zircons

Several potential source areas share DZ age peaks, causing ambiguity over source assignments. However, the ever-changing geologic history of North America reflects in various remarkable differences in age spectra of zircons, especially in the basement rocks. Many areas in the region yield "fingerprints", i.e., clusters of ages specific to just this area but not or not to the same extent to other areas, making them potential source indicators.

The youngest zircons may stem from the area under study itself. Volcanism in the Paleogene affected this (cf. fig. 4.1), including some of the most voluminous eruptions ever reconstructed. JENSEN ET AL. (2020) summarize the current knowledge: Various volcanic fields formed during different periods. Eruptions yielding the largest volumes successively increase in age from the south to the north. The activity started north of our studied sites mainly in Idaho at 45 – 51 Ma. At 39 – 43 Ma it reached the latitude of our Great Basin sample sites. RM is the site closest to one of the affected areas; the last eruptive activity is dated approximately 37 Ma there (COLGAN ET AL. 2010; LUND SNEE ET AL., 2016). 34 Ma is the youngest reported age at this latitude (JOHN ET AL. 2008; CANADA ET AL. 2019). Further south the eruptions occurred at 23 – 35 Ma (BEST ET AL. 2009) and 18 – 36 Ma (BEST ET AL. 2013) in large super-volcanic caldera complexes. DZ of 19 Ma are also frequent in fluvial deposits of the lower reach of the Colorado River (DICKINSON ET AL. 2014). In southernmost Nevada eruptive ages are even younger (8-18 Ma, COLOMBINI ET AL. 2011). This volcanism affected the Colorado Plateau as well. Some eroded laccoliths such as the La Sal Mountains (29 Ma, RØNNEVIK ET AL., 2017) and the Henry Mountains (23-31 Ma, WILSON ET AL. 2016), each close to sites of this study, belong to it. However, the most extreme volcanism has been reconstructed mainly in the west (22-30 Ma, HACKER ET AL. 2018) and east (28-37 Ma, MALFAIT ET AL. 2014) of the plateau. Even younger zircons may be derived from the hotspot spur of the Yellowstone volcanism starting in the western Snake River Plain (Oregon) at 17 Ma and ending with the Quaternary super volcanoes of the Yellowstone area (BERANEK ET AL. 2006). Zircons re-deposited from these eruptions have been detected in Quaternary river sediments of the Snake River and have already been used as accurate tracers of their respective sources (LINK ET AL. 2005).

The amalgamation of modern North America ended with the orogenic events that formed the continent's western margin. These started at around 380 Ma probably not connected with magmatism (non-collisional Antler orogen underlying parts of Nevada, BERANEK ET AL. 2016). The formation of the Cordilleran magmatic arc system started at around 250 Ma with a distinctive peak in the zircon record some 25 myr later; a second, even larger peak occurred at 165 Ma; finally, the main body of the Sierra Nevada batholith formed (LACKEY ET AL. 2012; PATERSON AND DUCEA 2015; DEGRAAFF SURPLESS ET AL. 2019). This latter, most productive phase peaked 85-95 Ma in the center of the mountain range (MEMETI ET AL. 2010). Further south zircon ages start at 135 Ma, peak at 105 – 110 Ma (MALKOWSKI ET AL. 2019), and last until 84 Ma (SALEEBY ET AL. 2008). The latter age also depicts the final stage of the batholith intrusion at the eastern margin of the Sierra (SALEEBY ET AL. 2008).

The eastern parts of the continent have been amalgamated around an Archean cratonic core during several orogenic phases adding belt by belt (WHITMEYER AND KARLSTROM 2007). The youngest of these is the multiphasic orogeny of the Appalachian mountain complex, which, including the Carolina-Suwanee terranes, lasted 260 – 735 Ma (HATCHER 2010).

The Grenville orogen is the next older element of the amalgamation of North America. It formed 1.0 – 1.2 Ga with a peak at 1.1 Ga (WHITMEYER AND KARLSTROM 2007). This was preceded by the accretion of the anorogenic Granite-Rhyolite Province 1.35 – 1.55 Ga (BICKFORD ET AL. 2015). Rocks of this age are found south of our studied area in Nevada, Arizona and Utah (ANDERSON AND BENDER, 1989; DOE ET AL., 2013). Before this, two series of terranes had become part of the continent, Mazatzal (1.6 – 1.69 Ga) and Yavapai (1.68 – 1.76 Ga) which also underlie the Colorado Plateau (Fig. 4.1; WHITMEYER AND KARLSTROM 2007). Zircons older than 1.8 Ga possibly are derived from North American cratons, whose formation ended with the Trans-Hudson orogeny. The closest outcrops are found in the north-east of the study area and are frequent southward until the Uinta Mountains (LAWTON ET AL. 2010).

Identifying possible cascades of transportation is urgently needed for provenance analyses (ZIEGER ET AL. 2020). It is known that DZ derived from the cratons to the north and the belts in the east and south east of our study area had been reworked by large river systems until they were deposited somewhere north of our study area (RAHL ET AL. 2003; DICKINSON ET AL. 2010). During Jurassic times aeolian processes deflated those deposits until they covered most of the Colorado Plateau (DICKINSON AND GEHRELS 2003, 2009b, 2009a). These aeolianites later were reworked into Cretaceous sediments (DICKINSON AND GEHRELS 2008a). The latter two cover much of the Colorado Plateau. This has the effect that all these DZ age groups from the eastern belts may be found there.

River deposits prone to deflation during dry seasons are supposed to be the major source of aeolian fines in many areas (BUGGLE ET AL. 2008; SMALLEY ET AL. 2009; NIE ET AL. 2015; 2018). The northern Great Basin consists of a system of nested endorheic basins. Most of them are not connected by any river system. Water mainly flows from the adjacent mountains into basins, which occasionally used to host lakes since the Neogene. Water may reach central parts of the Great Basin only by overflow from one basin into neighboring ones (BENSON ET AL. 1990; ADAMS 2010; REHEIS ET AL. 2014). This interrupts fluvial chains of transportation at each basin rim. Besides ancient Lake Lahontan this holds true for Lake Bonneville (Fig. 4.1; OVIATT 2015; SCHIDE ET AL. 2018). The southern segment of the latter, however, is the only one where DZ from the Colorado Plateau may have been deposited fluvially via the Sevier River.

Drainage has been oriented from north and north-east to the south-west at least since Eocene, possibly Paleocene times in the northern Colorado Plateau (DAVIS ET AL. 2010; ROSKOWSKI ET AL. 2010; WERNICKE2011) with the only exception of the Sevier River draining the westernmost border of the Colorado Plateau. Thus, fluvial transportation may carry northern DZ to the south, but not vice versa. Therefore, DZ evidence of southern sources would be indicative of southern, non-fluvial transportation directions in this area.

It has been doubted, whether the intrusion of the various generations of the Sierra Nevada batholith was accompanied by voluminous volcanism (DAVIS ET AL. 2012), contradicted by DUCEA ET AL. (2015). However, DZ from the Sierra Nevada batholith have been found even as far as the Colorado Plateau (DICKINSON AND GEHRELS 2010). Between this source area and the Colorado Plateau there was a high plateau at that time

(BEST ET AL. 2009; SNELL ET AL. 2014) forming a continuous divide (HENRY 2008), thus excluding transportation other than through the air.

4.3 Methods

4.3.1 End-member modelling of grainsize composition

Grainsizes were measured with a Retsch HORIBA LA-950® applying the Mie theory at SedLab, GfZ, Potsdam. Altogether 49 samples were measured from cover beds, dune sands and lake deposits, more than could be analysed for U-Pb dating. Only the PL profile was not included, which had no material left for grainsize analyses. Since these additional samples are from different sites and were only used for the model development but not yet for DZ analyses, they are not reproduced here.

End-member modeling is an option to decompose multi-modal grainsize distributions into sub-populations, so-called members. It is a tool to infer types of geomorphic genesis, especially transportation modes (DIETZE ET AL. 2012; DIETZE & DIETZE 2019). The different grainsizes require different energy levels, which allow drawing conclusions on transportation modes. Members are hypothetical grainsize distributions. All members together explain the entire given set derived from all samples of grainsize data; each particular member is interpreted to represent a different subpopulation in the grainsize dataset. Here we report a summary end-member model of most samples to examine the relative importance especially of aeolian contribution using the R-based tool EMMAgeo by DIETZE AND DIETZE (2019).

4.3.2 U-Pb dating

The complete sample preparation, SEM 3D backscattered electron (BSE) and cathodoluminescence (CL) images, Laser Ablation with Inductively Coupled Plasma Mass Spectrometry (LA-ICP-MS) U-Pb analyses, and age calculations were performed at the Geochronology Dept. of Senckenberg Naturhistorische Sammlungen Dresden (Germany). We collected around 1 kg of material for each sample and sieved each for the fraction 36-

400 µm. We accomplished the density separation of this fraction with LST (solution of lithium heteropolytungstates in water, d = 2.80 - 2.85 g/cm³). Then we used a Frantz isodynamic separator for the magnetic separation of the extracted heavy minerals. Single zircon grains were manually picked under a binocular microscope. We selected zircon grains of all grainsizes, colors, and morphological types and analyzed them regarding their morphology based on 3D BSE imaging using a Zeiss EVO50 SEM. Then we measured every single grain with Digital Image Processing System 2.9 ©. After this, the zircon grains were mounted in epoxy resin blocks (i.e., mounts) and then polished to half their thickness to expose their internal structure. We obtained CL images using the Zeiss EVO50 SEM coupled to a CL and a VPSE detector system. After these preinvestigations, the zircon grains were analyzed for U, Th, and Pb isotopes by LA ICP-MS, using a Thermo-Scientific Element 2 XR sector field ICP-MS coupled to an asi RESOlution 193 nm Excimer Laser System with laser spot sizes of 20-35 µm. During each analysis, 15 s of background acquisition were followed by 30 s of data acquisition. The signal was tuned for a maximum sensitivity for Pb and U, while keeping oxide production ($^{254}UO/^{238}U$) well below 1 %. Raw data were corrected for background signal, common-Pb, laser induced elemental fractionation, instrumental mass discrimination, time- and depth-dependent elemental fractionation of Pb/Th and Pb/U, using an Excel® macro developed by Dr. Axel Gerdes and Dr. Richard Albert Roper (Geosciences Inst., Goethe University Frankfurt a. M., Germany). For ages younger than 10 Ma, we corrected for ^{230}Th disequilibrium according to Horstwood et al. (2016). Reported uncertainties were propagated by quadratic addition of the external reproducibility obtained from the standard zircon GJ-1 (~0.6% and 0.5-1 % for $^{207}Pb/^{206}Pb$ and $^{206}Pb/^{238}U$, respectively) during individual analytical sessions and the within-run precision of each analysis. Concordia diagrams (2σ-error ellipses) were created and concordia ages (95 % confidence level) were calculated with Isoplot/Ex 2.49 (LUDWIG 2001). $^{207}Pb/^{206}Pb$ ages were used for zircons above 1.0 Ga, $^{206}Pb/^{238}U$ ages for younger ones. We applied this 1.0 Ga limit to data from other laboratories used for multidimensional scaling (see below) as well, even in cases when the original authors had used other limits (e.g., 1.2 Ga in BARTSCHI ET AL. 2018), just to keep the data as comparable as possible. We determined the age distribution of all grains that are concordant within ±10 %. Given the relatively young age of some zircons, only the

$^{206}Pb/^{238}U$ ratios provided reliable ages, whereas age calculations based on the $^{207}Pb/^{235}U$ ratios often could not be used due to the long decay period of ^{235}U. Therefore, the degree of concordance could not be calculated for the young zircon grains and is left clear in the table (Supplementary Material, SI1, chapter 6.3). Because ages below 120 Ma turned out to be especially valuable provenance indicators in this study, though they rarely were concordant, we included these ages and applied the requirement of concordance to older grains only. Nevertheless, all analyses resulting in a young age (<120 Ma) were calculated and interpreted considering that the majority of these ages are based solely on their $^{206}Pb/^{238}U$ ratio and the backup of these ages by a $^{207}Pb/^{235}U$ and/or $^{207}Pb/^{206}Pb$ age very often is not given. Without the possibility to calculate the degree of concordance, a control over the validity of the calculated age is not given. To minimize "wrong" ages, all analyses were disregarded for further interpretations, if inconsistent isotope signals were recorded. In addition, we relied only on young ages that are known to exist in the study area already. This knowledge is mainly based on rare findings of respective near-concordant young zircon ages in the unity of all analyzed sediments of this study and of the studies of KRAUTZ ET AL. (2018a; 2018b). Furthermore, repeated appearance of the same (young) age is another indication for the validity of this specific age.

4.3.3 Zircon dimensions and surfaces

The morphological classification of grains may be a correlation tool and yield insights into the dynamics of sediment transportation and recycling (SHAANAN AND ROSENBAUM 2018). The distance and/or intensity of transportation is indicated by the degree of roundness of the grains, whereas the number of steps of the transportation cascade is assumed to correlate with collision marks. Nevertheless, the surface and roundness of a DZ do not tell the number of transportation steps and their timing on their own. The combination of the size of a grain (energy level for transportation), its surface structures (degree of roundness, collision marks), and its age (probable proto source), allows solid provenance studies with zircons. We argue about grainsize with some caution because some zircons exhibit broken edges. This may have occurred to the grain not only before or during transportation to the final deposition but also thereafter due to weathering. The latter would render the size of that grain useless for estimating transportation energy. On the other hand, due to the physical robustness of zircons and the fact that unweathered rock fragments are frequent in our samples, we assume this effect to be small. Following GÄRTNER ET AL (2013) and ZIEGER ET AL. (2019), zircons may be differentiated into ten classes of roundness and 4 classes of collision marks. Roundness alone is not a biunique indication for transportation; by geochemical processes, e.g., recrystallization, may cause it as well.

4.3.4 Statistical and graphical representations

Provenance studies require graphical representations to illustrate and discuss the age distributions. Two types of relative-probability distributions are commonly plotted for visual comparison of age distributions. Probability density functions (PDF) are constructed from summing a set of distributions using as standard deviation the 1σ analytical uncertainty of the analysis (VERMEESCH 2012). Kernel density estimations (KDE) are computed similarly but use a fixed bandwidth rather than the analytical uncertainty (VERMEESCH 2012). It became clear in our study that neither variant was completely unambiguous, because PDF overemphasize measurements with a small analytical error, whereas KDE tend to envelop neighboring distinct peaks (cf. MCRIVETTE ET AL. 2019). Therefore, we use both KDE and PDF representations side by side, though there is some concern regarding the theoretical foundation of PDFs (VERMEESCH 2012). However, PDFs still are frequently used in the DZ literature especially in North America (recently e.g., BARTSCHI ET AL. 2018) and elsewhere (e.g., EIZENHÖFER 2020), so that our representations may be compared to others. Because we found huge differences between both areas under study, we had to plot the density functions in different ways: For age distributions in the northern Great Basin samples, we provide a split PDF for ages greater and smaller than 120 Ma. We applied an Epanechnikov kernel to the KDE and subjected the data to a square-root transformation (VERMEESCH 2012; SPENCER ET AL. 2017), using the software Radialplotter© (VERMEESCH 2009). For the Colorado Plateau data, we plot both types of density functions on a linear scale using Radialplotter© for the

PDF and the R-based tool Provenance for the KDE (VERMEESCH 2020). Both types of probability distributions were normalized as to the curve integrals.

Multidimensional Scaling (MDS) is an appropriate tool for evaluating and visualising how similar or dissimilar samples are. Pairwise distances are calculated, and the similarity of individual samples is depicted as points within an abstract Cartesian space (VERMEESCH 2013). We used the R-based tool Provenance for rendering (VERMEESCH 2020), applying the distance proposed by SIRCOMBE AND HAZELTON (2004). For comparison with our data from the Colorado Plateau, we produced composite datasets from published data derived from a single lab (Arizona LaserChron Center, https://sites.google.com/laser-chron.org/arizonalaserchroncenter) to reduce effects of analytical differences.

We illustrate relationships between age, axis length, roundness, and collision marks using scatter plots. Two types of plots are used. One depicts almost all measurements from a single sample and uses pinhead color as indicator of surficial collision marks. The other focuses on particular age ranges and includes all samples from either region. The latter indicates the region by the pinhead color and surface marks by the pinhead diameter.

4.4 Results and discussion

4.4.1 Aeolian contribution to cover beds

In earlier work based on texture and heavy-mineral analyses, KLEBER (1993, 2013) concluded that the cover beds contain aeolian compounds in remarkable quantities (also cf. CHADWICK AND DAVIS 1990; LAWRENCE ET AL. 2011). We used end-member-modeling of grainsizes of most samples to test these findings.

End-member modelling separates a convolute of samples into various members that are defined by the appearance of the different grainsizes. The members 1 and 2 (Fig. 4.3) are interpreted to represent lake deposits; but they also occur in samples close to former lakes as indicated by the secondary peaks which are in accordance to the maxima in the other members (grainsizes of individual samples are given in the Supplementary

Materials, chapter 6.3). The members 5 and 6 have their major peaks in the sand fraction and, thus, are assumed to be associated mainly with regional rocks or aeolian dune deposits but also have secondary contributions near the class 55. The modeling confirms that there is an important aeolian contribution to the samples within a narrow band of textures around class 55, i.e., 65 μm (end member 4 and secondary maxima of members

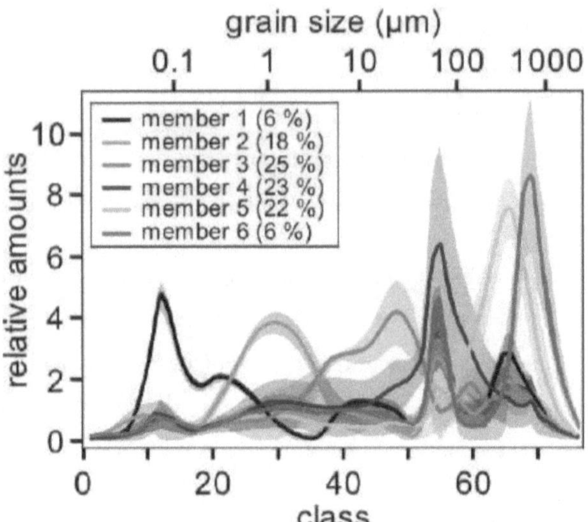

Figure 4.3 - Robust end-member loadings computed from fine-earth grainsize distributions of 49 samples from cover beds and lake deposits in the Great Basin and Colorado Plateau. Shading indicates error margins. Percentages depict the explained variances.

6 and 3) in the typical range of short-travelled, suspended aeolian matter. Furthermore, member 3 with a somewhat wider range around group 50 (~30 μm) is assumed to denote far-travelled aeolian matter because of the finer grainsizes. The modelled aeolian grainsize distribution is on par with measured dust grain sizes slightly east of our study area (LAWRENCE ET AL. 2011).

In a previous study, KRAUTZ ET AL. (2018b) analyzed cover beds of the Colorado Plateau, the underlying and upslope bedrock of which was Paleogene rhyolite, as to the DZ provenance of aeolian components. These sediments were dated older than 1.3 Ma. Accordingly, they are remarkably older than the sediments of the present study and, thus, had been subjected to weathering several times longer. Despite of this long time and the fact that clasts from the rhyolite were ubiquitous in the sediments, just one out of 175 grains had an age corresponding to the age of the local bedrock. All other DZ probably were allochthonous, i.e., aeolian. This finding suggests that even more than 1 myr of weathering was insufficient to separate zircons from rock clasts in relevant amounts, although HECKMAN AND RASMUSSEN (2011) showed that rhyolite weathers faster than other magmatic rocks under semi-arid climate. Most rocks that could have contributed zircons to the samples of the present study are more easily weathered (mainly sandstones). However, given the much shorter time available for weathering, we assume that most zircons of the present study have not been derived from local bedrock or gravel but are allochthonous as well, because all layers contain conspicuous amounts of unweathered bedrock clast. Furthermore, end members 5 and 6 are dominated by grainsizes beyond even the largest zircons in our data (see fig 4.4 and 4.6 below). Smaller grains would not at all appear in our analyses because of the grainsize separation during sample preparation with a lower limit of 36 µm.

An additional argument against a decisive share of *in-situ* zircons in our sediments comes from the results of DZ dating (see below): Due to the very different bedrock geologies of our sites, especially in the Great Basin, the variety of the zircon spectrum would have to differ much more from each other, if it were decisively local borne. Though, comparison to local bedrock geology should be in the scope of further research.

As shown by the end-member modeling, the grainsizes which are not air-borne but from the local substrates are finer (lake deposits) or coarser (sandstone, granite) than the members 3 and 4, and the secondary peaks of 6 and 5. For these non-aeolian grains, transportation-induced winnowing due to the different density of zircons compared to other minerals would not play a noticeable role. Members 3 and 4, assumed to be aeolian, are present in all individual samples (Supplementary Material, chapter 6.3). All samples also have major grainsize maxima derived from local materials. However, these latter are, as in the EMMA model, either coarser or finer than essentially all zircons in our analyses (see below).

4.4.2 Zircon morphology

We interpret the morphologic properties of DZ grains as follows: Idiomorphic crystals with few or no collision marks are supposed to be deposited close to their magmatic or metamorphic origination area. Their transportation distance is short. Grains that are well rounded and have many to numerous collision marks are regarded to indicate a long transportation pathway, which includes an aeolian transportation process. This is because aeolian transportation of small grains in suspension is more abrasive than fluvial: due to the buoyancy in water, mass of the grain is smaller than in the air, and kinetic energy is reduced by the resistance posed by the water (MENCKHOFF 2010). In short, we will refer to those grains as assumed aeolian grains. Grains of medium roundness and low collision marks may have been transported by wind as well as by water. In the study area to our knowledge no research has been conducted on morphological characteristics of bedrock zircons. Therefore, we cannot attribute the reconstructed pathways of the DZ to the last transportation processes to the current deposition site because there is no clue differentiating the grain properties from properties inherited from parent rocks.

Figure 4.4 depicts the relationships between morphological parameters and age (see next section) of Great Basin samples. PL almost exclusively contains young grains. Among these many are idiomorphic or nearly so (low roundness groups) and poor in collision marks. On average, these have a slightly greater size than the group with clear indication of transportation provided by collision marks and roundness. Most grains above roundness class 5 have many or numerous collision marks making an important contribution of aeolian transportation to their morphology likely. In CM, almost no grains, even the idiomorphic, are without collision marks, and assumed aeolian grains are numerous. We interpret this that all grains are far from their primary deposition. This holds true for all age classes, though the grains with probable aeolian properties are older on average. In the RM sample there are some assumed aeolian and some angular grains with no or little collision marks, but the majority has been transported without clear process attribution. In DDS there are only very few older grains, some of which appear aeolian. Among the group of young grains, there are only a few fully angular, but many show no or only a few marks. Only with higher roundness collision marks and, thus, indication of transportation increase.

Upper Wyemaha layer

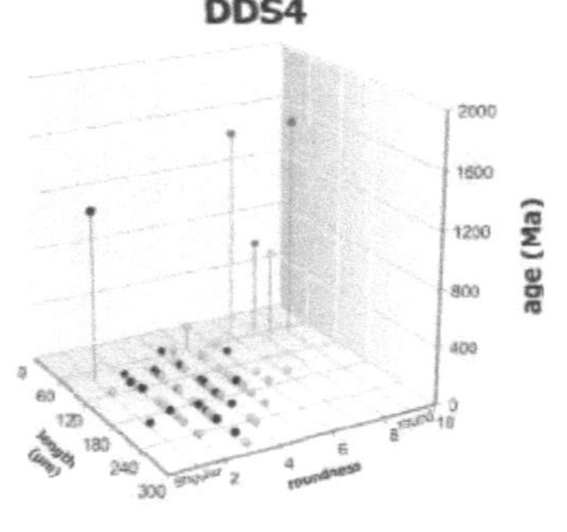

Lower Wyemaha and older layers

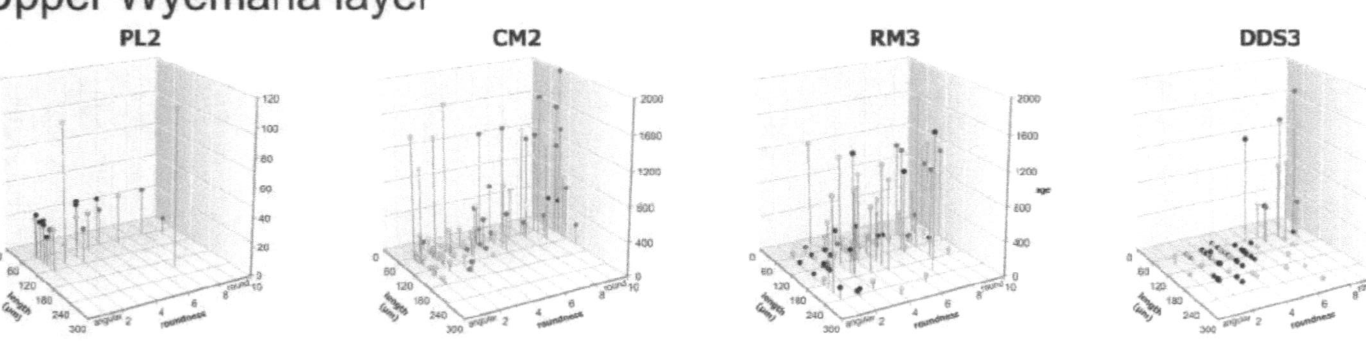

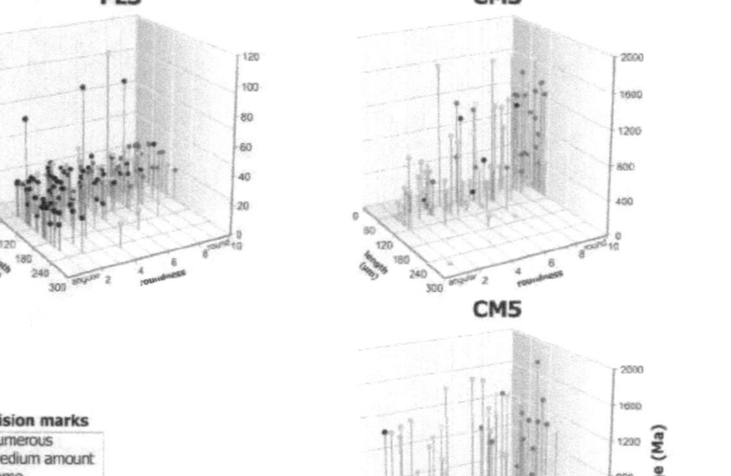

Figure 4.4 - Length, roundness, age and collision marks of Great Basin samples. Ages are cut at 2 Ga. Age scale for the PL samples is different from the others.

Most analysed DZ grains in the Great Basin, especially at the basin borders, have ages between 20 and 45 Ma (Figure 4.5), assumed to stem from volcanic sources. For comparison, we show the data from a tephra in primary deposition from the La Sal Mountains, Utah, some 400 km remote from its source in the Jemez Mountains, New Mexico (age data in KRAUTZ ET AL. 2018a). Its grains are less idiomorphic, smaller on average, and have less collision marks than the Great Basin tephra-derived grains. Thereafter, many of those DZ in the Great Basin may have had a long transportation pathway, though rounding is no certain indicator for this. The age group 80-120 Ma, supposed to have its roots in Sierra Nevada volcanic events, does not contain grains with presumed aeolian history but some idiomorphic and less chattered ones. Probably the

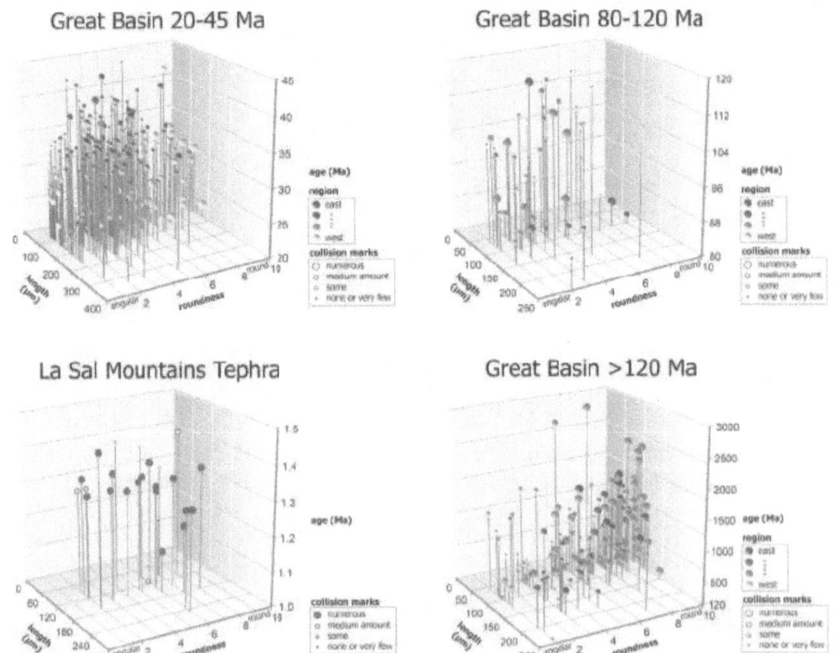

Figure 4.5 - Length, roundness, age, region and collision marks of Great Basin samples and of the La Sal Mountains tephra (Krautz et al., 2018a) for selected age groups.

DZ of this age group have a fluvial background and at least some, especially in the west-central site CM (orange pinheads), appear to be close to their primary deposition. The largest grains are found at RM (red) and sizes decrease into both directions. This indicates that increasing distance to the eruptive sources is not the decisive factor for zircon-grainsize differences but age, because almost all large grains are younger (80-90 Ma) than the smaller ones (around 100 Ma). This relationship between age and size holds also true within the grains at CM. Contrastingly, at all sites in the Great Basin the DZ older than 120 Ma have a strong bias towards a probable aeolian history, especially the grains with ages around 1.4 Ga.

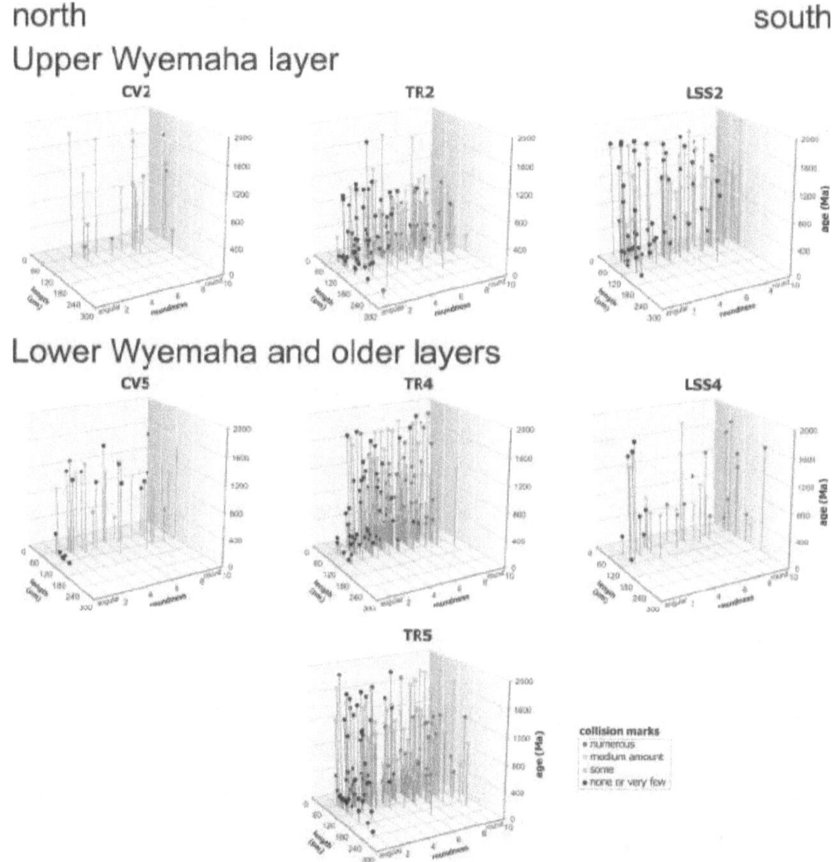

Figure 4.6 - Length, roundness, age and collision marks of Colorado Plateau samples.

The samples from CV yielded few concordant grains, but they show distinct differences between the two layers (Fig. 4.6). Collision marks are much more frequent in the upper layer, whereas the lower one has many unchattered grains and even some idiomorphic ones preferably in the youngest age group. This implies that both layers have different provenances with the upper one showing a greater encroachment by collisions, though this provenance difference is not evident in the grains size and the degree of roundness beyond the idiomorphic fraction. At the TR site grainsizes in the upper layer and the lower part of the bifid lower layer are remarkably larger than in the middle. The same difference is found in the portion of idiomorphic grains. However, rounding and encroachment by collision marks are similar in all layers. The portion of grains attributed aeolian is small in marks differ all layers. Again, the idiomorphic grains are mostly in the youngest age groups. Both layers at LSS show remarkable differences again. Although the average grainsizes are similar, the relative amounts of idiomorphic grains and, conspicuously, the amount of collision.

Figure 4.7 puts the focus on the major age groups of North American tectonic episodes. The group 20-35 Ma occurs at all sites, though seldom, and the grains are idiomorphic and almost scattered. Among them is the largest DZ grain of all samples. These findings indicate transportation cascades with few steps. DZ of that age range in the Great Basin on average are rounder and more damaged.

Most DZ grains in the age group of the Cordilleran magmatic arc ($\leq$250 Ma) are idiomorphic, and even better-rounded grains are often less chattered but do not reach the same lengths as the idiomorphic ones. We interpret these findings that most of these grains were volcanic and did not cascade much.

Contrastingly, DZ derived from the Appalachian orogenies (260 – 735 Ma) are rarely idiomorphic but mostly scattered. Angular grains are smaller than better-rounded ones. Some presumably indicate an aeolian background, but most are undefined as to dominant transportation modes. The next older age group (800 – 1350 Ma) shows a similar picture, but aeolian grains are more frequent, and there is a cluster of less-collided grains around 1300 Ma. The grains correlative to the Grenville orogeny (1 – 1.2 Ga) show similar properties as the Appalachian group but lack very angular grains. The age range encompassing the anorogenic granite province (1.35 – 1.6 Ga) contributes grains with many collision marks, equally distributed over all roundness classes, but less chattered ones prevail. The maximum grainsize is smaller than in the younger sets. The Yavapai/Mazatzal age range (1.6 – 1.8 Ga) is like the anorogenic granites range but with less unscathed grains. The North American cratons (>1.8 Ga) are represented by a few grains only with varying roundness and collision marks.

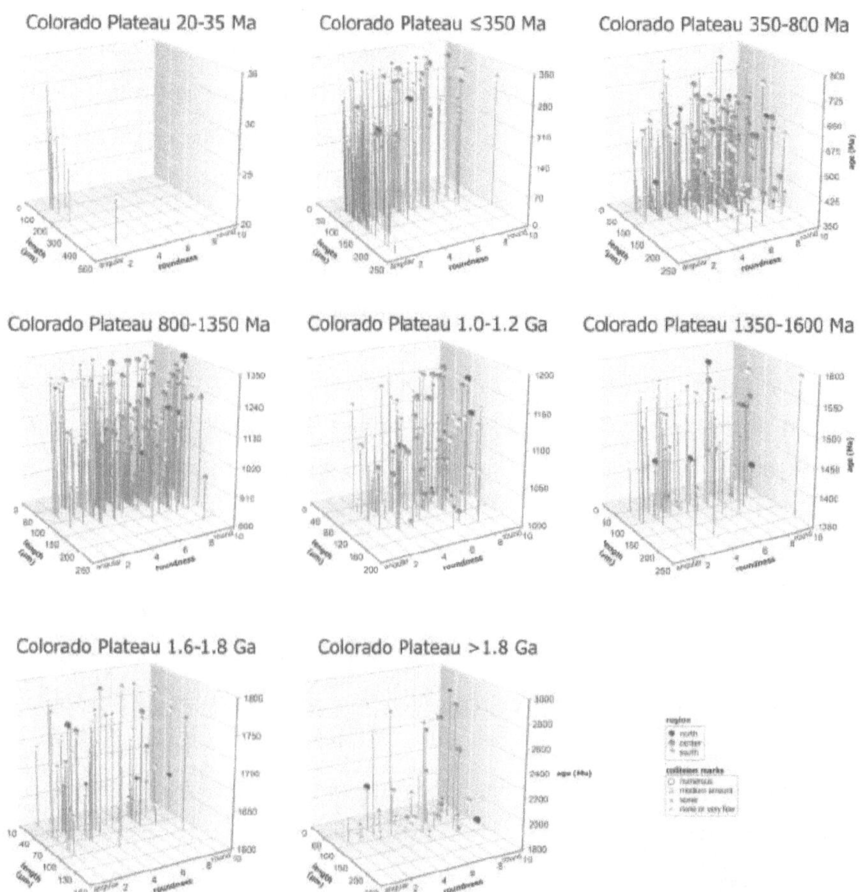

Figure 4.7 - Length, roundness, age, region and collision marks of Colorado Plateau samples for selected age groups.

Idiomorphism does not exclude long transportation pathways, because zircons with zonation may unshell their rim upon impact (GÄRTNER ET AL. 2013). Presumably, such zircons should have a smaller mean grainsize than others. Accordingly, this effect may be excluded for the DZ <250 Ma, but may apply to older DZ, especially those 260-375 Ma.

	age	surface	roundness
Great Basin			
length	-0.183	-0.204	-0.266
roundness	0.392	0.496	
surface	0.310		
Colorado Plateau			
length	-0.133	-0.013	0.045
roundness	0.233	0.608	
surface	0.142		

Values are correlation coefficients (r).

Due to the large number of cases almost all linear regressions one may calculate are statistically significant (Table 4.2). As may be expected, collision marks and roundness explain 25 % or more of one another, probably due to the fact that both are products of the transportation processes. Contrariwise, length of the crystal correlates only slightly with all other parameters, presumably because breaking during transportation and original size are stochastic.

Since there is not much knowledge of regional zircon properties other than age, it is unknown whether the differences found in the data are due to transportation from regional sources to the study sites, due to earlier transportation processes, or a combination of both.

4.4.3 Age distributions of detrital zircons

Since this is a pilot feasibility study, we put the focus on a broad spatial coverage of our analyses at the expense of sample sizes. Therefore, some samples yielded less than 60 DZ. This number has established as a standard for geological provenance studies since DODSON ET AL. (1988), who argue that 60 concordant ages suffice to recognize a single age element, which represents more than 5 % of the complete age population on a confidence level of 95 %. Therefore, for samples <60 DZ we draw conclusions from the existing data only and do not argue with respect to missing age ranges. VERMEESCH (2004) calculated that 117 grains per sample allow for reliably (i.e., with a probability of 95 %) achieving all relevant (>5 % fraction) constituents of a sample; according to EIZENHÖFER (2020), 97 DZ are sufficient for this 95 % confidence. For samples with numbers around these values, we assume that likely, all relevant age ranges are represented, and missing age ranges probably are not a relevant fraction of the entirety.

Possible sources of DZ, i.e., lake deposits, were analysed with even larger sets to receive clues on possible minor components and a more precise picture of the actual data distribution function. For the same reason, we produced larger datasets by combining published smaller datasets derived from one area to form composite sets (cf., e.g., DICKINSON AND GEHRELS 2008a).

Figure 4.8 depicts that the young age signals often frame subordinate age groups. Especially the CM and RM samples contain an even younger age group below 18 Ma, which may either be attributed to the latest and southernmost eruptions from the volcanic fields or to the early phase of the Yellowstone hotspot spur to the northwest. Similar DZ ages are found in the lake deposits. Ages older than 120 Ma are less common in most samples than younger ones. At all but the sites marginal to the Great Basin, PL and DDS, there is evidence of the older phases of the Cordilleran magmatic arc. The prevalent signals that may have been derived from the regional bedrocks (GEHRELS 2000; LINDE ET AL. 2014; CANADA ET AL. 2020), 1.4 – 1.5 Ga (anorogenic Antler orogeny) and 1.6 – 1.8 Ga (Yavapai/Mazatzal terranes) are of subordinate importance.

Although KDEs do not overemphasize young ages due to their higher analytical precision, Figure 4.9 shows that young ages dominate most DZ age spectra of Great-Basin samples. There are two pronounced maxima, one around 30 Ma, the other around 100 Ma. However, the samples differ in the precise location of the respective peaks and of their relative contribution. In both lake deposits from RPD the most pronounced peak centers at 96 Ma. This peak is repeated in the nearby CM profile, especially in the UWL. At best, it is only a minor part in the other samples. This signal is ascribed to the major magmatic episodes of the Sierra Nevada batholith.

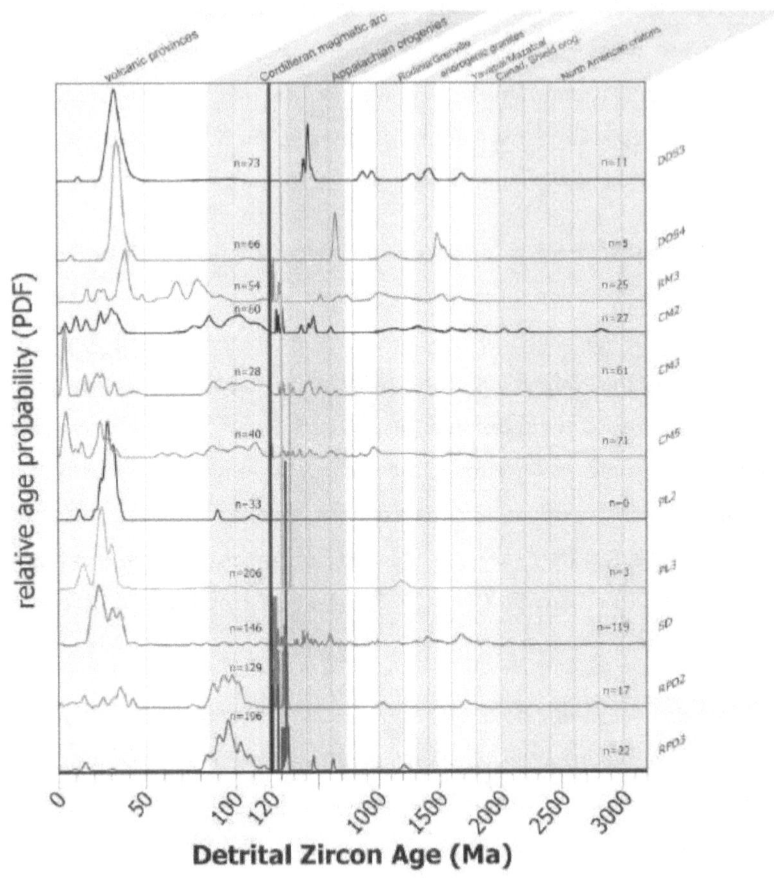

Figure 4.8 - Probability Density Functions (PDF) of samples from the Great Basin with the age scale split at 120 Ma. The left and the right functions have been normalized separately. Upper Wyemaha layer (UWL) in blue, lower Wyemaha layer (LWL) and deeper layers in orange, lake deposits in green. Some functions overlap and are drawn in different shades of green or orange, therefore.

The ages assigned to the volcanic fields peak at 15 and 34 Ma in the RPD lake deposits. Signals delimited by these ages occur in all other samples as well, representing the best-expressed peak in most of them. Ages of 34 Ma and older may be derived from the region, whereas by far most of the DZ found in lake deposits and cover beds must be ascribed to more southern sources.

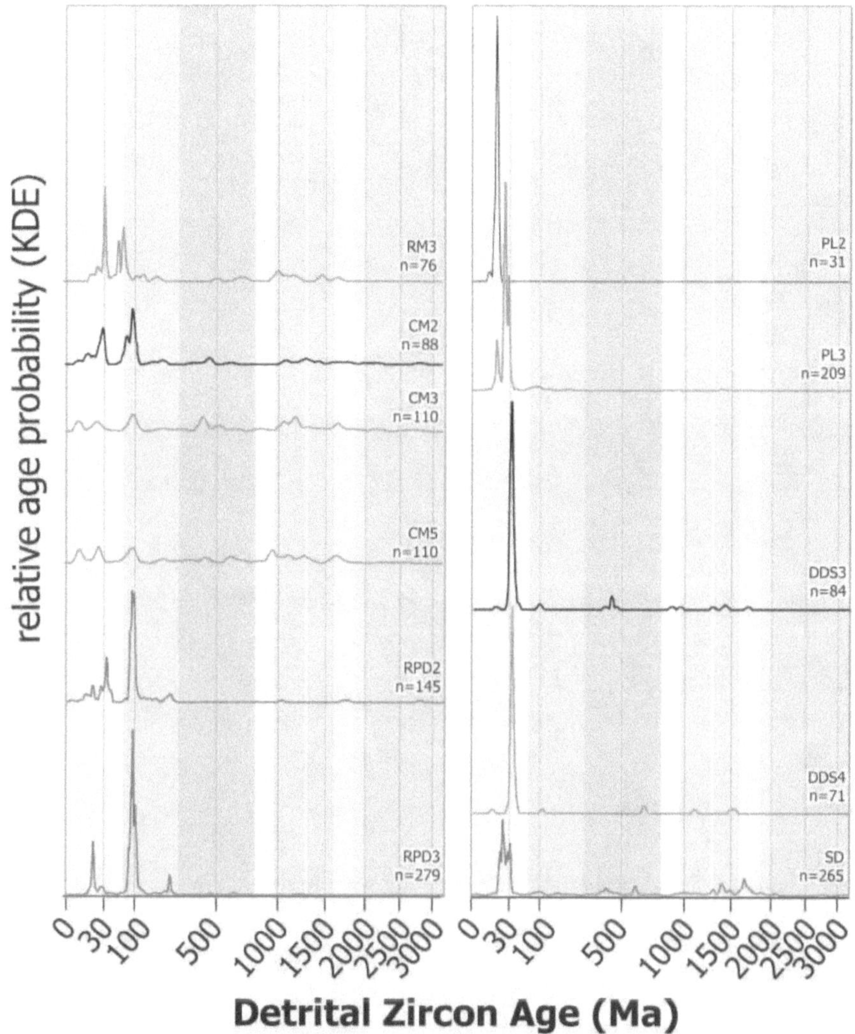

Figure 4.9 - Normalized Kernel Density Estimations (KDE) at logarithmic scale of samples from the Great Basin plotted with a bin width of 25 myr. UWL is in blue, LWL and deeper layers in orange, lake deposits in green. The samples are grouped after visual inspection of their similarities. See figure 4.8 for background colour coding.

On the Colorado Plateau, the two samples in the Castle Valley (CV) at the foot of the northern La Sal Mountains, located upon a river terrace assigned to the penultimate glaciation (KLEBER 1999), yielded rather different DZ age distributions (Figure 4.10). They tentatively support the

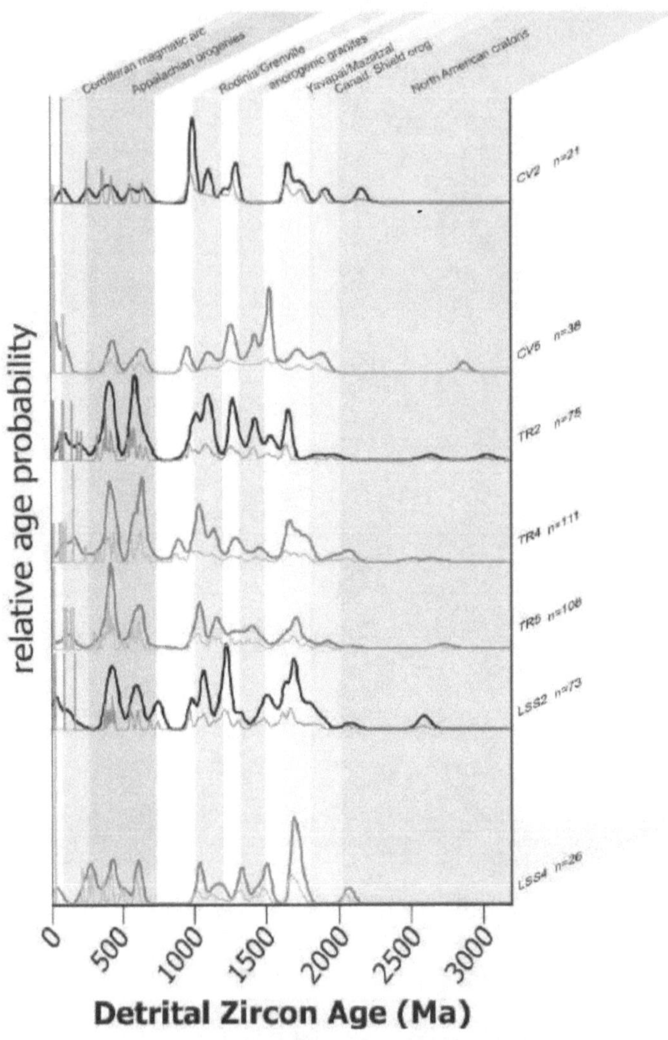

Figure 4.10 - Normalized KDE (thick lines) and PDE (thin dotted lines) of samples from the Colorado Plateau, the former plotted with a bin width of 25 myr; UWL in blue, LWL and older layers in orange.

field-based interpretation that both layers were influenced by different sources. The PDF of the upper layer shows a dominant Sierra Nevada signal of 90 Ma, which differs from the KDE. Within the latter there are the same peaks but the prominent one is at 1 Ga. The Appalachian orogenies show up (400-670 Ma), as well as the Grenville orogeny (1.1, 1.22, 1.3 Ga peaks). Yavapai/Mazatzal can be found (1.66 and 1.74 Ga), the same holds true for the Canadian Shield orogeny with 1.92 Ga and the North American cratons with 2.16 Ga. The differences to the lower layer are depicted in KDE and PDF. The volcanic province is prominent within this sample (30 Ma). Then the Sierra Nevada batholith follows with 100 Ma. Furthermore, there are strong signals of the anorogenic granites with 1.42 and 1.52 Ga. Older DZ show similar ages as in the upper layer.

In contrast, the site Trachyte Ranch (TR) located in the foothills of the Henry Mountains yielded rather similar spectra from three different layers, two of them assigned to the LWL, one to the UWL. The site is underlain by Mid-Jurassic rocks, upslope bedrock is Late Jurassic Morrison Formation (UTAH GEOLOGICAL SURVEY 2018). The latter contains sandstone layers that may have yielded DZ. Their age spectra (DICKINSON AND GEHRELS 2008a) are similar to those found in TR samples. Tephra layers may be another local source for zircons. The Cordilleran magmatic arc system is clearly visible in the KDE as well as in the PDF. The ages from the Appalachian orogenies are the most dominant in all three layers (380 – 700 Ma), followed by DZ from the Grenville orogeny (1 – 1.3 Ga), Yavapai/Mazatzal (1.6 – 1.72 Ga), and the anorogenic granite intrusions (1.32 – 1.54 Ga). Different peaks show up belonging to the Canadian Shield and other North American cratons.

The profile in the Abajo Mountains (LSS) is underlain by Cretaceous sandstone. Despite the different bedrock, its DZ age spectrum is similar to TR. The close relationship of the data from TR and LSS residing on different bedrock gives a hint on allochthonous impact. Furthermore, there are young sub-populations with age peaks around 30 and 90 Ma in two UWL samples. This profile is an example that both types of functions may give different important pieces of information. The PDF of the lower layer shows a dominant age peak (~30 Ma). Comparing this to the KDE of the same sample, the 30 Ma peak nearly disappears, but a Yavapai/Mazatzal peak dominates the curves.

Samples from Jurassic and Cretaceous rocks in Utah contain pronounced peaks at around 1.1 Ga (DICKINSON AND GEHRELS 2008a), which is not to the same extent reflected in our samples except for the UWL from CV and TR. The LSS samples, on the other hand, show a peak at around 1.7 Ga, which is typical for Mesozoic (and older) sources in the Four

Corners Region and New Mexico, i.e., to the south-east of this profile (DICKINSON AND GEHRELS 2008a).

The most remarkable difference between the two areas Great Basin and Colorado Plateau is the dominance of the ages from the volcanic province (18-45 Ma) only in the Great Basin samples and the palaeolake deposits. Similarly, the signals derived from the Cordilleran magmatic arc, especially from the Sierra Nevada batholith are much more frequent than in the Colorado Plateau. The majority of DZ of the Colorado-Plateau samples are older than about 300 Ma, assumed to be a mixture of grains from the Proterozoic to Early Phanerozoic orogens at the southern and south-eastern borders of North America with a few admixed DZ from North American cratons. Such a spectrum was originally transported to an area north of the Colorado Plateau by wind and relocated into Cretaceous sediments (cf. Potential sources of detrital zircons, chapter 4.3.4).

4.4.4 Multidimensional scaling (MDS)

Figure 4.11 shows that the profiles form three DZ facies defined by multidimensional scaling. At a scale involving all studied profiles, the Colorado Plateau samples (green circle) plot close together and are distant to all other samples. However, only larger sample sizes are assumed to yield trustworthy results by multidimensional scaling (MATTHEWS ET AL. 2017), therefore, the position of smaller samples has to be interpreted with caution.

The other two clusters of cover-bed samples are each grouped around lake deposits. Bonneville lake deposits located in the Sevier Desert (SD) plot in the middle of the second cluster (orange circle), which is the internally most dissimilar. This lake deposit was taken from the late glacial White Marl (OVIATT AND SHRODER 2016) of Lake Bonneville, from a marginal position of the Sevier subbasin. Thus, it is younger than the cover beds of this study. Therefore, the similarities to the cover-bed samples may be interpreted in various ways: The lake deposit may be an allochthonous product of a reworked older lake deposit. In this case, the Sevier basin was host to an earlier, separated lake, because Lake Bonneville probably has not reached this basin before the last major highstand (OVIATT 1988). Or the position marginal to the former lake has led to a contribution by reworked

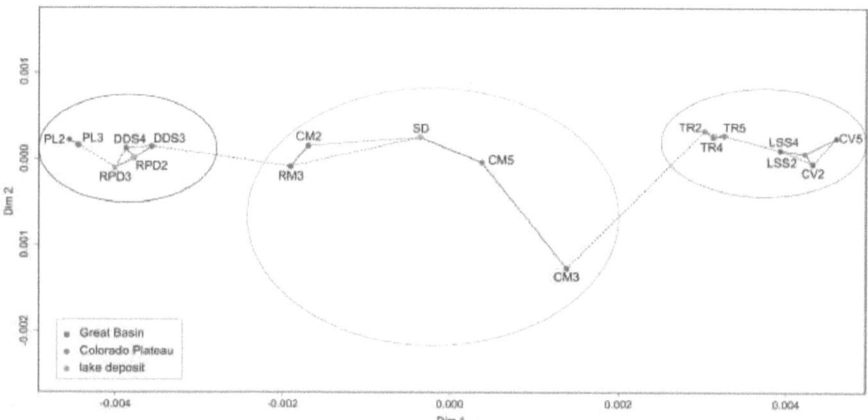

Figure 4.11 - Multidimensional scaling diagram using Sircombe-Hazelton distances between corresponding samples (Vermeesch, 2020) of all samples of this study. Solid lines correspond to close neighbors, dashed lines to the second closest neighbor. Three clusters are encircled (see text). Samples with less than 60 near-concordant DZ are marked by an 'x'.

cover beds from the nearby slopes, maybe connected with the dramatically dropping lake level during the Bonneville Flood (O'CONNOR 2016), which certainly has caused erosion around its shores because of the increased relief energy (KLEBER 1993). This cluster (orange circle) comprises all cover-bed samples located in the center of the Great Basin transect, Coal Mountains (CM) and Ruby Mountains (RM). The samples from the lower layer of the CM profile are closer to the Colorado Plateau samples than all others. This may be understood by influence of the nearby rocks, as CM is located upon fan deposits solely derived from Triassic to Jurassic upslope rocks (NEVADA BUREAU OF MINES AND GEOLOGY 2018), which might contain zircons of similar provenance as the contemporary deposits of the Colorado Plateau. However, the upper layer of the same profile plots much closer to the third cluster (blue circle), which is due to the much greater share of the two young age ranges. This indicates a remarked change in the DZ provenance between those layers. The CM upper layer sample plots closer to the RM sample than to the deeper layer of the CM profile. RM is located upon Paleogene felsic tuff (NEVADA BUREAU OF MINES AND GEOLOGY 2018) with zircon ages close to the youngest age range in the sample. However, a local derivation of the young age group seems unlikely, because these ages occur in all Great Basin samples, in most of them even with an even higher content. Many of those DZ show pronounced indications of transportation, indicating a long cascade of relocation (see below).

The third cluster (blue circle) plots farthest from the Colorado Plateau cluster (green circle). It is narrow within itself, indicating great similarity among cover beds and the lake deposits at Rye Patch Dam (RPD). Geographically, this is surprising because this comprises the two most distant of our Great Basin profiles with the Pyramid Lake site (PL) near the western, and the Dimple Dell (DDS) site at the eastern border of the Great Basin. On the other hand, the CM profile, though located close to the RPD lake deposits, has dissimilar DZ age compositions. There is evidence that substrates close to lakes were at least partially derived from those lakes through aeolian transportation (CHADWICK AND DAVIS 1990; KLEBER 1990; DIETZE ET AL. 2016). However, our findings indicate that this influence decreases with elevation (CM lies in a hilly area compared to RPD) and distance (RM is the most distant from larger paleolakes), whereas sites near the elevation of the former lakes such as PL and DDS plot very close to the palaeolakes. This is mainly due to the large contribution of the ~30 and the ~100 Ma signals found in the lake deposits as well as in the lower profiles. We speculate this could be due to the size of the studied grains, which are in the typical range of shorter-range aeolian transportation.

In Figure 4.12 similarities and dissimilarities of the Colorado Plateau samples are computed and compared to reported Mesozoic DZ ages from the literature, some of which were composed from several datasets from a particular region and the same stratigraphic age. The latter are colour-coded per region. There are two clusters. In the large one in the right of the figure all but one of our samples plot closer to composite datasets from the central or even the southern Colorado Plateau. The exception is one sample from the profile farthest north, Castle Valley (CV), which is similar to the northern composites, indicating a regional derivation. The upper layer from the same profile, however, plots quite distant to it. This indicates a more allochthonous source. However, these two samples have a low number of analyzed DZ, relativising this conclusion.

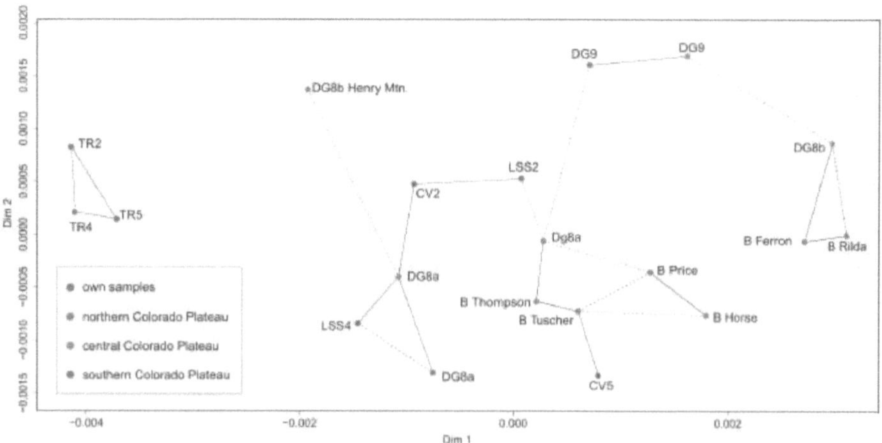

Figure 4.12 - Multidimensional scaling diagram of samples from the Colorado Plateau compared to published data. DG8a = Dickinson and Gehrels (2008a); DG8b = Dickinson and Gehrels (2008b); DG9 = Dickinson and Gehrels (2009b); B = Bartschi et al. (2018). Samples with less than 60 concordant DZ are marked by an 'x'.

The smaller cluster to the left comprises the Trachyte Ranch (TR) located in the Henry Mountains. This is comparably dissimilar to all other samples including the composite ones, indicating a different source of their DZ. TR is underlain by Mid-Jurassic rocks, upslope bedrock is Late Jurassic Morrison Formation (UTAH GEOLOGICAL SURVEY 2018). The latter contains sandstone layers that may have yielded DZ. Their age spectra (DG8b Henry Mountains in figure 4.11, DICKINSON AND GEHRELS 2008b,) are similar to those found in TR samples, except for the fact that data from the regional Morrison Formation share a pronounced peak at around 1100 Ma, which is only expressed in the UWL but missing in the other TR samples (cf. fig. 4.10). Accordingly, the TR DZ age spectra may not entirely be explained by local sources. Local tephra of 160 – 180 Ma (KOWALLIS ET AL. 2007) may be another source for zircons. In fact, idiomorphic zircons (cf. Zircon morphology, chapter 4.5.2) of that age were found in the TR cover beds. Furthermore, the TR samples contain grains with ages around 30 and 100 Ma, ages typical for the Great Basin samples, which are allochthonous as they are older than the underlying rocks. The location is prone to deposition from western sources because it lies in the lee of a gentle slope in the west, whereas steeper and higher slopes may have shielded it from northern or southern influences.

4.5 Conclusions

The most remarkable difference between the two areas Great Basin and Colorado Plateau is the dominance of the ages attributed to the volcanic province (18 – 45 Ma) only in the Great Basin samples and the palaeolake deposits. Similarly, the signals derived from the Cordilleran magmatic arc, especially from the Sierra Nevada batholith, are much more frequent than in the Colorado Plateau. The majority of DZ of the Colorado-Plateau samples are older than about 300 Ma, assumed to be a mixture of grains from the Proterozoic to Early Phanerozoic orogens at the southern and south-eastern margins of North America with a few admixed DZ from North American cratons. Such a spectrum was originally transported by rivers to an area north of the Colorado Plateau, re-allocated by wind in Jurassic times, and relocated into Cretaceous sediments. We assume that fluvial deposits derived from those Mesozoic sedimentary rocks are an intermediate sink of DZ. Ages above 120 Ma are of minor importance in most samples of the Great Basin, whereas they dominate all Colorado-Plateau samples.

Great Basin and Colorado Plateau were equally affected by the volcanic activities (cf. fig. 4.1). However, they show different contributions in the DZ record. Thus, we assume that the zircons of the volcanic province had been trapped within the Great-Basin palaeolakes, some of which have existed already in the Neogene (REHEIS ET AL. 2014). I.e., the lakes acted as intermediate reservoirs and eventually as dispersers of aeolian fines. This needs to be tested using additional samples from lake deposits. Paleogene deposits of the Rocky Mountains, i.e., more distant to the largest flare-ups but approximately contemporary to them, contain abundant zircons from this volcanic period (ZHU AND FAN 2018). Presumably, these young zircons originally had been deposited in the Colorado Plateau but have been denuded from slopes in the meanwhile, whereas in the Great Basin lakes acted as reservoirs. Regional winds played a major role in the distribution of DZ during the cover-bed formation. Apparently, the young DZ did not re-enter the Colorado Plateau to the same degree as in the Great Basin. Therefore, we render aeolian transportation from the west as a major pathway of DZ unlikely. Southern aeolian pathways are more likely as indicated by the southern provenance of the young DZ in the Great Basin and the results of the multidimensional scaling (MDS) of the Colorado Plateau samples compared to ages reported in the literature. This complies well with the southern to south-western provenance of recent dust in the Uinta Mountains north of the Colorado Plateau (MUNROE ET AL. 2019). MDS turned out to be a valuable tool to define clusters of age distributions.

The Cordilleran magmatic period may have been more eruptive than previously thought. Although western winds as well as fluvial transportation are unlikely, the age peak derived from the Cordilleran arc is obvious in our samples. Having a closer look at the surfaces of these zircons frequently shows idiomorphic grains without or with just a few collision marks. We interpret this as indicating strong kinetic energy of this volcanism without many steps in a transportation cascade. We recommend in further provenance studies not to rely only on ages rather having a closer look at DZ surfaces. However, there is no specific type of collision marks to differentiate fluvial from aeolian transportation, length, surface and age together may give a more precise picture of a zircon's pathway. Yet, very well rounded DZ with many marks are interpreted as having a long transportation history including aeolian transportation, whereas idiomorphic grains with only a few marks indicate a cascade with not many sedimentation cycles and transportation modes. Especially the assignment of morphological parameters to particular age groups turned out of great value, because there are obvious morphological disparities due to different transportation histories in different age ranges.

Before applying our approach to another area or research question, one should consider possible recent and past transportation pathways, cascades, surrounding geomorpho-logic processes and geologic structures. Figure 4.13 gives an overview of reconstructed transportation pathways. Since we studied cover beds, we knew that the last step of transportation happened during the cover-bed deposition. In our semiarid study area with one dominant fluvial system (the Colorado River) and a few minor rivers, we considered fluvial transportation first. Past river systems have not changed much

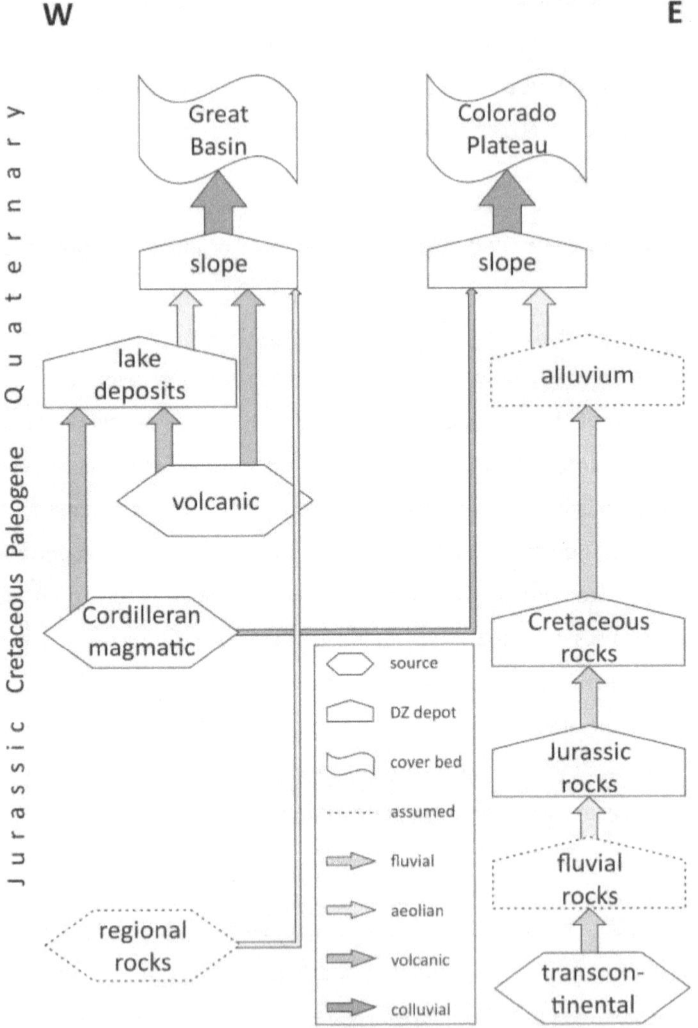

Fig. 4.13 - *Scheme of probable transportation cascades of DZ. Arrows indicate transportation directions; thin arrows denote minor contributions. Except for the regional rocks, which may decisively vary in age, the chronological order is from bottom upwards, with the major chronostratigraphic systems at the left. The main cascades of DZ are: (1) a transcontinental river system, whose grains were wind-blown to the studied area in Jurassic times, were reworked into Cretaceous rocks and probably by the southward river system during the Quaternary. From all these depots, grains may have been blown to the slopes and then reworked during the formation of cover beds. (2) Cretaceous and Paleogene volcanic eruptions, whose grains were deposited on the slopes directly or indirectly via intermediate depots in palaeolakes. We render the local rocks as DZ sources unlikely for most samples, because there is evidence that weathering did not release quantities of zircons from rocks, the grainsizes typically for most rocks underlying the sites are smaller or larger than the DZ in our analyses, and because several analyses yielded DZ ages much younger than the local rocks.*

compared to modern ones, and there was a divide in the Great Basin between areas (HENRY, 2008), which now yield similar DZ spectra. Glacial transportation in principle is another possible process. However, within our study area only the Ruby Mountains developed large mountain glaciers (LAABS ET AL. 2013), whereas none were affected by the Cordilleran Ice Sheet (BOOTH ET AL. 2003). Some age groups can be traced back to volcanic events. Volcanic flare-ups may have had sufficient kinetic energy to reach the study sites directly. Some DZ are essentially idiomorphic despite of a distant source area, indicating such a direct path. Finally, aeolian transportation is another possible process. Our study sites were affected by the west-wind drift at least during the Quaternary (ALDER AND HOSTETLER 2019). Nevertheless, because of their specific weight, zircons probably have not reached the jet stream height except by a volcanic eruption. Furthermore, our results render western winds as main drivers of transportation unlikely. Within the current regional wind systems, strong winds able to carry sand are mainly from approximately northern directions (JEWELL AND NICOLL 2011). Some source areas of probable wind-borne grains are far distant. Therefore, we assume that DZ were transported to the sites with several intermediate sinks, which eventually became sources for the next aeolian impulse. Such a pathway is known for a transcontinental river system disposing matter for Jurassic aeolianites of the Colorado Plateau, which again was reworked into regional Cretaceous deposits. Furthermore, we are aware that some provenances cannot be unequivocally assigned to a specific source with our current approach alone, which especially holds true for the Colorado Plateau, where the spectra of the Mesozoic bedrocks are similar over greater distances.

By using cover beds for provenance analysis, we attempt opening a new field for environ-mental reconstructions of the Quaternary. Beyond classical approaches to provenance which mainly depend on age distributions, here we provide a pilot study how to also use morphological properties of DZ for reconstructing the genesis of this particular type of deposits: Cover beds are ideally suited for regional studies because they widely spread many slopes (KLEBER AND TERHORST 2013) and may be chronostratigraphically classified in some areas (KLEBER ET AL. 2013).

Other questions about, e.g., humidity, which was higher as seen from the size and expansion of the palaeolakes, other humidity archives, and the maturity of the palaeosols, must remain open, but be in the scope of further studies.

Appendix

The Supplemental Material provides the results of all U-Pb measurements, charts of all profiles and their stratigraphic correlation, grain size diagrams of most samples and the raw data of the grain-morphological analyses.

The data that supports the findings of this study are available in the supplementary material of this article (chapter 6.3).

[SI1 U-Pb ratios and calculated ages; SI2 Charts of profiles and stratigraphic connections (see chapter 5.1, fig. 5.1); SI3 Grainsize distributions; SI4 Zircon morphology data]

4.6 References

Adams, K.D. (2010) Lake levels and sedimentary environments during deposition of the Trego Hot Springs and Wono Tephras in the Lake Lahontan Basin, Nevada, USA. Quaternary Research, 73(01), 118–129. Available from: doi:10.1016/j.yqres.2009.08.001.

Alder, J.R. & Hostetler, S.W. (2019) Applying the Community Ice Sheet Model to evaluate PMIP3 LGM climatologies over the North American ice sheets. Climate Dynamics, 53(5-6), 2807–2824. Available from: doi:10.1007/s00382-019-04663-x.

Aleinikoff, J.N., Muhs, D.R., Bettis III, E.A., Johnson, W.C., Fanning, C.M. & Benton, R. (2008) Isotopic evidence for the diversity of late Quaternary loess in Nebraska: Glaciogenic and nonglaciogenic sources. Geological Society of America Bulletin, 120(11-12), 1362–1377. Available from: doi:10.1130/B26222.1.

Amorosi, A., Bruno, L., Cleveland, D.M., Morelli, A. & Hong, W. (2017) Paleosols and associated channel-belt sand bodies from a continuously subsiding late Quaternary system (Po Basin, Italy): New insights into continental sequence stratigraphy. Geological Society of America Bulletin, 129(3-4), 449–463. Available from: doi:10.1130/B31575.1.

Anderson, J. & Bender, E. (1989) Nature and origin of Proterozoic A-type granitic magmatism in the southwestern United States of America. Lithos, 23(1-2), 19–52. Available from: doi:10.1016/0024-4937(89)90021-2.

Ayers, J.C., Liu, X., Lasley, C., Wang, X., Hyun Nam, J. & Katsiaficas, N.J. (2017) U-Pb zircon geochronology for determining soil provenance in a limestone terrane, Middle Tennessee, United States. Available from: http://get.iedadata.org/doi/100733 [Accessed 30 December 2018].

Baars, D.L. (2000) The Colorado Plateau: A geologic history. University of New Mexico Press: Albuquerque.

Bartschi, N.C., Saylor, J.E., Lapen, T.J., Blum, M.D., Pettit, B.S. & Andrea, R.A. (2018) Tectonic controls on Late Cretaceous sediment provenance and stratigraphic architecture in the Book Cliffs, Utah. Geological Society of America Bulletin, 130(11-12), 1763–1781. Available from: doi:10.1130/B31927.1.

Bayat, O., Karimzadeh, H., Karimi, A., Eghbal, M.K. & Khademi, H. (2017) Paleoenvironment of geomorphic surfaces of an alluvial fan in the eastern Isfahan, Iran, in the light of micromorphology and clay mineralogy. Arabian Journal of Geosciences, 10(4). Available from: doi:10.1007/s12517-017-2848-9.

Bellosi, E.S. & González, M.G. (2010) Paleosols of the middle Cenozoid Sarmiento Formation, central Patagonia. In: Madden, R.H., Carlini, A.A., Vucetich, M.G. & Kay, R.F. (Eds.) The paleontology of Gran Barranca: Evolution and environmental change through the middle Cenozoic of Patagonia. Cambridge University Press: New York, pp. 293–305.

Benson, L.V., Currey, D.R., Dorn, R.I., Lajoie, K.R., Oviatt, C.G., Robinson, S.W., Smith, G.I. & Stine, S. (1990) Chronology of expansion and contraction of four Great Basin lake systems during the past 35,000 years. Palaeogeography, Palaeoclimatology, Palaeoecology, 78, 241–286.

Beranek, L.P., Link, P.K. & Fanning, C.M. (2006) Miocene to Holocene landscape evolution of the western Snake River Plain region, Idaho: Using the SHRIMP detrital zircon provenance record to track eastward migration of the Yellowstone hotspot. Geological Society of America Bulletin, 118(9-10), 1027–1050. Available from: doi:10.1130/B25896.1.

Beranek, L.P., Link, P.K. & Fanning, C.M. (2016) Detrital zircon record of mid-Paleozoic convergent margin activity in the northern U.S. Rocky Mountains: Implications for the Antler orogeny and early evolution of the North American Cordillera. Lithosphere, 8(5), 533–550. Available from: doi:10.1130/L557.1.

Berger, G.W. (1991) The use of glass for dating volcanic ash by thermoluminescence. Journal of Geophysical Research, 96(B12), 19705–19720. Available from: doi:10.1029/91JB01899.

Best, M.G., Barr, D.L., Christiansen, E.H., Gromme, S., Deino, A.L. & Tingey, D.G. (2009) The Great Basin Altiplano during the middle Cenozoic ignimbrite flareup: Insights from volcanic rocks. International Geology Review, 51(7-8), 589–633. Available from: doi:10.1080/00206810902867690.

Best, M.G., Christiansen, E.H. & Gromme, S. (2013) Introduction: The 36-18 Ma southern Great Basin, USA, ignimbrite province and flareup: Swarms of subduction-related supervolcanoes. Geosphere, 9(2), 260–274. Available from: doi:10.1130/GES00870.1.

Bickford, M.E., van Schmus, W.R., Karlstrom, K.E., Mueller, P.A. & Kamenov, G.D. (2015) Mesoproterozoic-trans-Laurentian magmatism: A synthesis of continent-wide age distributions, new SIMS U–Pb ages, zircon saturation temperatures, and Hf and Nd isotopic compositions. Precambrian Research, 265, 286–312. Available from: doi:10.1016/j.precamres.2014.11.024.

Bird, A., Steven, T.A., Rittner, M., Vermeesch, P., Carter, A., Andò, S., Garzanti, E., Lu, H., Nie, J., Zeng, L., Zhang, H. & Xu, Z. (2015) Quaternary dust source variation across the Chinese Loess Plateau. Palaeogeography, Palaeoclimatology, Palaeoecology, 435, 254–264. Available from: doi:10.1016/j.palaeo.2015.06.024.

Booth, D.B., Troost, K.G., Clague, J.J. & Waitt, R.B. (2003) The Cordilleran Ice Sheet. In: Gillespie, A.R., Porter, S.C. & Atwater, B.F. (Eds.) The Quaternary Period in the United States. Elsevier: Amsterdam, pp. 17–43.

Bouchard, D.P., Kaufman, D.S., Hochberg, A. & Quade, J. (1998) Quaternary history of the Thatcher Basin, Idaho, reconstructed from the 87Sr/86Sr and amino acid composition of lacustrine fossils: implications for the diversion of the Bear River into the Bonneville Basin. Palaeogeography, Palaeoclimatology, Palaeoecology, 141(1-2), 95–114. Available from: doi:10.1016/S0031-0182(98)00005-4.

Bronger, A. (2003) Correlation of loess–paleosol sequences in East and Central Asia with SE Central Europe: towards a continental Quaternary pedostratigraphy and paleoclimatic history. Quaternary International, 106-107, 11–31. Available from: doi:10.1016/S1040-6182(02)00159-3.

Buggle, B., Glaser, B., Zöller, L., Hambach, U., Marković, S.B., Glaser, I. & Gerasimenko, N. (2008) Geochemical characterization and origin of Southeastern and Eastern European loesses (Serbia, Romania, Ukraine). Quaternary Science Reviews, 27(9-10), 1058–1075. Available from: doi:10.1016/j.quascirev.2008.01.018.

Canada, A.S., Cassel, E.J., McGrew, A.J., Smith, M.E., Stockli, D.F., Foland, K.A., Jicha, B.R. & Singer, B.S. (2019) Eocene exhumation and extensional basin formation in the Copper Mountains, Nevada, USA. Geosphere, 15(5), 1577–1597. Available from: doi:10.1130/GES02101.1.

Canada, A.S., Cassel, E.J., Stockli, D.F., Smith, M.E., Jicha, B.R. & Singer, B.S. (2020) Accelerating exhumation in the Eocene North American Cordilleran hinterland: Implications from detrital zircon (U-Th)/(He-Pb) double dating. Geological Society of America Bulletin, 132(1-2), 198–214. Available from: doi:10.1130/B35160.1.

Catuneanu, O. (2019) Model-independent sequence stratigraphy. Earth-Science Reviews, 188, 312–388. Available from: doi:10.1016/j.earscirev.2018.09.017.

Chadwick, O.A. & Davis, J.O. (1990) Soil-forming intervals caused by eolian sediment pulses in the Lahontan basin, northwestern Nevada. Geology, 18(3), 243. Available from: doi:10.1130/0091-7613(1990)018<0243:SFICBE>2.3.CO;2.

Colgan, J.P., Howard, K.A., Fleck, R.J. & Wooden, J.L. (2010) Rapid middle Miocene extension and unroofing of the southern Ruby Mountains, Nevada. Tectonics, 29(6), n/a-n/a. Available from: doi:10.1029/2009TC002655.

Colombini, L.L., Miller, C.F., Gualda, G.A.R., Wooden, J.L. & Miller, J.S. (2011) Sphene and zircon in the Highland Range volcanic sequence (Miocene, southern Nevada, USA): elemental partitioning, phase relations, and influence on evolution of silicic magma. Mineralogy and Petrology, 102(1-4), 29–50. Available from: doi:10.1007/s00710-011-0177-3.

Conroy, J., Guentner, W. & Grimley, D. (2016) Chronologies and mechanisms of last glacial loess deposition in the Central Lowlands of North America: National Science Foundation, Award Abstract #1637481.

Crowley, J.L., Schmitz, M.D., Bowring, S.A., Williams, M.L. & Karlstrom, K.E. (2006) U–Pb and Hf isotopic analysis of zircon in lower crustal xenoliths from the Navajo volcanic field: 1.4 Ga mafic magmatism and metamorphism beneath the Colorado Plateau. Contributions to Mineralogy and Petrology, 151(3), 313–330. Available from: doi:10.1007/s00410-006-0061-z.

Davis, S.J., Dickinson, W.R., Gehrels, G.E., Spencer, J.E., Lawton, T.F. & Carroll, A.R. (2010) The Paleogene California River: Evidence of Mojave-Uinta paleodrainage from U-Pb ages of detrital zircons. Geology, 38(10), 931–934. Available from: doi:10.1130/G31250.1.

DeGraaff Surpless, K., Clemens-Knott, D., Barth, A.P. & Gevedon, M. (2019) A survey of Sierra Nevada magmatism using Great Valley detrital zircon trace-element geochemistry: View from the forearc. Lithosphere, 11(5), 603–619. Available from: doi:10.1130/L1059.1.

Deutz, P., Montañez, I.P., Monger, H.C. & Morrison, J. (2001) Morphology and isotope heterogeneity of Late Quaternary pedogenic carbonates: Implications for paleosol carbonates as paleoenvironmental proxies. Palaeogeography, Palaeoclimatology, Palaeoecology, 166(3-4), 293–317. Available from: doi:10.1016/S0031-0182(00)00214-5.

Dickinson, W.R. & Gehrels, G.E. (2003) U–Pb ages of detrital zircons from Permian and Jurassic eolian sandstones of the Colorado Plateau, USA: paleogeographic implications. Sedimentary Geology, 163(1-2), 29–66. Available from: doi:10.1016/S0037-0738(03)00158-1.

Dickinson, W.R. & Gehrels, G.E. (2008a) Sediment delivery to the Cordilleran foreland basin: insights from U-Pb ages of detrital zircons in upper Jurassic and Cretaceous strata of the Colorado Plateau. American Journal of Science, 308, 1041–1082.

Dickinson, W.R. & Gehrels, G.E. (2008b) U-Pb Ages of Detrital Zircons in Relation to Paleogeography: Triassic Paleodrainage Networks and Sediment Dispersal Across Southwest Laurentia. Journal of Sedimentary Research, 78(12), 745–764. Available from: doi:10.2110/jsr.2008.088.

Dickinson, W.R. & Gehrels, G.E. (2009a) U-Pb ages of detrital zircons in Jurassic eolian and associated sandstones of the Colorado Plateau: Evidence for transcontinental dispersal and intraregional recycling of sediment. Geological Society of America Bulletin, 121(3-4), 408–433. Available from: doi:10.1130/B26406.1.

Dickinson, W.R. & Gehrels, G.E. (2009b) Use of U–Pb ages of detrital zircons to infer maximum depositional ages of strata: A test against a Colorado Plateau Mesozoic database. Earth and Planetary Science Letters, 288(1-2), 115–125. Available from: doi:10.1016/j.epsl.2009.09.013.

Dickinson, W.R. & Gehrels, G.E. (2010) Insights into North American Paleogeography and Paleo-tectonics from U–Pb ages of detrital zircons in Mesozoic strata of the Colorado Plateau, USA. International Journal of Earth Sciences, 99(6), 1247–1265. Available from: doi:10.1007/s00531-009-0462-0.

Dickinson, W.R., Gehrels, G.E. & Marzolf, J.E. (2010) Detrital zircons from fluvial Jurassic strata of the Michigan basin: Implications for the transcontinental Jurassic paleoriver hypothesis. Geology, 38(6), 499–502. Available from: doi:10.1130/G30509.1.

Dickinson, W.R., Karlstrom, K.E., Hanson, A.D., Gehrels, G.E., Pecha, M., Cather, S.M. & Kimbrough, D.L. (2014) Detrital-zircon U-Pb evidence precludes paleo–Colorado River sediment in the ex-posed Muddy Creek Formation of the Virgin River depression. Geosphere, 10(6), 1123–1138. Available from: doi:10.1130/GES01097.1.

Dietze, E. & Dietze, M. (2019) Grain-size distribution unmixing using the R package EMMAgeo. E&G Quaternary Science Journal, 68(1), 29–46. Available from: doi:10.5194/egqsj-68-29-2019.

Dietze, E., Hartmann, K., Diekmann, B., IJmker, J., Lehmkuhl, F., Opitz, S., Stauch, G., Wünnemann, B. & Borchers, A. (2012) An end-member algorithm for deciphering modern detrital processes from lake sediments of Lake Donggi Cona, NE Tibetan Plateau, China. Sedimentary Geology, 243-244, 169–180. Available from: doi:10.1016/j.sedgeo.2011.09.014.

Dietze, M., Bartel, S., Lindner, M. & Kleber, A. (2012) Formation mechanisms and control factors of vesicular soil structure. Catena, 99, 83–96. Available from: doi:10.1016/j.catena.2012.06.011.

Dietze, M., Dietze, E., Lomax, J., Fuchs, M., Kleber, A. & Wells, S.G. (2016) Environmental history recorded in aeolian deposits under stone pavements, Mojave Desert, USA. Quaternary Re-search, 85(1), 4–16. Available from: doi:10.1016/j.yqres.2015.11.007.

Dodson, M.H., Compston, W., Williams, I.S. & Wilson, J.F. (1988) A search for ancient detrital zir-cons in Zimbabwean sediments. Journal of the Geological Society, 145(6), 977–983. Available from: doi:10.1144/gsjgs.145.6.0977.

Doe, M.F., Jones, J.V., Karlstrom, K.E., Dixon, B., Gehrels, G.E. & Pecha, M. (2013) Using detrital zir-con ages and Hf isotopes to identify 1.48–1.45Ga sedimentary basins and fingerprint sources of exotic 1.6–1.5Ga grains in southwestern Laurentia. Precambrian Research, 231, 409–421. Available from: doi:10.1016/j.precamres.2013.03.002.

Ducea, M.N., Paterson, S.R. & DeCelles, P.G. (2015) High-Volume Magmatic Events in Subduction Systems. Elements, 11(2), 99–104. Available from: doi:10.2113/gselements.11.2.99.

Eizenhöfer, P.R. (2020) Subduction and closure of the Palaeo-Asian Ocean along the Solonker Su-ture Zone: Constraints from an integrated sedimentary provenance analysis. Springer: Singa-pore.

Fenn, K., Steven, T.A., Bird, A., Limonta, M., Rittner, M., Vermeesch, P., Andò, S., Garzanti, E., Lu, H., Zhang, H. & Lin, Z. (2018) Insights into the provenance of the Chinese Loess Plateau from joint zircon U-Pb and garnet geochemical analysis of last glacial loess. Quaternary Research, 89(3), 645–659. Available from: doi:10.1017/qua.2017.86.

Fillmore, R. (2011) Geological Evolution of the Colorado Plateau of Eastern Utah and Western Colorado. The University of Utah Press: Salt Lake City.

Gärtner, A., Linnemann, U. & Hofmann, M. (2014) The provenance of northern Kalahari Basin sediments and growth history of the southern Congo Craton reconstructed by U–Pb ages of zircons from recent river sands. International Journal of Earth Sciences, 103(2), 579–595. Available from: doi:10.1007/s00531-013-0974-5.

Gärtner, A., Linnemann, U., Sagawe, A., Hofmann, M., Ullrich, B. & Kleber, A. (2013) Morphology of zircon crystal grains in sediments – characteristics, classifications, definitions // Morphologie von Zirkonen in Sedimenten – Merkmale, Klassifikationen, Definitionen. Geologica Saxonia, 59, 65–73 [Accessed 3 December 2015].

Gehrels, G.E. (2000) Introduction to detrital zircon studies of Paleozoic and Triassic strata in western Nevada and northern California. Geological Society of America Special Papers, 347, 1–17. Available from: doi:10.1130/0-8137-2347-7.1.

Gehrels, G.E., Blakey, R., Karlstrom, K.E., Timmons, J.M., Dickinson, B. & Pecha, M. (2011) Detrital zircon U-Pb geochronology of Paleozoic strata in the Grand Canyon, Arizona. Lithosphere, 3(3), 183–200. Available from: doi:10.1130/L121.1.

Geslin, J.K., Link, P.K. & Fanning, C.M. (1999) High-precision provenance determination using detrital-zircon ages and petrography of Quaternary sands on the eastern Snake River Plain, Idaho. Geology, 27(4), 295–298. Available from: doi:10.1130/0091-7613(1999)027<0295:HPPDUD>2.3.CO;2.

Hacker, D.B., Rowley, P.D. & Biek, R.F. (2018) Catastrophic Collapse Features in Volcanic Terrains: Styles and Links to Subvolcanic Magma Systems. In: Breitkreuz, C. & Rocchi, S. (Eds.) Physical Geology of Shallow Magmatic Systems: Dykes, Sills and Laccoliths. Springer International Publishing: Cham, pp. 215–248.

Hall, S.A. & Goble, R.J. (2012) Berino paleosol, Late Pleistocene argillic soil development on the mescalero sand sheet in New Mexico. The Journal of Geology, 120(3), 333–345. Available from: doi:10.1086/664777.

Hatcher, R.D. (2010) The Appalachian orogen: A brief summary. In: Tollo, R.P., Bartholomew, M.J., Hibbard, J.P. & Karabinos, P.M. (Eds.) From Rodinia to Pangea: The Lithotectonic Record of the Appalachian Region. Geological Society of America, pp. 1–19.

Heckman, K. & Rasmussen, C. (2011) Lithologic controls on regolith weathering and mass flux in forested ecosystems of the southwestern USA. Geoderma, 164(3-4), 99–111. Available from: doi:10.1016/j.geoderma.2011.05.003.

Henry, C.D. (2008) Ash-flow tuffs and paleovalleys in northeastern Nevada: Implications for Eocene paleogeography and extension in the Sevier hinterland, northern Great Basin. Geosphere, 4(1), 1. Available from: doi:10.1130/GES00122.1.

Horstwood, M.S.A., Košler, J., Gehrels, G., Jackson, S.E., McLean, N.M., Paton, C., Pearson, N.J., Sircombe, K., Sylvester, P., Vermeesch, P., Bowring, J.F., Condon, D.J. & Schoene, B. (2016) Community-Derived Standards for LA-ICP-MS U-(Th-)Pb Geochronology - Uncertainty Propagation, Age Interpretation and Data Reporting. Geostandards and Geoanalytical Research, 40(3), 311–332. Available from: doi:10.1111/j.1751-908X.2016.00379.x.

Huang, C.C., Pang, J., Chen, S. & Zhang, Z. (2003) Holocene dust accumulation and the formation of polycyclic cinnamon soils (luvisols) in the Chinese Loess Plateau. Earth Surface Processes and Landforms, 28(12), 1259–1270. Available from: doi:10.1002/esp.512.

Jacobs, P.M. & Mason, J.A. (2007) Late Quaternary climate change, loess sedimentation, and soil profile development in the central Great Plains: A pedosedimentary model. Geological Society of America Bulletin, 119(3-4), 462–475. Available from: doi:10.1130/B25868.1.

Jensen, M., Kowallis, B., Christiansen, E., Webb, C., Dorais, M., Sprinkel, D. & Jicha, B. (2020) Fallout tuffs in the Eocene Duchesne River Formation, northeastern Utah: ages, compositions, and likely source. Geology of the Intermountain West, 7, 1–27. Available from: doi:10.31711/giw.v7.pp1.

Jewell, P.W. & Nicoll, K. (2011) Wind regimes and aeolian transport in the Great Basin, U.S.A. Geomorphology, 129(1-2), 1–13. Available from: doi:10.1016/j.geomorph.2011.01.005.

John, D.A., Henry, C.D. & Colgan, J.P. (2008) Magmatic and tectonic evolution of the Caetano caldera, north-central Nevada: A tilted, mid-Tertiary eruptive center and source of the Caetano Tuff. Geosphere, 4(1), 75–106. Available from: doi:10.1130/GES00116.1.

Karlstrom, K.E., Hagadorn, J., Gehrels, G.E., Matthews, W., Schmitz, M., Madronich, L., Mulder, J., Pecha, M., Giesler, D. & Crossey, L. (2018) Cambrian Sauk transgression in the Grand Canyon region redefined by detrital zircons. Nature Geoscience, 11(6), 438. Available from: doi:10.1038/s41561-018-0131-7.

Katsiaficas, N.J. (2014) Provenance of modern soils of middle Tennessee assessed using zircon U-Pb geochronology and element mass fluxes, Master book, Nashville, TN. Available from: https://etd.library.vanderbilt.edu/available/etd-11192014-145347/unrestricted/Katsiaficas.pdf [Accessed 30 December 2018].

Kimbrough, D.L., Grove, M., Gehrels, G.E., Dorsey, R.J., Howard, K.A., Lovera, O., Aslan, A., House, P.K. & Pearthree, P.A. (2015) Detrital zircon U-Pb provenance of the Colorado River: A 5 m.y. record of incision into cover strata overlying the Colorado Plateau and adjacent regions. Geosphere, 11(6), 1719–1748. Available from: doi:10.1130/GES00982.1.

Kleber, A. (1990) Upper Quaternary sediments and soils in the Great Salt Lake area, USA. Zeitschrift für Geomorphologie Neue Folge, 34, 271–281.

Kleber, A. (1993) A stratigraphy of slope deposits and soils in the northeastern Great Basin and its vicinity. Z. Geomorphol. N.F. Suppl.-vol. (Zeitschrift für Geomorphologie N.F. Supplementband), 92, 173–188.

Kleber, A. (1997) Cover-beds as soil parent materials in midlatitude regions. Catena, 30(2-3), 197–213. Available from: doi:10.1016/S0341-8162(97)00018-0.

Kleber, A. (1999) Cover-beds as relative-dating tools - examples from the western U.S.A. Zeitschrift für Geomorphologie N. F., 43(1), 51–59.

Kleber, A. (2000) Compound soil horizons with mixed calcic and argillic properties — examples from the northern Great Basin, USA. Catena, 41(1-3), 111–131. Available from: doi:10.1016/S0341-8162(00)00111-9.

Kleber, A. (2013) Heavy-mineral analysis as a tool in tephrochronology, with an example from the La Sal Mountains, Utah, U.S.A. Geologos, 19(1-2), 87–94. Available from: doi:10.2478/logos-2013-0006.

Kleber, A., Leopold, M., Vonlanthen, C. & Völkel, J. (2013) Transferring the concept of cover beds. In: Kleber, A. & Terhorst, B. (Eds.) Mid-latitude slope deposits (cover beds). Elsevier: Amsterdam, pp. 171–228.

Kleber, A. & Terhorst, B. (2013) Introduction. In: Kleber, A. & Terhorst, B. (Eds.) Mid-latitude slope deposits (cover beds). Elsevier: Amsterdam, pp. 1–8.

Kowallis, B.J., Britt, B.B., Greenhalgh, B.W. & Sprinkel, D.A. (2007) New U-Pb zircon ages from an ash bed in the Brushy Basin Member of the Morrison Formation near Hanksville, Utah. UGA Publication, 36, 75–80. Available from: http://archives.datapages.com/data/uga/data/079/079001/pdfs/75.pdf [Accessed 30 December 2018].

Krautz, J., Hofmann, M., Gärtner, A., Linnemann, U. & Kleber, A. (2018a) Capability of U–Pb dating of zircons from Quaternary tephra: Jemez Mountains, NM, and La Sal Mountains, UT, USA. E&G Quaternary Science Journal, 67(1), 7–16. Available from: doi:10.5194/egqsj-67-7-2018.

Krautz, J., Gärtner, A., Hofmann, M., Linnemann, U. & Kleber, A. (2018b) Cover beds older than the mid-pleistocene revolution and the provenance of their eolian components, La Sal Mountains, Utah, USA. Quaternary Science Reviews, 185, 1–8. Available from: doi:10.1016/j.quascirev.2018.01.012.

Laabs, B.J., Munroe, J.S., Best, L.C. & Caffee, M.W. (2013) Timing of the last glaciation and subsequent deglaciation in the Ruby Mountains, Great Basin, USA. Earth and Planetary Science Letters, 361, 16–25. Available from: doi:10.1016/j.epsl.2012.11.018.

Lackey, J.S., Cecil, M.R., Windham, C.J., Frazer, R.E., Bindeman, I.N. & Gehrels, G.E. (2012) The Fine Gold Intrusive Suite: The roles of basement terranes and magma source development in the Early Cretaceous Sierra Nevada batholith. Geosphere, 8(2), 292–313. Available from: doi:10.1130/GES00745.1.

LaMaskin, T.A. (2012) Detrital zircon facies of Cordilleran terranes in western North America. GSA Today, 22 (3), 4–11. Available from: doi:10.1130/GSATG142A.1.

Lawrence, C.R., Neff, J.C. & Farmer, G.L. (2011) The accretion of aeolian dust in soils of the San Juan Mountains, Colorado, USA. Journal of Geophysical Research, 116(F2), 355. Available from: doi:10.1029/2010JF001899.

Lawton, T.F., Hunt, G.J. & Gehrels, G.E. (2010) Detrital zircon record of thrust belt unroofing in Lower Cretaceous synorogenic conglomerates, central Utah. Geology, 38(5), 463–466. Available from: doi:10.1130/G30684.1.

Licht, A., Pullen, A., Kapp, P., Abell, J. & Giesler, N. (2016) Eolian cannibalism: Reworked loess and fluvial sediment as the main sources of the Chinese Loess Plateau. Geological Society of America Bulletin, 128(5-6), 944–956. Available from: doi:10.1130/B31375.1.

Linde, G.M., Cashman, P.H., Trexler, J.H. & Dickinson, W.R. (2014) Stratigraphic trends in detrital zircon geochronology of upper Neoproterozoic and Cambrian strata, Osgood Mountains, Nevada, and elsewhere in the Cordilleran miogeocline: Evidence for early Cambrian uplift of the Transcontinental Arch. Geosphere, 10(6), 1402–1410. Available from: doi:10.1130/GES01048.1.

Link, P.K., Fanning, C.M. & Beranek, L.P. (2005) Reliability and longitudinal change of detrital-zircon age spectra in the Snake River system, Idaho and Wyoming: An example of reproducing the bumpy barcode. Sedimentary Geology, 182(1-4), 101–142. Available from: doi:10.1016/j.sedgeo.2005.07.012.

Link, P.K., Mahon, R.C., Beranek, L.P., Campbell-Stone, E.A. & Lynds, R. (2014) Detrital zircon provenance of Pennsylvanian to Permian sandstones from the Wyoming craton and Wood River Basin, Idaho, U.S.A. Rocky Mountain Geology, 49(2), 115–136. Available from: doi:10.2113/gsrocky.49.2.115.

Linnemann, U., Ouzegane, K., Drareni, A., Hofmann, M., Becker, S., Gärtner, A. & Sagawe, A. (2011) Sands of West Gondwana: An archive of secular magmatism and plate interactions — A case study from the Cambro-Ordovician section of the Tassili Ouan Ahaggar (Algerian Sahara) using U–Pb–LA-ICP-MS detrital zircon ages. Lithos, 123(1-4), 188–203. Available from: doi:10.1016/j.lithos.2011.01.010.

Ludwig, R.K. (2001) User Manual for Isoplot/Ex rev. 2.49. Berkeley Geochronology Center Special Publication, 1a, 1–58. Available from: doi:10.1016/S1471-3918(01)80062-9.

Lund Snee, J.-E., Miller, E.L., Grove, M., Hourigan, J.K. & Konstantinou, A. (2016) Cenozoic paleogeographic evolution of the Elko Basin and surrounding region, northeast Nevada. Geosphere, 12(2), 464–500. Available from: doi:10.1130/GES01198.1.

Malfait, W.J., Seifert, R., Petitgirard, S., Perrillat, J.-P., Mezouar, M., Ota, T., Nakamura, E., Lerch, P. & Sanchez-Valle, C. (2014) Supervolcano eruptions driven by melt buoyancy in large silicic magma chambers. Nature Geoscience, 7, 122-125. Available from: doi:10.1038/ngeo2042.

Malkowski, M.A., Sharman, G.R., Johnstone, S.A., Grove, M.J., Kimbrough, D.L. & Graham, S.A. (2019) Dilution and propagation of provenance trends in sand and mud: Geochemistry and detrital zircon geochronology of modern sediment from central California (U.S.A.). American Journal of Science, 319(10), 846–902. Available from: doi:10.2475/10.2019.02.

Matthews, W., Guest, B. & Madronich, L. (2017) Latest Neoproterozoic to Cambrian detrital zircon facies of western Laurentia. Geosphere, 14(1), 243–264. Available from: doi:10.1130/GES01544.1.

McCoy, W.D. (1987) Quaternary aminostratigraphy of the Bonneville Basin, western United States. Geological Society of America Bulletin, 98(1), 99–112. Available from: doi:10.1130/0016-7606(1987)98<99:QAOTBB>2.0.CO;2.

McRivette, M.W., Yin, A., Chen, X. & Gehrels, G.E. (2019) Cenozoic basin evolution of the central Tibetan plateau as constrained by U-Pb detrital zircon geochronology, sandstone petrology, and fission-track thermochronology. Tectonophysics, 751, 150–179. Available from: doi:10.1016/j.tecto.2018.12.015.

Memeti, V., Paterson, S., Matzel, J., Mundil, R. & Okaya, D. (2010) Magmatic lobes as "snapshots" of magma chamber growth and evolution in large, composite batholiths: An example from the Tuolumne intrusion, Sierra Nevada, California. Geological Society of America Bulletin, 122(11-12), 1912–1931. Available from: doi:10.1130/B30004.1.

Menckhoff, K. (2010) Mikro-Oberflächenstrukturen auf detritischen Quarzkörnern als Paläo-klimaindikator. PhD book: Hamburg.

Morrison, R.B. (1964) Lake Lahontan: Geology of southern Carson Desert, Nevada. U.S. Geological Survey Professional Paper, 401(401).

Muhs, D.R., Bettis III, E.A., Roberts, H.M., Harlan, S.S., Paces, J.B. & Reynolds, R.L. (2013) Chronology and provenance of last-glacial (Peoria) loess in western Iowa and paleoclimatic implications. Quaternary Research, 80(03), 468–481. Available from: doi:10.1016/j.yqres.2013.06.006.

Muhs, D.R., Budahn, J.R., Johnson, D.L., Reheis, M., Beann, J., Skipp, G., Fisher, E. & Jones, J.A. (2008) Geochemical evidence for airborne dust additions to soils in Channel Islands National Park, California. Geological Society of America Bulletin, 120(1-2), 106–126. Available from: doi:10.1130/B26218.1.

Munroe, J.S., Norris, E.D., Carling, G.T., Beard, B.L., Satkoski, A.M. & Liu, L. (2019) Isotope fingerprinting reveals western North American sources of modern dust in the Uinta Mountains, Utah, USA. Aeolian Research, 38, 39–47. Available from: doi:10.1016/j.aeolia.2019.03.005.

Nevada Bureau of Mines and Geology (2018) Interactive geologic map of Nevada. Available from: https://gisweb.unr.edu/NevadaGeology/ [Accessed 26 May 2020].

Nie, J., Pullen, A., Garzione, C.N., Peng, W. & Wang, Z. (2018) Pre-Quaternary decoupling between Asian aridification and high dust accumulation rates. Science Advances, 4(2), eaao6977. Available from: doi:10.1126/sciadv.aao6977.

Nie, J., Steven, T.A., Rittner, M., Stockli, D.F., Garzanti, E., Limonta, M., Bird, A., Andò, S., Vermeesch, P., Saylor, J., Lu, H., Breecker, D., Hu, X., Liu, S., Resentini, A., Vezzoli, G., Peng, W., Carter, A., Ji, S. & Pan, B. (2015) Loess Plateau storage of Northeastern Tibetan Plateau-derived Yellow River sediment. Nature Communications, 6, 8511. Available from: doi:10.1038/ncomms9511.

Niemi, N.A. (2013) Detrital zircon age distributions as a discriminator of tectonic versus fluvial transport: An example from the Death Valley, USA, extended terrane. Geosphere, 9(1), 126–137. Available from: doi:10.1130/GES00820.1.

O'Connor, J. (2016) The Bonneville Flood—A Veritable Débâcle. In: Oviatt, C.G. & Shroder, J.F. (Eds.) Lake Bonneville: A scientific update. Elsevier: Amsterdam, Netherlands, pp. 105–126.

Oviatt, C.G. (1988) Late Pleistocene and Holocene lake fluctuations in the Sevier Lake basin, Utah, USA. Journal of Paleolimnology, 1(1), 9–21. Available from: doi:10.1007/BF00202190.

Oviatt, C.G. (2015) Chronology of Lake Bonneville, 30,000 to 10,000 yr B.P. Quaternary Science Reviews, 110, 166–171. Available from: doi:10.1016/j.quascirev.2014.12.016.

Oviatt, C.G. (2016) The Snowplow, Lake Bonneville, Utah. Available from: https://serc.carleton.edu/39953 [Accessed 26 May 2020].

Oviatt, C.G. & Shroder, J.F. (2016) Introduction. In: Oviatt, C.G. & Shroder, J.F. (Eds.) Lake Bonneville: A scientific update. Elsevier: Amsterdam, Netherlands, pp. xxv–xxxvi.

Paterson, S.R. & Ducea, M.N. (2015) Arc Magmatic Tempos: Gathering the Evidence. Elements, 11(2), 91–98. Available from: doi:10.2113/gselements.11.2.91.

Pavich, M.J. & Chadwick, O.A. (2003) Soils and the Quaternary climate system. In: Gillespie, A.R., Porter, S.C. & Atwater, B.F. (Eds.) The Quaternary Period in the United States. Elsevier: Amsterdam, pp. 311–330.

Pierce, K.L. (2003) Pleistocene glaciations of the Rocky Mountains. In: Gillespie, A.R., Porter, S.C. & Atwater, B.F. (Eds.) The Quaternary Period in the United States. Elsevier: Amsterdam, pp. 63–76.

Presley, D.R., Hartley, P.E. & Ransom, M.D. (2010) Mineralogy and morphological properties of buried polygenetic paleosols formed in late quaternary sediments on upland landscapes of the central plains, USA. Geoderma, 154(3-4), 508–517. Available from: doi:10.1016/j.geoderma.2009.03.015.

Raab, T., Leopold, M. & Völkel, J. (2007) Character, age, and ecological significance of Pleistocene periglacial slope deposits in Germany. Physical Geography, 28(6), 451–473. Available from: doi:10.2747/0272-3646.28.6.451.

Raeside, J. D. (1964) Loess Deposits of the South Island, New Zealand, and Soils Formed on them, New Zealand Journal of Geology and Geophysics, 7(4), 811-838. Available from: doi: 10.1080/00288306.1964.10428132

Rahl, J.M., Reiners, P.W., Campbell, I.H., Nicolescu, S. & Allen, C.M. (2003) Combined single-grain (U-Th)/He and U/Pb dating of detrital zircons from the Navajo Sandstone, Utah. Geology, 31(9), 761. Available from: doi:10.1130/G19653.1.

Reheis, M.C., Adams, K.D., Oviatt, C.G. & Bacon, S.N. (2014) Pluvial lakes in the Great Basin of the western United States—a view from the outcrop. Quaternary Science Reviews, 97, 33–57. Available from: doi:10.1016/j.quascirev.2014.04.012.

Reynolds, R.L., Reheis, M.C., Neff, J.C., Goldstein, H. & Yount, J. (2006) Late Quaternary eolian dust in surficial deposits of a Colorado Plateau grassland: Controls on distribution and ecologic effects. Catena, 66(3), 251–266. Available from: doi:10.1016/j.catena.2006.02.003.

Rønnevik, C., Ksienzyk, A.K., Fossen, H. & Jacobs, J. (2017) Thermal evolution and exhumation history of the Uncompahgre Plateau (northeastern Colorado Plateau), based on apatite fission track and (U-Th)-He thermochronology and zircon U-Pb dating. Geosphere, 13(2), 518–537. Available from: doi:10.1130/GES01415.1.

Roskowski, J.A., Patchett, P.J., Spencer, J.E., Pearthree, P.A., Dettman, D.L., Faulds, J.E. & Reynolds, A.C. (2010) A late Miocene-early Pliocene chain of lakes fed by the Colorado River: Evidence from Sr, C, and O isotopes of the Bouse Formation and related units between Grand Canyon and the Gulf of California. Geological Society of America Bulletin, 122(9-10), 1625–1636. Available from: doi:10.1130/B30186.1.

Sack, D. (2020) Reconstructing the chronology of Lake Bonneville: An historical review. Binghamton Geomorphology Symposium, 19, 223–256.

Saleeby, J.B., Ducea, M.N., Busby, C.J., Nadin, E.S. & Wetmore, P.H. (2008) Chronology of pluton emplacement and regional deformation in the southern Sierra Nevada batholith, California. In: Wright, J.E. & Shervais, J.W. (Eds.) Ophiolites, arcs, and batholiths: a tribute to Cliff Hopson. Geological Society of America, pp. 397–427.

Schide, K.H., Jewell, P.W., Oviatt, C.G., Jol, H.M. & Larsen, C.F. (2018) Transgressive-phase barriers as indicators of basin-wide lake-level changes in late Pleistocene Lake Bonneville, Utah, USA. Geomorphology, 318, 390–403. Available from: doi:10.1016/j.geomorph.2018.07.007.

Shaanan, U. & Rosenbaum, G. (2018) Detrital zircons as palaeodrainage indicators: insights into southeastern Gondwana from Permian basins in eastern Australia. Basin Research, 30, 36–47. Available from: doi:10.1111/bre.12204.

Shang, Y., Prins, M.A., Beets, C.J., Kaakinen, A., Lahaye, Y., Dijkstra, N., Rits, D.S., Wang, B., Zheng, H. & van Balen, R.T. (2018) Aeolian dust supply from the Yellow River floodplain to the Pleistocene loess deposits of the Mangshan Plateau, central China: Evidence from zircon U-Pb age spectra. Quaternary Science Reviews, 182, 131–143. Available from: doi:10.1016/j.quascirev.2018.01.001.

Sircombe, K.N. & Hazelton, M.L. (2004) Comparison of detrital zircon age distributions by kernel functional estimation. Sedimentary Geology, 171(1-4), 91–111. Available from: doi:10.1016/j.sedgeo.2004.05.012.

Smalley, I., O'Hara-Dhand, K., Wint, J., Machalett, B., Jary, Z. & Jefferson, I. (2009) Rivers and loess: The significance of long river transportation in the complex event-sequence approach to loess deposit formation. Quaternary International, 198(1-2), 7–18. Available from: doi:10.1016/j.quaint.2008.06.009.

Snell, K.E., Koch, P.L., Druschke, P., Foreman, B.Z. & Eiler, J.M. (2014) High elevation of the 'Nevadaplano' during the Late Cretaceous. Earth and Planetary Science Letters, 386, 52–63. Available from: doi:10.1016/j.epsl.2013.10.046.

Soil Survey Staff (2014) Keys to Soil Taxonomy, 12th edn. USDA-Natural Res. Conserv. Serv.: Washington, DC.

Spencer, C.J., Yakymchuk, C. & Ghaznavi, M. (2017) Visualising data distributions with kernel density estimation and reduced chi-squared statistic. Geoscience Frontiers, 8(6), 1247–1252. Available from: doi:10.1016/j.gsf.2017.05.002.

Stevens, T., Palk, C., Carter, A., Lu, H. & Clift, P.D. (2010) Assessing the provenance of loess and desert sediments in northern China using U-Pb dating and morphology of detrital zircons. Geological Society of America Bulletin, 122(7-8), 1331–1344. Available from: doi:10.1130/B30102.1.

Sun, J., Ding, Z., Xia, X., Sun, M. & Windley, B.F. (2018) Detrital zircon evidence for the ternary sources of the Chinese Loess Plateau. Journal of Asian Earth Sciences, 155, 21–34. Available from: doi:10.1016/j.jseaes.2017.10.012.

Újvári, G. & Klötzli, U. (2015) U–Pb ages and Hf isotopic composition of zircons in Austrian last glacial loess: constraints on heavy mineral sources and sediment transport pathways. International Journal of Earth Sciences, 104(5), 1365–1385. Available from: doi:10.1007/s00531-014-1139-x.

Újvári, G., Klötzli, U., Kiraly, F. & Ntaflos, T. (2013) Towards identifying the origin of metamorphic components in Austrian loess: insights from detrital rutile chemistry, thermometry and U–Pb geochronology. Quaternary Science Reviews, 75, 132–142. Available from: doi:10.1016/j.quascirev.2013.06.002.

Újvári, G., Varga, A., Ramos, F.C., Kovács, J., Németh, T. & Steven, T.A. (2012) Evaluating the use of clay mineralogy, Sr–Nd isotopes and zircon U–Pb ages in tracking dust provenance: An example from loess of the Carpathian Basin. Chemical Geology, 304-305, 83–96. Available from: doi:10.1016/j.chemgeo.2012.02.007.

Utah Geological Survey (2018) Geologic Map Portal. Available from: geology.utah.gov/apps/intgeomap/ [Accessed 30 December 2018].

Vermeesch, P. (2004) How many grains are needed for a provenance study? Earth and Planetary Science Letters, 224(3-4), 441–451. Available from: doi:10.1016/j.epsl.2004.05.037.

Vermeesch, P. (2009) Radialplotter: a java application for fission track, luminescence and other radial plots. Radiation Measurements, 44(4), 409–410. Available from: doi:10.1016/j.radmeas.2009.05.003.

Vermeesch, P. (2012) On the visualisation of detrital age distributions. Chemical Geology, 312-313, 190–194. Available from: doi:10.1016/j.chemgeo.2012.04.021.

Vermeesch, P. (2013) Multi-sample comparison of detrital age distributions. Chemical Geology, 341, 140–146. Available from: doi:10.1016/j.chemgeo.2013.01.010.

Vermeesch, P. (2020) Provenance: Statistical toolbox for sedimentary provenance analysis (Version 2.4). Available from: http://provenance.london-geochron.com/ [Accessed 26 May 2020].

Wernicke, B. (2011) The California River and its role in carving Grand Canyon. Geological Society of America Bulletin, 123(7-8), 1288–1316. Available from: doi:10.1130/B30274.1.

Whitmeyer, S.J. & Karlstrom, K.E. (2007) Tectonic model for the Proterozoic growth of North America. Geosphere, 3(4), 220. Available from: doi:10.1130/GES00055.1.

Wilson, P.I., McCaffrey, K.J., Wilson, R.W., Jarvis, I. & Holdsworth, R.E. (2016) Deformation structures associated with the Trachyte Mesa intrusion, Henry Mountains, Utah: Implications for sill and laccolith emplacement mechanisms. Journal of Structural Geology, 87, 30–46. Available from: doi:10.1016/j.jsg.2016.04.001.

Zhu, L. & Fan, M. (2018) Detrital zircon provenance record of middle Cenozoic landscape evolution in the southern Rockies, USA. Sedimentary Geology, 378, 1–12. Available from: doi:10.1016/j.sedgeo.2018.10.003.

Zieger, J., Rothe, J., Hofmann, M., Gärtner, A. & Linnemann, U. (2019) The Permo-Carboniferous Dwyka Group of the Aranos Basin (Namibia) – How detrital zircons help understanding sedimentary recycling during a major glaciation. Journal of African Earth Sciences, 158, 103555. Available from: doi:10.1016/j.jafrearsci.2019.103555.

Zieger, J., Stutzriemer, M., Hofmann, M., Gärtner, A., Gerdes, A., Marko, L. & Linnemann, U. (2020) The evolution of the southern Namibian Karoo-aged basins: implications from detrital zircon geochronologic and geochemistry data. International Geology Review, 1–24. Available from: doi:10.1080/00206814.2020.1795732.

5 Extended summary

5.1 Synthesis

I) Reliable datability of DZ below the established age limit (>= 100 Ma)

According to chapter 2 we showed that it is possible to get reliable zircon ages even with such young zircons if you are aware of the fragility of the results. The Jemez tephras are 1.513 ± 0.021 Ma and 1.316 ± 0.012 Ma, the La Sal Mountains tephras (2 samples) are 1.327 ± 0.017 Ma (2013) and 1.341 ± 0.059 Ma (2014) old. Our results built on the mean age of the youngest cluster of grain ages may be seen as groups of analyses resulting in ages close together, thereby validating each other even without a reliable Pb-Pb age. We knew the ages from published ^{40}Ar/^{39}Ar datings; being aware of the different closure temperatures of the dated minerals (and therefore the slightly different ages), we could show reliable ages of four Pleistocene tephra samples. Thus, we can encourage giving the U–Pb method – which is available in a variety of labs worldwide – a try for dating and even distinguishing volcanic material of undisclosed age even if the assumed age is Quaternary. Since zircons from the same source (only eruptions differed in time), which are difficult or impossible to distinguish because of the same inherited material using methods such as glass chart analysis alone, U-Pb dating offers an additional possibility of validation. (cf. chapter 2.1, 2.5 and 2.6)

II) Distinguishing palaeosol parent materials and their aeolian components

In the same profile sampled for the La Sal Mountains tephra layer, several cover beds (CB) separated by palaeosols follow beneath. This situation offers the first evidence of cover-bed ages older than the mid-Pleistocene revolution. We tried to distinguish three of these layers by looking at their chemical composition and at detrital-zircon ages. We found that the aeolian matter varied remarkably between these layers, although they are in the same location. Furthermore, we found older zircons in the uppermost studied layer that should occur as well in the underlying layers, which is not the case, if they had received constant influx from similar sources. The zircon ages are consistent with the field-derived evidence from palaeosols that the CB underlying the LSM tephra layer comprise several CB of different ages, as the distributions of zircon U-Pb ages vary remarkably. Thus, there must have been different contributions between the different time slots.

On the other hand, we show that physical and chemical properties of CB and especially of soils older than the mid-Pleistocene Revolution were similar to those formed after the revolution. The amount of time for soil formation must have been quite similar to the Late Pleistocene, despite the different durations of glacial-interglacial cycles. Contrary to this, the oldest exposed soil is remarkably more mature than the younger ones. Stage four or higher carbonate-enriched horizons have already been assumed to need more than 500 kyr to form (cf. MACHETTE 1985). This mature soil was formed possibly even before the Quaternary, at least during a long time of geomorphic stability under environmental conditions different from those that formed the overlying soils. The most remarkable finding was the high accuracy with which the provenance of the aeolian components of some of the CB appears to be discernible in the three layers in the La Sal Mountain profile, which indicates the region around the Grand Canyon as the source region. Grains from underlying and up-slope younger rocks do essentially not contaminate them.

III) Test of the applicability of zircon analysis for area-wide provenance study

The findings from a single profile encouraged expanding the zircon analyses supraregionally. We could show that the two different physiogeographical regions Great Basin (GB) and Colorado Plateau (CP) have different zircon patterns (cf. chapter 5.2). The GB profiles are dominated by the tephra zircons of the various volcanic activities between Eocene and Oligocene. Surprisingly, these volcanic zircons appear only marginally in the CP profiles. In addition, the Cordilleran input (Cretaceous) is clearly visible within the GB profiles. This difference is possibly due to the palaeolakes Lahonton and Bonneville, which apparently intercepted and enclosed the zircons and thus prevented further aeolian transportation. (cf. fig. 4.1, chapters 4.3.3 & 4.6)

The CP profiles, even with different underlying bedrocks (Quaternary, Cretaceous and Jurassic), still have similar zircon age patterns. We discussed the missing dominance of the bedrock and assume this is due to fundamental aeolian input and only a faint influence from weathering of local rocks. Though the area eventually must have been more humid, the weathering of the bedrock did not increase proportionally. (cf. chapters 4.5.1, 4.5.3 & 2.5). This is also indicated by the fresh appearance of the La Sal Mountains tephra layer despite of its age although it is embedded into a developed soil (chapter 2, Fig. 2.1). Thus, we compared our data to the extensive database of detrital zircon ages in the US: Multidimensional scaling (MDS) shows – compared to published ages from possible sources - first regional similarities. At this scale involving all studied profiles, this statistical approach shows how close or distant the samples are to each other. Again, to increase the accuracy one should increase the profile number. (cf. chapter 4.5.4)

Because of the small amount of young DZ, which dominate the GB, detected in the CP, we exclude a westerly wind direction for the CP profiles. The wind for the aeolian distribution must have come either from the north or from the south. We suspect the latter due to the moisture needed for the thickness of the soils, which most probably occurred during El Niño-like atmospheric circulation conditions when precipitation reached farther north.

We still cannot name a transportation direction for the GB. Many of the young (≤120 Ma) zircons are idiomorphic and show no or only traces of collisions (see below). Therefore, these young zircon ages with a high degree of probability come from Cretaceous and Paleogene volcanic flare-ups, were deposited close to the studied sites and had no long transportation pathway. However, there are too many other open questions, which definitely should be concerned in further research.

In figure 4.2, we demonstrate the development of the soils in the study area between the lake level highs (Lahonton and Bonneville basins). The palaeosol between the Bonneville and Provo high stand is developed in the so-called Wyemaha formation. It consists of the Upper and the Lower Wyemaha layers, and the palaeosol is divided in two clearly discernible soils as well. We can show the presence of the Upper and the Lower Wyemaha layers in the whole region: Fig. 5.1 (originally from chapter 4, Appendix, SI 2) connects all profiles of chapter 4. The stratigraphic approach primarily is based on the interfingering

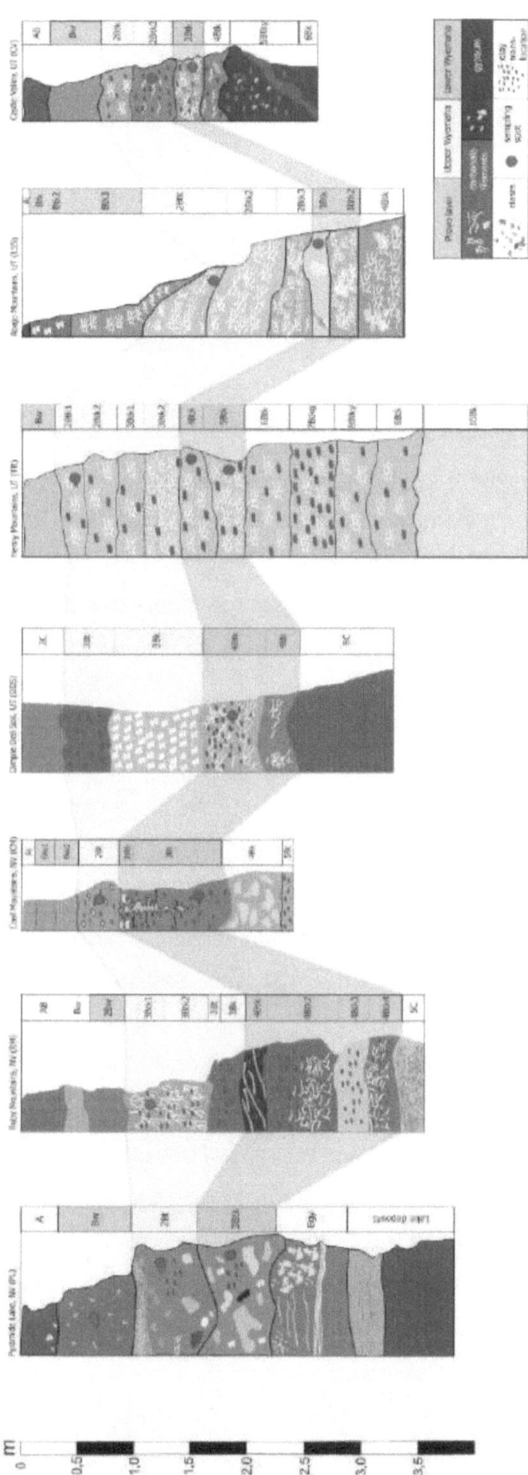

Figure 5.1 - Charts of profiles and stratigraphic connections.
The profile sketches reflect relative hardness under field conditions and the colours are an abstraction of the natural colour.

of CB with lake deposits and tills (fig. 4.2), tracing the unconformity between the Wyemaha layers is based on palaeopedological means.

The grainsize end-member modelling confirms that there is an important aeolian contribution to the samples within a narrow band of textures around 65 µm, in the typical range of short-travelled, suspended aeolian matter. Furthermore, far-travelled aeolian matter has finer grainsizes, which cluster with a somewhat wider range around 30 µm. The grainsizes, which are not air-borne, do not play an important role within our samples because they are mainly not within the range of sizes of the studied zircons. The modelled aeolian grain-size distribution is on par with measured dust grain sizes slightly east of our study area, which keeps the discussion open about the differences between recent environmental conditions and those during soil formation. (cf. chapter 4.5.1)

Zircon morphology is currently still underrated in the regional literature and databases of bedrock zircon morphology are missing to be able to make comparisons. Therefore, at the present time, we cannot draw any conclusions about the provenance only through the zircon surface. But we can show that most of our GB and CP zircons have a non-idiomorphic surface in contrast to the examined zircons of the two Jemez tephras (cf. Chapter 2), which are on average smaller but also more idiomorphic than the zircons of the GB and CP profiles. Thereafter, many of those DZ in the GB and CP may have had a long transportation pathway, though rounding is no certain indicator for this. However, idiomorphism does not exclude long transportation pathways, because zircons with zonation may unshell their rim upon impact. We calculated the linear regressions between zircon morphological parameters, and they are all statistically significant (Table 4.2). Not very surprisingly, collision marks and roundness correlate with each other with the highest coefficient, probably because both are products of the transportation processes. Having a closer look at the surfaces of, e.g., probable Cordilleran zircons shows idiomorphic grains without or with just a few collision marks. We interpret this as indicating strong kinetic energy of this volcanism without many steps in a transportation cascade (see above). Because there was quite a lot of volcanic activity, direct transport in the study area should actually be considered more frequently. (cf. fig 4.13).

5.2 Regional differences and similarities

If one compares the age distribution of the CP profiles with those of the GB, clear differences become apparent. (cf. Chapters 4.3.4 & 4.5.3)

Young signals (≤ 120 Ma) dominate the age distributions of the GB samples. We found ages possibly linked to the latest and southernmost volcanic eruptions (~ 18 – 33 Ma) or even the early eruptions of the Yellowstone hot spot spur. The further west the profiles are located, the higher is the input from the Cordilleran magmatic arc (~ 100 Ma). Following the same direction, the ages connected to the Appalachian orogenies increase. Findings of the Granite Rhyolite Province (cf. fig. 1.1 & 4.8) cannot be explained with

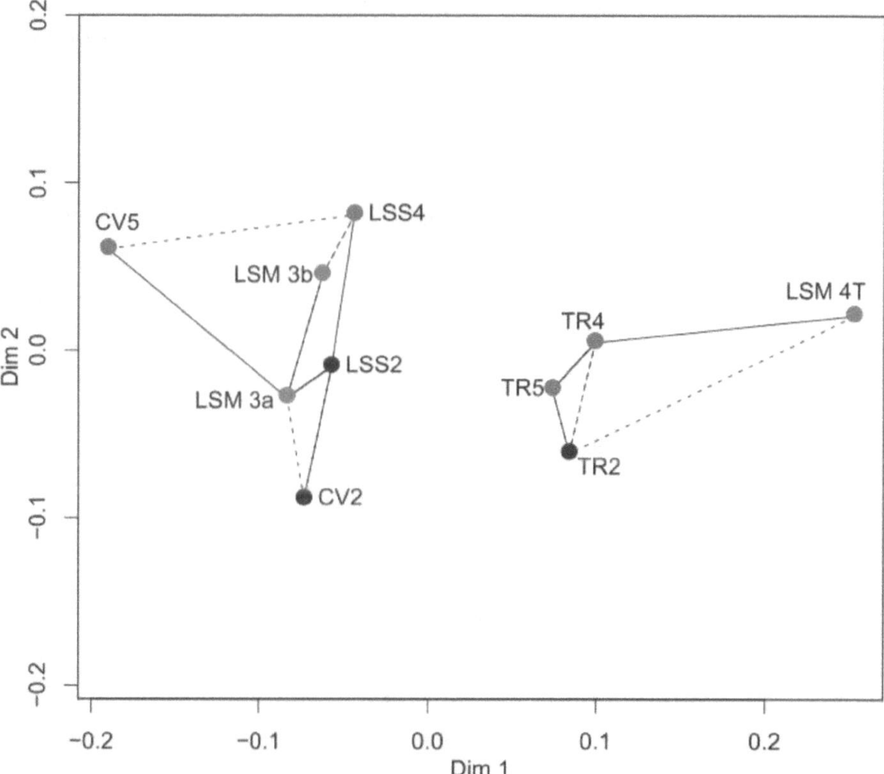

Figure 5.2 – MDS (Kolmogorow-Smirnow distance) of all CP samples including LSM (chapter 3); green = whole LSM profile, blue = the upper Wyemaha layer of the profiles, red = lower Wyemaha layer.

underlying bedrock. The input must stem from the southeast, whereas a direct transportation without stopovers is unlikely here. (cf. fig. 4.13)

The CP profile distribution is very different to GB. However, the CP profiles are similar to each other despite there are over 100 km between them.

Here I show three new figures bringing together all CP samples from chapter 4 and the La Sal Mountains (LSM) profile from chapter 3. The selection of displayed ages is the same for all samples as outlined in chapter 4 (including discordant ages ≤120 Ma). Having a closer look at the MDS of these profiles (fig. 5.2), one can clearly see two groups of similarity. 4T (which is the deepest in the LSM profile) is very distant to all other samples, interestingly it is most similar to TR, which is in the SW. This could mean that the input during sedimentation of 4T was similar to the TR profile, supporting the conclusion in chapter 3 of a more south-western provenance of this layer. LSM 3a and 3b, on the other hand, are more like LSS to the south (and the nearby CV). This could be a confirmation of the assumption of chapter 3 regarding the southern direction of the input to these two layers. Altogether the LSM samples depict the same degree of dissimilarity as the whole other data set dispersed over the northern CP. This may be interpreted that environmental conditions during the formation of the layers changed more decisively than either later (during the time of upper and lower Wyemaha layers) or at lower elevations.

Conversely, especially the TR samples are very close to each other, so hardly anything has changed here over time (possibly nothing at all because the samples represent only a subpopulation of the total DZ population). In contrast, CV and LSS show more pronounced dissimilarities within the same profile.

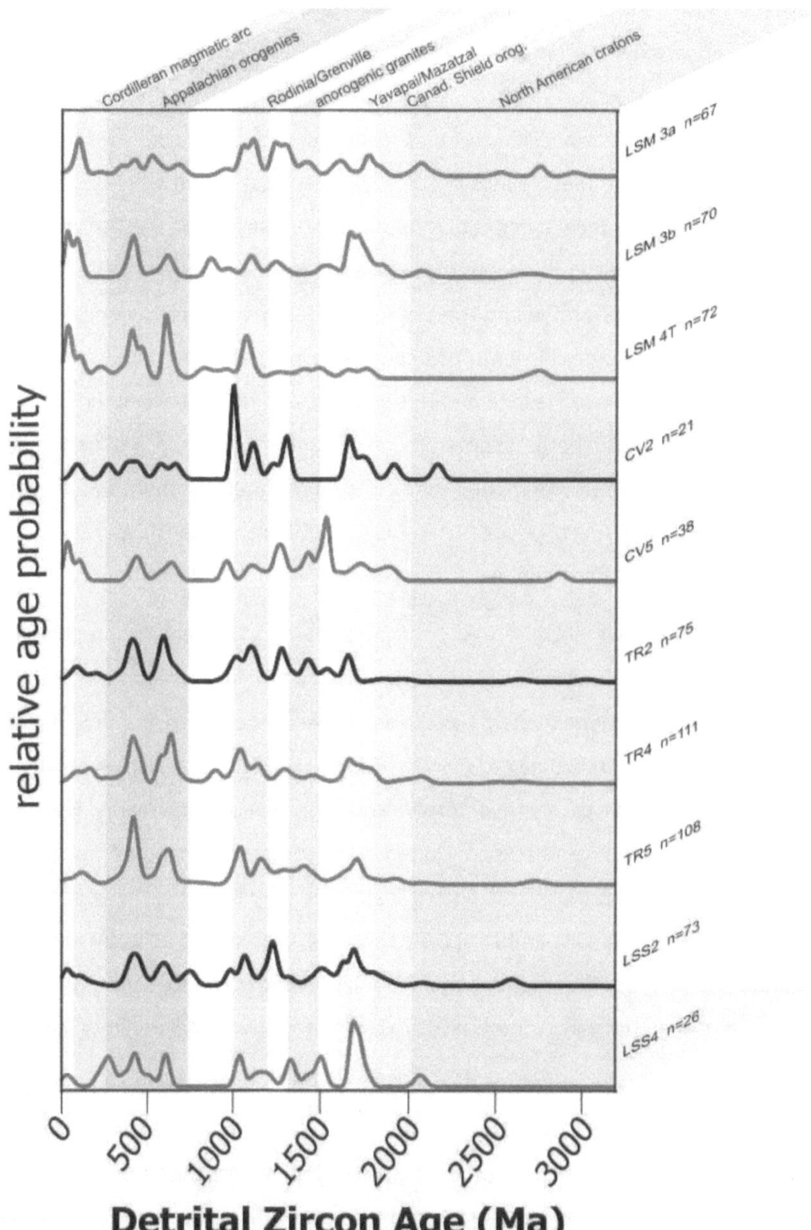

Figure 5.3 – KDE of all CP samples including the upper La Sal Mountains samples. Bin width = 25 myr. Green = whole LSM profile, blue = upper Wyemaha layer of the profiles, red = lower Wyemaha layer.

Figure 5.3 shows relative age probabilities of all these samples (CP including LSM). There is a main Appalachian input within all CP profiles as well as from Rodinia/Grenville orogenies. Fluvial transportation by a transcontinental river system and an relocation during the Jurassic to the CP by wind, and during the Cretaceous within the CP, is assumed. One can see these peaks but as well how dissimilar they look in detail. The Granite Rhyolite Province (from the southeastern margin of the North American continent) is only shown in the upper LSM (3a) and TR2 and the lower CV5 and LSS4. It is remarkable that the TR profile and the LSS2 show nearly the same age peaks. The only layer that actually distances itself in this comparison is the upper Wyemaha in the CV, which is amazing, because the LSM profile is so close, and none of the other layers shows such an age distribution there. However, if one considers the temporal aspect, namely at least one change in the environmental conditions, since otherwise there would have been no stopping and restarting of pedogenesis and the stratigraphic layer separation, changes within profiles are not necessary but likely.

All three LSM samples show a few younger zircons (≤120 Ma) connected with the Cordilleran magmatic arc and the Paleogene volcanic provinces (separated in the figure by colour coding), but none of them has the same age as the underlying La Sal mountains laccolith, which is well U-Pb dated at ~ 29 Ma (RØNNEVIK ET AL. 2017) (cf. fig. 5.4). All young zircons are idiomorphic or near idiomorphic and show few collision marks. This indicates rather direct input by volcanism as concluded for CP samples in chapter 4. Comparing this to the profile down the valley, CV, shows that there only the lower layer shows young zircons (however, the small number of DZ in these samples should be taken into account). All CP samples with a sufficient number of DZ analyses show a minor contribution of these age groups, mainly from the southern GB volcanics and the Cordilleran magmatic arc.

Only ~ 2 km (beeline) and ~ 260 m (height difference) separate the two sampling points LSM (cf. fig. 3.3 & 5.2) and CV (cf. fig. 5.1 & 5.2; Tab. 4.1) from each other. They have not been included or compared to the other CP profiles yet because the focus changed slightly after chapter 3 to analysing more profiles and samples. There are many similarities, but also some very distinct differences between these two profiles as even in the LSM profile itself. The Appalachian signal decreases sharply the younger the soil is. Also the three LSM layers seem to have different distributions within each layer. However, maybe this is because of the Canyon structure of the CP. We already assume palaeolakes trapping the most of the young volcanic and Cordilleran DZ. maybe the same happened with the three stand-alone peaks of the La Sal Mountains. Thus, they got the high variety, the mesas could not catch them, thus, they were removed by aeolian or fluvial transportation and the 'rest' ended up in the canyons. We do have unpublished data of almost 800 DZ of the Little Sahara dunes, UT, eastern GB, which show an astonishing strong signal similar to the southwestern CP profiles.

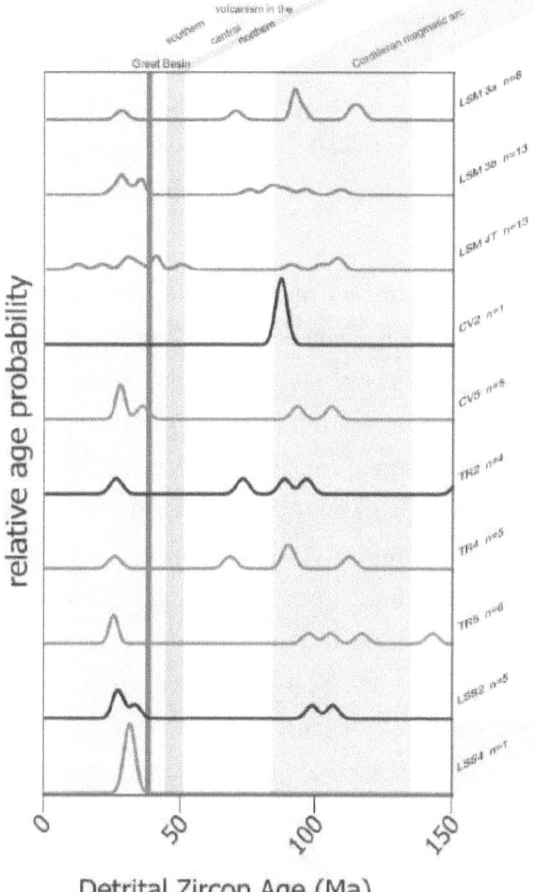

Figure 5.4 - KDE of all CP samples including the upper La Sal Mountains samples (LSM) limited to ages ≤150 Ma. Bin width = 2 myr. The main provenances in this age range are colour-coded as well as the U-Pb age of the LSM laccolith at ~29 Ma (red perpendicular line).

5.3 Outlook

This new approach gives the opportunity for many different disciplines to work together. We demonstrated the feasibility of provenance research even on Quaternary aeolian substrates using a combination of U-Pb dating and morphological analysis of DZ. However, this approach has to be tested in other regions. Are the answers more unambiguous than in this region? Is it a useful tool for other substrates, e.g., a dune within surrounding mountain ranges? Or for distinguishing two dune fields close to each other?

The answer of the question about the palaeoenvironmental conditions during soil formation are still unclear. Again there should be try out other methods e.g., biomarkers or pollen. Even if there is a risk that there is little that can be evaluated in the dry soils of this semi-arid climate (LERCH 2017, LERCH ET AL. 2018). The questions whether the shorter periods of the glacial-interglacial cycles have been more humid and colder or hotter or if soil formation in drier but hotter or more humid but colder within the same time range. It may be assumed that the soil formation in the studied layers took place under environmental conditions different from those that formed the top soils and, on the other hand, that the oldest studied soil formed under even more extreme conditions, possibly before the Quaternary. Such an intensity of pedogenesis is possible only during a long time of low geomorphic activity, especially because the preservation of calcic horizons indicates that precipitation was not exceptionally high during soil formation. Frequent climate changes would have interrupted soil formation and would have induced its start-over in a new deposit. This long duration also points to a period preceding the frequent changes of conditions typical for the Quaternary. Thus, we should try out to combine other methods for getting an answer.

Using U-Pb dating as a stand-alone tool is of limited use for aeolian (re-allocated) material. There are too many variables, which prevent clear answers. There have to be further investigations which methods helping each other to support answers. Obviously, condensing the database for morphology of autochthon DZ is nessesary as well as get more information about the chemical composition. Furthermore, the question arose about 'How many grains are enough for aeolian DZ provenance in such an admixed substrate?' It is obvious from our findings that larger sample sizes than were feasible in this pilot study probably would improve the conclusions.

Nevertheless, the understanding of the changes of wind directions, humidity and necessities of forming such huge soils within such arid regions will lead to the understanding of durations of climate changes measured in human life.

The lake deposits of the palaeolakes Lahonton and Bonneville may be very helpful together with biomarker analyses. Further sampling and analysing to get a denser network are necessary to testify this approach, also in other study areas.

To verify the sequence-stratigraphic approach it is urgently needed to carry out OSL, better IRSL, dating and here, too, more samples have to be examined and the analyses of the samples already prepared in the luminescence laboratory have to be processed.

6 Supplementary Information

6.1 Supplementary material chapter 2 'Capability of U-Pb dating of zircons from Quaternary tephra: Jemez Mountains, NM, and La Sal Mountains, UT, USA'

The Supplement related to this article is also available online at https://doi.org/10.5194/egqsj-67-7-2018-supplement.

6.1.1 Raw data electron microprobe analyses of glass shards from tephra layers

a)   Raw data (weight-percent oxide), sample 2014-NM-Gu

	SiO_2	TiO_2	Al_2O_3	Fe_2O_3	MnO	MgO	CaO	Na_2O	K_2O	P_2O_5	BaO	SO_3	Cl	Total
3_34	74.37	.05	11.63	1.37	.08	.02	.24	3.70	4.74	.00	.01	.01	.22	96.43
3_35	72.30	.08	11.31	1.37	.11	.03	.24	3.58	4.92	.01	.03	.00	.21	94.18
3_38	74.46	.08	11.79	1.46	.11	.00	.23	3.75	4.65	.00	.00	.02	.23	96.77
3_06	73.91	.05	11.71	1.30	.09	.01	.25	3.82	4.26	.01	.04	.00	.23	95.68
3_13	74.22	.03	11.76	1.41	.11	.01	.24	3.84	4.15	.00	.00	.00	.24	96.01
3_14	72.24	.03	11.77	1.50	.10	.19	.35	3.62	4.46	.01	.02	.00	.22	94.51
3_28	74.60	.02	11.79	1.38	.08	.00	.24	3.97	4.39	.00	.01	.00	.22	96.69
3_31	74.06	.03	11.82	1.37	.08	.01	.24	3.57	4.79	.00	.02	.00	.24	96.24
3_32	72.10	.04	11.17	1.42	.12	.09	.29	2.69	4.74	.00	.00	.00	.24	92.90
3_21	75.30	.08	12.00	1.26	.10	.03	.29	3.54	4.66	.00	.04	.02	.12	97.41
3_22	75.94	.06	12.19	1.48	.09	.03	.28	3.81	4.61	.00	.00	.00	.11	98.60
3_40	74.25	.06	11.68	1.39	.09	.02	.23	3.78	4.82	.01	.03	.00	.24	96.57
3_41	74.26	.05	11.72	1.39	.09	.01	.26	3.83	4.34	.01	.05	.01	.23	96.25
3_42	74.17	.06	11.77	1.36	.09	.01	.24	3.94	4.30	.00	.00	.01	.22	96.16
3_47	75.00	.05	12.13	1.14	.07	.02	.29	3.70	4.57	.00	.00	.05	.12	97.14
3_48	73.81	.03	11.64	1.36	.07	.01	.26	3.46	5.16	.00	.01	.03	.22	96.04
3_50	74.62	.06	11.71	1.39	.08	.01	.22	4.00	4.23	.00	.00	.00	.22	96.54
3_54	73.11	.03	12.96	1.29	.09	.01	.23	4.01	5.36	.04	.04	.00	.17	97.34
3_56	74.40	.04	11.69	1.39	.07	.02	.21	3.86	4.62	.00	.00	.02	.25	96.58
3_58	73.86	.03	12.28	1.28	.07	.01	.21	3.90	4.92	.00	.01	.01	.20	96.77
3_60	73.08	.04	11.72	1.27	.06	.01	.23	3.48	4.87	.00	.00	.00	.22	94.98
3_61	73.87	.04	11.67	1.36	.06	.02	.23	3.29	5.24	.00	.01	.01	.23	96.03
3_64	73.39	.04	11.77	1.32	.09	.01	.23	3.45	4.92	.01	.00	.01	.23	95.45

	SiO$_2$	TiO$_2$	Al$_2$O$_3$	Fe$_2$O$_3$	MnO	MgO	CaO	Na$_2$O	K$_2$O	P$_2$O$_5$	BaO	SO$_3$	Cl	Total
3_65	74.00	.06	11.84	1.35	.05	.01	.24	3.73	4.81	.00	.04	.01	.21	96.34
3_66	73.89	.04	11.69	1.33	.10	.01	.27	3.83	4.56	.00	.00	.02	.20	95.94
3_67	73.64	.04	11.81	1.36	.07	.01	.28	3.73	4.60	.01	.02	.01	.23	95.82
3_68	74.35	.05	11.76	1.40	.07	.02	.25	3.96	4.73	.00	.02	.04	.30	96.95
3_69	74.05	.07	11.69	1.41	.11	.01	.23	3.86	4.25	.01	.00	.00	.23	95.92
3_71	74.22	.04	11.79	1.45	.06	.01	.25	3.89	4.57	.00	.00	.00	.21	96.48
3_72	73.74	.04	11.81	1.43	.08	.02	.21	3.46	5.04	.01	.00	.00	.22	96.05
3_73	72.57	.04	11.34	1.35	.09	.02	.24	3.20	4.96	.01	.00	.00	.21	94.03
3_74	73.85	.03	11.73	1.34	.08	.01	.25	3.49	5.20	.01	.02	.00	.25	96.25

b)　Raw data (weight-percent oxide), sample 2014-LSM-T

	SiO$_2$	TiO$_2$	Al$_2$O$_3$	Fe$_2$O$_3$	MnO	MgO	CaO	Na$_2$O	K$_2$O	P$_2$O$_5$	BaO	SO$_3$	Cl	Total
2_01	74.23	.06	11.62	1.52	.04	.02	.27	3.71	4.04	.01	.00	.02	.14	95.69
2_03	71.50	.06	11.80	1.60	.08	.01	.29	4.30	3.77	.05	.00	.00	.31	93.78
2_04	75.15	.10	11.79	1.41	.07	.03	.27	3.84	4.32	.02	.01	.01	.14	97.15
2_06	75.16	.07	11.79	1.50	.07	.02	.27	3.95	3.97	.00	.01	.00	.16	96.97
2_07	74.62	.07	11.60	1.58	.05	.01	.26	4.16	4.10	.00	.03	.01	.31	96.80
2_09	74.53	.08	11.72	1.39	.03	.03	.28	3.88	3.98	.00	.02	.00	.12	96.07
2_10	74.57	.13	11.77	1.39	.08	.05	.27	3.47	4.61	.02	.01	.02	.11	96.49
2_11	73.82	.04	11.72	1.59	.08	.02	.26	3.84	3.91	.02	.00	.00	.18	95.49
2_12	74.55	.07	11.77	1.41	.09	.03	.27	3.89	4.26	.01	.00	.00	.15	96.50
2_13	74.28	.07	11.73	1.37	.07	.03	.30	3.81	4.02	.01	.00	.02	.15	95.86
2_14	74.29	.05	11.60	1.48	.08	.02	.26	3.93	3.97	.02	.02	.01	.08	95.79
2_17	74.39	.03	11.76	1.46	.07	.03	.28	4.02	4.21	.01	.00	.00	.05	96.32
2_18	74.18	.09	11.82	1.49	.08	.02	.26	3.84	4.11	.01	.00	.03	.19	96.10
2_20	74.80	.09	11.75	1.41	.06	.02	.26	3.82	3.99	.00	.00	.00	.14	96.33
2_21	75.02	.09	11.78	1.53	.06	.02	.28	4.02	4.04	.02	.00	.01	.18	97.05
2_23	74.80	.08	12.04	.89	.07	.03	.35	3.70	4.05	.01	.00	.02	.14	96.17

	SiO_2	TiO_2	Al_2O_3	Fe_2O_3	MnO	MgO	CaO	Na_2O	K_2O	P_2O_5	BaO	SO_3	Cl	Total
2_24	74.69	.10	11.79	1.47	.06	.03	.27	3.73	3.98	.01	.06	.02	.13	96.35
2_25	74.94	.06	11.84	1.51	.06	.02	.26	3.97	4.21	.00	.02	.01	.19	97.10
2_26	74.78	.08	11.81	1.43	.08	.08	.33	4.10	4.46	.00	.03	.08	.42	97.68
2_27	74.96	.06	11.86	1.44	.08	.03	.30	3.93	3.75	.00	.03	.00	.17	96.62
2_28	74.54	.07	11.78	1.50	.07	.02	.24	3.89	4.13	.00	.00	.02	.15	96.41
2_29	74.54	.09	11.90	1.45	.07	.02	.25	3.75	4.65	.00	.00	.00	.12	96.84
2_30	74.82	.08	11.82	1.48	.11	.01	.27	3.67	4.57	.00	.03	.00	.18	97.03

c)	Raw data (weight-percent oxide), sample 2013-LSM-T

	SiO_2	TiO_2	Al_2O_3	Fe_2O_3	MnO	MgO	CaO	Na_2O	K_2O	P_2O_5	BaO	SO_3	Cl	Total
6_01	74.66	.14	11.85	1.36	.05	.05	.28	3.60	4.74	.01	.04	.11		96.88
6_02	71.60	.13	11.21	1.33	.05	.04	.30	3.31	4.24	.00	.02	.12		92.34
6_03	75.18	.10	11.78	1.29	.07	.05	.29	3.66	4.53	.03	.00	.10		97.07
6_05	74.63	.14	11.80	1.36	.06	.05	.29	3.63	4.53	.00	.02	.09		96.58
6_06	74.82	.06	11.83	1.42	.05	.02	.27	3.93	4.24	.03	.02	.15		96.83
6_07	75.00	.07	11.87	1.42	.06	.04	.27	4.00	4.26	.00	.01	.17		97.15
6_08	75.19	.06	11.88	1.46	.07	.02	.27	3.94	4.25	.00	.03	.15		97.31
6_09	73.52	.10	11.75	1.46	.09	.03	.31	3.82	4.59	.00	.02	.14		95.83
6_10	75.16	.09	11.86	1.43	.05	.02	.26	3.92	4.29	.01	.00	.18		97.26
6_11	75.07	.15	11.85	1.44	.08	.04	.30	3.67	4.59	.03	.02	.11		97.36
6_12	75.23	.14	11.84	1.46	.02	.05	.29	3.84	4.50	.03	.00	.11		97.51
6_13	74.52	.05	11.77	1.56	.10	.01	.24	3.99	3.88	.00	.03	.24		96.39
6_14	75.25	.12	11.85	1.44	.07	.03	.26	3.63	4.45	.00	.00	.10		97.20
6_16	74.40	.06	11.77	1.61	.11	.01	.26	4.03	4.05	.01	.02	.29		96.60
6_17	74.34	.10	11.80	1.44	.04	.04	.28	3.61	4.66	.00	.03	.14		96.47
6_18	75.31	.07	11.84	1.56	.10	.02	.27	3.90	4.42	.01	.00	.18		97.67
6_19	75.12	.09	11.79	1.51	.07	.04	.25	3.81	4.40	.01	.00	.13		97.22
6_20	75.03	.09	11.82	1.50	.04	.03	.26	3.89	4.20	.00	.00	.11		96.96

	SiO_2	TiO_2	Al_2O_3	Fe_2O_3	MnO	MgO	CaO	Na_2O	K_2O	P_2O_5	BaO	SO_3	Cl
6_21	75.20	.08	11.72	1.45	.10	.02	.27	4.02	4.47	.00	.00	.18	97.51
6_22	75.46	.08	11.65	1.48	.08	.02	.26	3.85	4.27	.00	.03	.14	97.33
6_23	75.11	.09	11.73	1.45	.07	.03	.27	3.78	4.25	.01	.02	.12	96.93
6_24	74.86	.13	11.86	1.43	.06	.03	.30	3.60	4.57	.01	.04	.10	96.97
6_25	72.38	.06	11.57	1.52	.08	.01	.26	3.94	4.47	.00	.00	.28	94.57
6_27	75.22	.11	11.77	1.48	.05	.03	.27	3.66	4.56	.03	.00	.12	97.29
6_28	75.20	.07	11.79	1.47	.05	.03	.29	3.98	4.09	.01	.04	.17	97.19
6_30	74.81	.07	11.75	1.45	.04	.02	.26	3.93	3.93	.00	.00	.17	96.43
6_31	75.18	.10	11.90	1.43	.08	.04	.29	3.70	4.13	.01	.00	.12	96.98
6_32	74.03	.05	11.58	1.66	.08	.01	.23	3.74	4.34	.00	.00	.32	96.05
6_33	75.16	.07	11.78	1.48	.08	.03	.26	3.99	3.83	.00	.00	.18	96.86
6_34	75.29	.07	11.88	1.46	.05	.03	.24	3.94	4.04	.00	.01	.19	97.19
6_35	74.44	.06	11.60	1.42	.08	.02	.28	3.73	4.21	.00	.00	.17	96.03
6_36	75.00	.06	11.79	1.37	.07	.02	.26	3.90	4.42	.00	.00	.21	97.10
6_37	75.19	.09	11.87	1.55	.05	.02	.26	4.10	4.01	.01	.01	.18	97.32
6_38	75.10	.07	11.80	1.51	.06	.03	.26	3.96	4.25	.04	.00	.17	97.25
6_39	75.15	.11	11.82	1.47	.08	.04	.27	3.63	4.62	.02	.01	.15	97.34
6_40	74.35	.16	11.77	1.35	.07	.07	.31	3.55	4.51	.02	.00	.09	96.25
6_41	74.57	.10	11.86	1.04	.07	.04	.30	3.33	4.37	.00	.00	.13	95.80
6_39	74.28	.08	11.84	.97	.07	.04	.34	3.31	4.46	.00	.03	.10	95.51
6_40	74.70	.09	11.93	.92	.08	.04	.29	3.21	4.47	.01	.00	.11	95.85
6_41	73.76	.06	11.89	.96	.08	.04	.31	3.27	4.61	.01	.05	.12	95.16
6_43	74.52	.06	11.73	1.57	.07	.01	.24	4.28	4.14	.01	.02	.31	96.94
6_44	74.32	.12	11.87	1.37	.04	.06	.31	3.55	4.72	.00	.04	.11	96.49
6_46	75.01	.07	11.87	1.48	.09	.02	.28	4.14	4.19	.00	.03	.15	97.32
6_47	74.63	.11	11.79	1.45	.04	.04	.29	3.77	4.61	.00	.00	.13	96.84
6_48	74.47	.08	11.85	1.47	.07	.02	.23	4.02	4.36	.00	.00	.19	96.75

d)   Measured standards

	SiO$_2$	TiO$_2$	Al$_2$O$_3$	Fe$_2$O$_3$	MnO	MgO	CaO	Na$_2$O	K$_2$O	P$_2$O$_5$	Total
ATHO-G-gb	74.665	.229	12.315	3.677	.092	.088	1.762	3.548	2.715	.089	99.18
ATHO-G-gb	74.788	.259	12.342	3.513	.155	.116	1.865	3.568	2.613	.023	99.242
ATHO-G-gb	75.385	.254	12.436	3.404	.134	.083	1.703	3.749	2.751	.000	99.899
ATHO-G-gb	75.683	.237	12.425	3.326	.163	.078	1.697	3.617	2.606	.089	99.921
ATHO-G-gb	75.404	.217	12.417	3.516	.106	.104	1.689	3.696	2.621	.011	99.781
StHs-gb	62.874	.727	17.797	4.521	.092	1.915	5.312	4.613	1.299	.199	99.349
StHs-gb	63.206	.651	17.812	4.521	.084	1.890	5.264	4.556	1.203	.144	99.331
StHs-gb	63.241	.667	18.080	4.581	.134	1.884	5.365	4.691	1.318	.167	100.128
StHs-gb	63.385	.674	17.632	4.463	.071	1.890	5.487	4.667	1.338	.199	99.806
StHs-gb	63.564	.727	17.806	4.523	.028	1.950	5.453	4.619	1.282	.167	100.119
BCR-2G-gb	54.533	2.315	13.447	12.196	.216	3.505	7.132	3.061	1.697	.378	98.48
BCR-2G-gb	54.028	2.317	13.190	12.200	.292	3.444	7.041	3.029	1.807	.367	97.715
BCR-2G-gb	54.017	2.168	13.706	11.954	.230	3.469	7.255	3.170	1.732	.442	98.143
BCR-2G-gb	54.267	2.350	13.409	12.296	.187	3.615	7.437	3.040	1.756	.486	98.843
BCR-2G-gb	54.062	2.277	13.209	12.888	.214	3.701	7.011	3.077	1.756	.541	98.736

Average values:

	SiO$_2$	TiO$_2$	Al$_2$O$_3$	Fe$_2$O$_3$	MnO	MgO	CaO	Na$_2$O	K$_2$O	P$_2$O$_5$	Total
ATHO-G	75.19	.24	12.39	3.49	.13	.09	1.74	3.64	2.66	.04	99.60
StHs	63.25	.69	17.83	4.52	.08	1.91	5.38	4.63	1.29	.18	99.75
BCR-2G	54.18	2.29	13.39	12.31	.23	3.55	7.18	3.08	1.75	.44	98.38

std dev.		SiO$_2$	TiO$_2$	Al$_2$O$_3$	Fe$_2$O$_3$	MnO	MgO	CaO	Na$_2$O	K$_2$O	P$_2$O$_5$
std dev.	ATHO-G	.44	.02	.05	.13	.03	.02	.07	.09	.07	.04
std dev.	StHs	.26	.04	.16	.04	.04	.03	.09	.05	.05	.02
std dev.	BCR-2G	.22	.07	.21	.35	.04	.11	.17	.06	.04	.07

target value:

	SiO$_2$	TiO$_2$	Al$_2$O$_3$	Fe$_2$O$_3$	MnO	MgO	CaO	Na$_2$O	K$_2$O	P$_2$O$_5$	Total
ATHO-G	75.6	.255	12.2	3.27	.106	.103	1.70	3.75	2.64	.025	99.649

119

	SiO_2	TiO_2	Al_2O_3	Fe_2O_3	MnO	MgO	CaO	Na_2O	K_2O	P_2O_5	Total
StHs6/80	63.7	.703	17.8	4.37	.076	1.970	5.28	4.44	1.29	.164	99.793
BCR-2G	54.4	2.270	13.4	12.40	.190	3.56	7.06	3.23	1.74	.370	98.62

normalized values:

	SiO_2	TiO_2	Al_2O_3	Fe_2O_3	MnO	MgO	CaO	Na_2O	K_2O	P_2O_5	Total
ATHO-G	75.48	0.24	12.44	3.5	0.13	0.09	1.75	3.65	2.67	0.04	99.99
StHs	63.42	0.69	17.87	4.53	0.08	1.91	5.39	4.64	1.29	0.18	100
BCR-2G	55.07	2.32	13.61	12.51	0.23	3.61	7.29	3.13	1.78	0.45	100

target value:

	SiO_2	TiO_2	Al_2O_3	Fe_2O_3	MnO	MgO	CaO	Na_2O	K_2O	P_2O_5	Total
ATHO-G	75.87	0.26	12.24	3.28	0.11	0.1	1.71	3.76	2.65	0.03	100.01
StHs6/80	63.83	0.7	17.84	4.38	0.08	1.97	5.29	4.45	1.29	0.16	99.99
BCR-2G	55.16	2.3	13.59	12.57	0.19	3.61	7.16	3.28	1.76	0.38	100

6.1.2 Raw data U-Pb ratios and calculated ages for all samples.

Data shaded in yellow represent the measurements used to calculate the Concordia ages of the tephras.

[a] within-run background-corrected mean ^{207}Pb signal in counts per second.

[b] U and Pb content and Th/U ratio were calculated relative to GJ-1 and are accurate to approximately 10%.

[c] corrected for background. mass bias. laser-induced U-Pb fractionation and common Pb (if detectable. see Materials and Methods Section) using STACEY & KRAMERS' (1975) model-Pb composition. ^{207}Pb/^{235}U calculated using ^{207}Pb/^{206}Pb/(^{238}U/^{206}Pb × 1/137.88). Errors are propagated by quadratic addition of within-run errors (2 σ) and the reproducibility of GJ-1 (2 σ).

[d] ρ is the error correlation defined as err^{206}Pb/^{238}U/err^{207}Pb/^{235}U.

La Sal Mountains tephra sampled in 2013

Sample and spot number	^{207}Pb^a (cps)	U^b (ppm)	Pb^b (ppm)	Th^b U	$\frac{^{206}Pb^c}{^{204}Pb}$	$\frac{^{206}Pb^c}{^{238}U}$	2 σ %	$\frac{^{207}Pb^c}{^{235}U}$	2 σ %	$\frac{^{207}Pb^c}{^{206}Pb}$	2 σ %	ρ^d	$\frac{^{206}Pb}{^{238}U}$	2 σ (Ma)	$\frac{^{207}Pb}{^{235}U}$	2 σ (Ma)	$\frac{^{207}Pb}{^{206}Pb}$	2 σ (Ma)	Conc %
2013-LSM-T-seq1-a01	52620	401	73	0.51	1676	0.17114	2.5	1.90093	3.1	0.080559	1.8	0.82	1018.39	23.79	1081	21	1211	35	84
2013-LSM-T-seq1-a02	204	1267	0	0.07	417	0.00020472	2.6	0.0012507	20.3	0.048198	20.2	0.13	1.32	0.03	-	-	-	-	-
2013-LSM-T-seq1-a03	878	133	0	0.8	158	0.00029149	17.3	0.021515	20.3	0.56028	10.6	0.85	1.88	0.33	-	-	-	-	-
2013-LSM-T-seq1-a04	346	1751	0	0.06	219	0.00021089	6.8	0.0018057	18.8	0.067393	17.5	0.36	1.36	0.09	-	-	-	-	-
2013-LSM-T-seq1-a05	-	-	-	-	-	-	-	-	-	-	-	-	-	-	-	-	-	-	-
2013-LSM-T-seq1-a06	56051	63	20	0.28	20431	0.31569	2.1	4.66362	2.4	0.10714	1.2	0.88	1768.66	33.34	1761	21	1751	21	**101**
2013-LSM-T-seq1-a07	4560	2018	1	0.07	46	0.00015481	10.6	0.007229	20.5	0.37919	17.6	0.52	1	0.11	-	-	-	-	-
2013-LSM-T-seq1-a08	10969	10	6	5.07	102	0.23579	3.7	7.47014	7	0.22977	6	0.52	1364.79	45.16	2169	65	3050	96	45
2013-LSM-T-seq1-a09	874	432	2	0.91	1665	0.0041189	2.7	0.030137	8.5	0.053066	8.1	0.32	26.5	0.72	-	-	-	-	-
2013-LSM-T-seq1-a10	-	-	-	-	-	-	-	-	-	-	-	-	-	-	-	-	-	-	-
2013-LSM-T-seq1-a11	1717	13	0	3.62	64	0.011926	3.3	0.72799	6.1	0.44271	5.1	0.53	76.43	2.47	-	-	-	-	-
2013-LSM-T-seq1-a12	16971	569	28	0.61	669	0.04469	3.3	0.32839	6.4	0.053295	5.5	0.52	281.84	9.15	288	16	341	124	83
2013-LSM-T-seq1-a13	1353	754	1	0.27	84	0.00035114	6.8	0.01494	12.5	0.32282	10.5	0.54	2.26	0.15	-	-	-	-	-
2013-LSM-T-seq1-a14	24183	164	31	1.25	1223	0.16316	3	1.82276	4	0.081024	2.7	0.74	974.32	26.93	1054	27	1222	53	80
2013-LSM-T-seq1-a15	-	-	-	-	-	-	-	-	-	-	-	-	-	-	-	-	-	-	-
2013-LSM-T-seq1-a16	4405	75	7	1.52	511	0.074108	2.2	1.13165	9.1	0.11075	8.8	0.24	460.86	9.73	769	50	1812	160	25

Sample and spot number	$^{207}Pb^a$ (cps)	U^b (ppm)	Pb^b (ppm)	Th^b U	$\frac{^{206}Pb^c}{^{204}Pb}$	$\frac{^{206}Pb^c}{^{238}U}$	2 σ %	$\frac{^{207}Pb^c}{^{235}U}$	2 σ %	$\frac{^{207}Pb^c}{^{206}Pb}$	2 σ %	ρ^d	$\frac{^{206}Pb}{^{238}U}$	2 σ (Ma)	$\frac{^{207}Pb}{^{235}U}$	2 σ (Ma)	$\frac{^{207}Pb}{^{206}Pb}$	2 σ (Ma)	Conc %
2013-LSM-T-seq1-a17	112	1439	0	0.44	238	0.00020832	3.9	0.0012805	28.3	0.04798	28	0.14	1.34	0.05	-	-	-	-	-
2013-LSM-T-seq1-a18	8152	70	1	0.72	23	0.0036939	6	0.17118	59.4	0.33611	59.1	0.1	23.77	1.43	-	-	-	-	-
2013-LSM-T-seq1-a19	32148	398	39	0.91	3724	0.084139	2.4	0.67625	3.3	0.058292	2.3	0.72	520.78	11.79	525	14	541	50	**96**
2013-LSM-T-seq1-a20	246990	979	125	0.12	252	0.10392	2.1	2.12475	2.4	0.14828	1.1	0.89	637.36	12.91	1157	17	2326	19	27
2013-LSM-T-seq2-b01	1066	94	0	0.66	20	0.0010504	19.1	0.08679	22.1	0.60656	11	0.87	6.77	1.29	-	-	-	-	-
2013-LSM-T-seq2-b02	4440	735	1	0.44	23	0.00046805	6.2	0.03631	8.3	0.58074	5.4	0.75	3.02	0.19	-	-	-	-	-
2013-LSM-T-seq2-b03	3939	2	1	5.02	23	0.1684	2.6	10.34267	30.5	0.44544	30.4	0.09	1003.29	24.54	2466	331	4070	452	25
2013-LSM-T-seq2-b04	6354	68	5	1.75	11707	0.052314	2.4	0.39322	3.7	0.054515	2.8	0.64	328.71	7.58	337	11	392	63	84
2013-LSM-T-seq2-b05	37958	305	24	0.49	639	0.076676	1.9	0.5799	3	0.054852	2.4	0.61	476.25	8.55	464	11	406	54	117
2013-LSM-T-seq2-b06	2521	181	0	1.07	30	0.00043677	16.8	0.025658	38.1	0.43768	34.2	0.44	2.81	0.47	-	-	-	-	-
2013-LSM-T-seq2-b07	-	-	-	-	-	-	-	-	-	-	-	-	-	-	-	-	-	-	-
2013-LSM-T-seq2-b08	104021	608	122	0.69	14714	0.17996	2.9	2.52307	3.9	0.10168	2.5	0.76	1066.77	28.75	1279	28	1655	47	64
2013-LSM-T-seq2-b09	-	-	-	-	-	-	-	-	-	-	-	-	-	-	-	-	-	-	-
2013-LSM-T-seq2-b10	5661	456	6	0.67	1199	0.011071	1.8	0.09948	8.2	0.065172	8	0.23	70.97	1.3	-	-	-	-	-
2013-LSM-T-seq2-b11	-	-	-	-	-	-	-	-	-	-	-	-	-	-	-	-	-	-	-
2013-LSM-T-seq2-b12	-	-	-	-	-	-	-	-	-	-	-	-	-	-	-	-	-	-	-
2013-LSM-T-seq2-b13	-	-	-	-	-	-	-	-	-	-	-	-	-	-	-	-	-	-	-
2013-LSM-T-seq2-b14	-	-	-	-	-	-	-	-	-	-	-	-	-	-	-	-	-	-	-
2013-LSM-T-seq2-b15	31426	499	25	0.61	654	0.045584	3.3	0.3386	4.2	0.053874	2.6	0.78	287.35	9.2	296	11	366	59	79
2013-LSM-T-seq2-b16	34642	647	51	0.21	973	0.075984	2.3	0.62528	3.3	0.059683	2.4	0.69	472.11	10.56	493	13	592	52	80
2013-LSM-T-seq2-b17	4852	140	14	0.99	1960	0.093302	2.2	0.83302	3.9	0.064753	3.2	0.56	575.04	11.98	615	18	766	68	75
2013-LSM-T-seq2-b18	-	-	-	-	-	-	-	-	-	-	-	-	-	-	-	-	-	-	-
2013-LSM-T-seq2-b19	-	-	-	-	-	-	-	-	-	-	-	-	-	-	-	-	-	-	-
2013-LSM-T-seq2-b20	271	94	0	0.81	14	0.00085473	12.2	0.064946	30.7	0.55915	28.1	0.4	5.51	0.67	-	-	-	-	-
2013-LSM-T-seq2-b21	-	-	-	-	-	-	-	-	-	-	-	-	-	-	-	-	-	-	-
2013-LSM-T-seq2-b22	-	-	-	-	-	-	-	-	-	-	-	-	-	-	-	-	-	-	-
2013-LSM-T-seq2-b23	8158	5	5	5.56	19	0.30033	4.8	17.79081	36.8	0.42964	36.5	0.13	1692.93	71.3	2978	435	4016	545	42

Sample and spot number	$^{207}Pb^a$ (cps)	U^b (ppm)	Pb^b (ppm)	Th^b U	$\frac{^{206}Pb^c}{^{204}Pb}$	$\frac{^{206}Pb^c}{^{238}U}$	2 σ %	$\frac{^{207}Pb^c}{^{235}U}$	2 σ %	$\frac{^{207}Pb^c}{^{206}Pb}$	2 σ %	ρ^d	$\frac{^{206}Pb}{^{238}U}$ (Ma)	2 σ (Ma)	$\frac{^{207}Pb}{^{235}U}$	2 σ (Ma)	$\frac{^{207}Pb}{^{206}Pb}$	2 σ (Ma)	Conc %
2013-LSM-T-seq2-b24	-	-	-	-	-	-	-	-	-	-	-	-	-	-	-	-	-	-	-
2013-LSM-T-seq2-b25	45952	6	5	2	21	0.26175	3.3	10.38629	44.8	0.28778	44.7	0.07	1498.81	44	2470	533	3406	695	44
2013-LSM-T-seq2-b26	3236	663	2	0.7	38	0.00079126	6.3	0.0522	26.7	0.48641	26	0.24	5.1	0.32	-	-	-	-	-
2013-LSM-T-seq2-b27	82	89	0	0.87	47	0.00019897	7.3	0.0049005	44	0.19082	43.4	0.17	1.28	0.09	-	-	-	-	-
2013-LSM-T-seq2-b28	185	626	0	0.12	384	0.000211	2.9	0.0012392	25.8	0.046148	25.6	0.11	1.36	0.04	-	-	-	-	-
2013-LSM-T-seq2-b29	51	58	0	0.59	41	0.00020031	14.5	0.0036449	60.4	0.14193	58.6	0.24	1.29	0.19	-	-	-	-	-
2013-LSM-T-seq2-b30	-	-	-	-	-	-	-	-	-	-	-	-	-	-	-	-	-	-	-
2013-LSM-T-seq2-b31	875	1569	0	0.06	531	0.0002513	3.1	0.0052868	12.9	0.16333	12.5	0.24	1.62	0.05	-	-	-	-	-
2013-LSM-T-seq2-b32	-	-	-	-	-	-	-	-	-	-	-	-	-	-	-	-	-	-	-
2013-LSM-T-seq2-b33	133	362	0	0.46	221	0.00020535	4.8	0.0016824	25.5	0.063994	25.1	0.19	1.32	0.06	-	-	-	-	-
2013-LSM-T-seq2-b34	10410	1	2	4.37	18	0.48413	8.7	31.24638	115	0.4681	114.7	0.08	2545.24	185.01	3527	-	4144	1700	61
2013-LSM-T-seq2-b35	4969	3	1	3.94	20	0.15742	3.8	15.9816	7.2	0.73632	6.1	0.52	942.41	33.16	2876	71	4803	88	20
2013-LSM-T-seq2-b36	-	-	-	-	-	-	-	-	-	-	-	-	-	-	-	-	-	-	-
2013-LSM-T-seq2-b37	-	-	-	-	-	-	-	-	-	-	-	-	-	-	-	-	-	-	-
2013-LSM-T-seq2-b38	-	-	-	-	-	-	-	-	-	-	-	-	-	-	-	-	-	-	-
2013-LSM-T-seq2-b39	-	-	-	-	-	-	-	-	-	-	-	-	-	-	-	-	-	-	-
2013-LSM-T-seq2-b40	229	239	0	0.67	201	0.00020512	6.8	0.0031077	18.5	0.11772	17.2	0.37	1.32	0.09	-	-	-	-	-
2013-LSM-T-seq2-b41	-	-	-	-	-	-	-	-	-	-	-	-	-	-	-	-	-	-	-
2013-LSM-T-seq2-b42	-	-	-	-	-	-	-	-	-	-	-	-	-	-	-	-	-	-	-
2013-LSM-T-seq2-b43	5449	147	1	1.14	17	0.0027636	3.3	0.29177	5	0.7657	3.8	0.65	17.79	0.58	-	-	-	-	-
2013-LSM-T-seq2-b44	-	-	-	-	-	-	-	-	-	-	-	-	-	-	-	-	-	-	-
2013-LSM-T-seq2-b45	100	58	0	0.61	24	0.00032865	17.9	0.015705	96.8	0.36183	95.2	0.18	2.12	0.38	-	-	-	-	-
2013-LSM-T-seq2-b46	322	252	0	0.65	196	0.00020144	8.4	0.003996	24.2	0.15443	22.7	0.35	1.3	0.11	-	-	-	-	-
2013-LSM-T-seq2-b47	93478	389	49	0.28	803	0.11838	2.1	1.40703	2.3	0.086203	1	0.9	721.23	14.27	892	14	1343	19	54
2013-LSM-T-seq2-b48	89	71	0	0.6	59	0.00021237	12.7	0.0044232	34.4	0.16172	32	0.37	1.37	0.17	-	-	-	-	-
2013-LSM-T-seq2-b49	93708	91	29	0.58	9441	0.2936	1.9	4.20655	2.2	0.10391	1	0.9	1659.49	28.27	1675	18	1695	18	**98**
2013-LSM-T-seq2-b50	619	1363	0	0.04	539	0.00022415	2.3	0.0017162	7.1	0.059985	6.8	0.32	1.44	0.03	-	-	-	-	-

Sample and spot number	$^{207}Pb^a$ (cps)	U^b (ppm)	Pb^b (ppm)	Th^b / U	$^{206}Pb^c$ / ^{204}Pb	$^{206}Pb^c$ / ^{238}U	2 σ %	$^{207}Pb^c$ / ^{235}U	2 σ %	$^{207}Pb^c$ / ^{206}Pb	2 σ %	ρ^d	^{206}Pb / ^{238}U	2 σ (Ma)	^{207}Pb / ^{235}U (Ma)	2 σ (Ma)	^{207}Pb / ^{206}Pb (Ma)	2 σ (Ma)	Conc %
2013-LSM-T-seq2-b51	29663	2	3	4.17	18	0.90867	2.9	30.49485	42	0.2434	41.9	0.07	4167.01	91.08	3503	530	3142	666	133
2013-LSM-T-seq2-b52	-	-	-	-	-	-	-	-	-	-	-	-	-	-	-	-	-	-	-
2013-LSM-T-seq2-b53	9118	2	2	2.86	20	0.39215	3.8	15.38208	84.5	0.28449	84.4	0.04	2132.78	69.35	2839	1601	3388	1316	63
2013-LSM-T-seq2-b54	4914	39	3	0.62	6267	0.063204	2.5	0.48691	4.2	0.055873	3.4	0.6	395.08	9.77	403	14	447	75	88
2013-LSM-T-seq2-b55	2083	3	2	6.41	44	0.14684	5.8	12.42169	6.7	0.61352	3.4	0.86	883.24	47.92	2637	65	4540	49	19
2013-LSM-T-seq2-b56	-	-	-	-	-	-	-	-	-	-	-	-	-	-	-	-	-	-	-
2013-LSM-T-seq2-b57	83	1207	0	0.5	171	0.000201	7.2	0.0011891	36.7	0.046231	36	0.2	1.3	0.09	-	-	-	-	-
2013-LSM-T-seq2-b58	1601	3	1	3.2	64	0.14884	6.6	11.68284	11.5	0.56927	9.4	0.58	894.48	55.52	2579	113	4431	137	20
2013-LSM-T-seq2-b59	5424	152	3	0.59	10380	0.016542	2.3	0.11996	5.8	0.052596	5.3	0.4	105.76	2.45	115	6	311	122	34
2013-LSM-T-seq2-b60	1522	3	1	4.41	33	0.15259	4.2	11.09837	14.9	0.52751	14.3	0.28	915.47	35.62	2531	149	4320	210	21
2013-LSM-T-seq3-c01	-	-	-	-	-	-	-	-	-	-	-	-	-	-	-	-	-	-	-
2013-LSM-T-seq3-c02	-	-	-	-	-	-	-	-	-	-	-	-	-	-	-	-	-	-	-
2013-LSM-T-seq3-c03	3057	13	2	5.45	19	0.056121	3.2	4.24662	21.6	0.5488	21.4	0.15	351.99	11.08	1683	196	4378	313	8
2013-LSM-T-seq3-c04	3040	2	1	3.13	21	0.19677	4.5	4.58115	75.3	0.16885	75.1	0.06	1157.95	48.1	1746	977	2546	1259	45
2013-LSM-T-seq3-c05	2313	12	0	3.8	120	0.0097485	5.3	0.58259	15.8	0.43343	14.9	0.33	62.54	3.27	-	-	-	-	-
2013-LSM-T-seq3-c06	797	647	0	0.51	159	0.0001534	39.2	0.0097472	40.6	0.50879	10.6	0.97	0.99	0.39	-	-	-	-	-
2013-LSM-T-seq3-c07	216	1480	0	0.05	447	0.00020238	3	0.00126	19.2	0.049187	19	0.15	1.3	0.04	-	-	-	-	-
2013-LSM-T-seq3-c08	806	498	0	0.7	251	0.00028857	10.2	0.012255	15.2	0.32309	11.2	0.68	1.86	0.19	-	-	-	-	-
2013-LSM-T-seq3-c09	26976	154	52	0.63	22847	0.30538	2.3	4.42134	3.1	0.10501	2	0.76	1717.94	35.31	1716	26	1714	37	100
2013-LSM-T-seq3-c10	-	-	-	-	-	-	-	-	-	-	-	-	-	-	-	-	-	-	-
2013-LSM-T-seq3-c11	2888	98	11	1.15	4923	0.093282	3.4	0.76285	6.4	0.059312	5.4	0.53	574.92	18.73	576	29	579	118	99
2013-LSM-T-seq3-c12	31	253	0	0.6	67	0.00020529	5	0.0012344	35.5	0.046798	35.1	0.14	1.32	0.07	-	-	-	-	-
2013-LSM-T-seq3-c13	115	1026	0	0.33	261	0.00020434	3.4	0.0012124	32.3	0.046504	32.2	0.1	1.32	0.04	-	-	-	-	-
2013-LSM-T-seq3-c14	-	-	-	-	-	-	-	-	-	-	-	-	-	-	-	-	-	-	-
2013-LSM-T-seq3-c15	405	102	0	0.46	42	0.00029355	18.3	0.010666	32.1	0.27731	26.3	0.57	1.89	0.35	-	-	-	-	-
2013-LSM-T-seq3-c16	2733	671	0	0.08	73	0.00020266	6.5	0.0099202	15.6	0.38639	14.2	0.42	1.31	0.09	-	-	-	-	-
2013-LSM-T-seq3-c17	39974	187	51	0.57	42687	0.24934	2.2	3.24237	3	0.094313	2.1	0.73	1435.07	28.81	1467	24	1514	39	95

Sample and spot number	$^{207}Pb^a$	U^b	Pb^b	Th^b	$\frac{^{206}Pb^c}{^{204}Pb}$	$\frac{^{206}Pb^c}{^{238}U}$	2 σ	$\frac{^{207}Pb^c}{^{235}U}$	2 σ	$\frac{^{207}Pb^c}{^{206}Pb}$	2 σ	ρ^d	$\frac{^{206}Pb}{^{238}U}$	2 σ	$\frac{^{207}Pb}{^{235}U}$	2 σ	$\frac{^{207}Pb}{^{206}Pb}$	2 σ	Conc
	(cps)	(ppm)	(ppm)	U	^{204}Pb	^{238}U	%	^{235}U	%	^{206}Pb	%		^{238}U	(Ma)	^{235}U	(Ma)	^{206}Pb	(Ma)	%
2013-LSM-T-seq3-c18	30760	81	13	0.7	22971	0.15198	1.9	1.55758	2.1	0.074332	1	0.88	912.03	15.89	954	13	1050	20	87
2013-LSM-T-seq3-c19	-	-	-	-	-	-	-	-	-	-	-	-	-	-	-	-	-	-	-
2013-LSM-T-seq3-c20	8233	167	17	0.65	13615	0.095652	2.5	0.7918	4.3	0.060037	3.5	0.59	588.88	14.33	592	19	605	75	97
2013-LSM-T-seq3-c21	3281	164	5	1.14	1029	0.028311	3	0.23356	5.5	0.059834	4.6	0.55	179.97	5.35	213	11	598	99	30
2013-LSM-T-seq3-c22	-	-	-	-	-	-	-	-	-	-	-	-	-	-	-	-	-	-	-
2013-LSM-T-seq3-c23	98	1472	0	0.36	210	0.00020734	4.9	0.0012735	26.3	0.048055	25.9	0.19	1.34	0.07	-	-	-	-	-
2013-LSM-T-seq3-c24	7305	222	28	1.39	11878	0.10239	2.3	0.86682	4.1	0.0614	3.4	0.56	628.4	13.81	634	20	653	73	96
2013-LSM-T-seq3-c25	202	806	0	0.36	229	0.0002458	6.5	0.0027869	24.7	0.087621	23.9	0.26	1.58	0.1	-	-	-	-	-
2013-LSM-T-seq3-c26	32988	184	31	0.53	6314	0.16351	2.1	1.66738	3.3	0.07396	2.6	0.63	976.24	18.97	996	21	1040	52	94
2013-LSM-T-seq3-c27	1110	117	1	0.22	1066	0.0044985	2	0.039248	8.3	0.063278	8.1	0.23	28.93	0.56	-	-	-	-	-
2013-LSM-T-seq3-c28	1502	17	0	2.34	65	0.0041971	15.9	0.23344	18.1	0.40338	8.8	0.88	27	4.28	-	-	-	-	-
2013-LSM-T-seq3-c29	-	-	-	-	-	-	-	-	-	-	-	-	-	-	-	-	-	-	-
2013-LSM-T-seq3-c30	8644	17	1	6.07	26	0.021087	6.4	0.58077	72.9	0.19975	72.6	0.09	134.52	8.58	465	316	2824	1185	5
2013-LSM-T-seq3-c31	1365	291	1	0.55	22	0.0010998	13.7	0.083597	14.6	0.55791	5	0.94	7.09	0.97	-	-	-	-	-
2013-LSM-T-seq3-c32	-	-	-	-	-	-	-	-	-	-	-	-	-	-	-	-	-	-	-
2013-LSM-T-seq3-c33	277	1059	0	0.05	603	0.00018653	3.3	0.0010855	13.3	0.046337	12.9	0.25	1.2	0.04	-	-	-	-	-
2013-LSM-T-seq3-c34	-	-	-	-	-	-	-	-	-	-	-	-	-	-	-	-	-	-	-
2013-LSM-T-seq3-c35	34214	284	43	0.75	546	0.12092	3.6	1.91581	5.3	0.11491	3.9	0.68	735.84	25.21	1087	36	1879	71	39
2013-LSM-T-seq3-c36	5983	320	14	0.66	10369	0.040899	2.3	0.29024	3.5	0.051467	2.6	0.66	258.41	5.9	259	8	262	60	99
2013-LSM-T-seq3-c37	-	-	-	-	-	-	-	-	-	-	-	-	-	-	-	-	-	-	-
2013-LSM-T-seq3-c38	1007	112	0	0.95	16	0.00043538	6.8	0.033153	11.1	0.56816	8.9	0.61	2.81	0.19	-	-	-	-	-
2013-LSM-T-seq3-c39	1233	1028	5	0.16	2651	0.0047637	1.9	0.030683	6.5	0.046714	6.2	0.3	30.64	0.59	-	-	-	-	-
2013-LSM-T-seq3-c40	1677	39	1	1.38	3418	0.026535	2.2	0.18126	4.7	0.049542	4.2	0.46	168.82	3.59	169	7	174	98	97
Standard: GJ1																			
GJ1-1_seq a	26272	235	22	0.02	43988	0.0994958	2.1	0.82588	2.6	0.060202	1.5	0.81	611.45	12.38	611	12	611	33	100
GJ1-2_seq a	26928	243	22	0.02	45022	0.098259	2	0.81606	2.2	0.060235	0.9	0.91	604.2	11.51	606	10	612	20	99

125

Sample and spot number	$^{207}Pb^a$	U^b	Pb^b	Th^b	$\frac{^{206}Pb^c}{^{204}Pb}$	$\frac{^{206}Pb^c}{^{238}U}$	2 σ	$\frac{^{207}Pb^c}{^{235}U}$	2 σ	$\frac{^{207}Pb^c}{^{206}Pb}$	2 σ	ρ^d	$\frac{^{206}Pb}{^{238}U}$	2 σ	$\frac{^{207}Pb}{^{235}U}$	2 σ	$\frac{^{207}Pb}{^{206}Pb}$	2 σ	Conc
	(cps)	(ppm)	(ppm)	U			%		%		%			(Ma)		(Ma)		(Ma)	%
GJ1-3_seq a	25643	241	22	0.02	42953	0.0989509	2	0.82083	2.4	0.060163	1.4	0.82	608.26	11.53	609	11	609	30	**100**
GJ1-4_seq a	26275	244	22	0.02	54925	0.0969665	2.1	0.8039	2.4	0.060128	1.3	0.85	596.61	11.72	599	11	608	27	**98**
GJ1-5_seq a	26390	249	23	0.02	128966	0.098907	2.1	0.82156	2.4	0.060244	1.1	0.88	608	12.43	609	11	612	25	**99**
GJ1-6_seq a	26389	253	23	0.02	12197	0.0985451	2.2	0.81887	2.5	0.060267	1.3	0.85	605.88	12.53	607	12	613	29	**99**
GJ1-7_seq a	26511	252	23	0.02	44400	0.0987556	2.1	0.81963	2.4	0.060194	1.3	0.85	607.11	12.08	608	11	611	27	**99**
GJ1-8_seq a	26056	249	23	0.02	19232	0.0993428	2	0.82412	2.3	0.060166	1.1	0.88	610.56	11.89	610	11	610	23	**100**
GJ1-9_seq a	21843	212	19	0.02	36492	0.098088	2	0.81275	2.6	0.060095	1.6	0.77	603.19	11.28	604	12	607	35	**99**
GJ1-10_seq a	25621	253	23	0.02	42944	0.0980944	2.1	0.81323	2.6	0.060127	1.6	0.78	603.23	11.9	604	12	608	35	**99**
GJ1-11_seq a	25106	253	23	0.02	14719	0.0991943	2.2	0.82279	2.5	0.060159	1.2	0.89	609.69	12.82	610	11	609	25	**100**
GJ1-1_seq b	25264	237	22	0.02	42207	0.0984323	1.9	0.81761	2.2	0.060243	1.1	0.86	605.21	11.05	607	10	612	24	**99**
GJ1-2_seq b	25562	241	22	0.02	16144	0.0986553	1.9	0.81841	2.2	0.060166	1.1	0.87	606.52	11.29	607	10	610	24	**100**
GJ1-3_seq b	25206	243	22	0.02	15035	0.097622	1.9	0.81074	2.2	0.060232	1.2	0.85	600.46	10.96	603	10	612	25	**98**
GJ1-4_seq b	25023	243	22	0.02	41867	0.0981425	2.1	0.81507	2.6	0.060233	1.4	0.83	603.51	12.32	605	12	612	31	**99**
GJ1-5_seq b	25478	248	23	0.02	25781	0.0997443	1.9	0.82601	2.2	0.060062	1.1	0.87	612.91	11.31	611	10	606	24	**101**
GJ1-7_seq b	25237	244	22	0.02	10100	0.0992448	1.9	0.82311	2.3	0.060152	1.3	0.81	609.98	10.9	610	11	609	29	**100**
GJ1-8_seq b	25693	249	23	0.02	52447	0.0990976	2	0.82196	2.3	0.060157	1.1	0.88	609.12	11.89	609	11	609	24	**100**
GJ1-9_seq b	25445	251	23	0.02	64897	0.0981442	2	0.81397	2.5	0.060151	1.6	0.79	603.52	11.46	605	12	609	34	**99**
GJ1-10_seq b	24768	238	22	0.02	11917	0.099642	2.1	0.8283	2.5	0.06029	1.3	0.86	612.31	12.56	613	12	614	28	**100**
GJ1-11_seq b	25430	248	22	0.02	35625	0.0976661	2	0.81022	2.4	0.060167	1.4	0.81	600.72	11.33	603	11	610	31	**99**
GJ1-12_seq b	25191	248	23	0.02	42174	0.098114	2	0.81415	2.4	0.060183	1.5	0.8	603.35	11.26	605	11	610	32	**99**
GJ1-13_seq b	24843	247	22	0.02	41577	0.098059	2.2	0.81327	2.5	0.060152	1.1	0.89	603.02	12.85	604	11	609	24	**99**
GJ1-14_seq b	25018	247	23	0.02	11821	0.0988467	2.3	0.82033	2.6	0.06019	1.3	0.87	607.65	13.1	608	12	610	28	**100**
GJ1-15_seq b	25349	247	22	0.02	84751	0.0981923	2.2	0.81602	2.5	0.060273	1.2	0.87	603.81	12.69	606	12	613	27	**98**
GJ1-16_seq b	24440	241	22	0.02	40893	0.09907	2	0.82153	2.4	0.060143	1.3	0.83	608.96	11.44	609	11	609	29	**100**
GJ1-17_seq b	24666	241	22	0.02	41293	0.098245	1.9	0.81444	2.3	0.060124	1.3	0.84	604.12	11.19	605	11	608	27	**99**
GJ1-18_seq b	24836	245	23	0.02	38708	0.0995968	2.1	0.82669	2.9	0.0602	1.9	0.75	612.05	12.49	612	13	611	41	**100**
GJ1-19_seq b	22594	226	21	0.02	37832	0.0992514	2	0.82354	2.4	0.060179	1.4	0.82	610.02	11.43	610	11	610	29	**100**

Sample and spot number	$^{207}Pb^a$	U^b	Pb^b	Th^b	$\frac{^{206}Pb^c}{^{204}Pb}$	$\frac{^{206}Pb^c}{^{238}U}$	2σ	$\frac{^{207}Pb^c}{^{235}U}$	2σ	$\frac{^{207}Pb^c}{^{206}Pb}$	2σ	ρ^d	$\frac{^{206}Pb}{^{238}U}$	2σ	$\frac{^{207}Pb}{^{235}U}$	2σ	$\frac{^{207}Pb}{^{206}Pb}$	2σ	Conc
	(cps)	(ppm)	(ppm)	U			%		%		%			(Ma)		(Ma)		(Ma)	%
GJ1-20_seq b	24774	247	22	0.02	41487	0.0982221	2	0.81381	2.3	0.060092	1.2	0.85	603.98	11.34	605	11	607	27	100
GJ1-21_seq b	24926	246	22	0.02	41716	0.0983686	1.9	0.81604	2.3	0.060166	1.4	0.8	604.84	10.76	606	11	610	30	99
GJ1-22_seq b	25172	244	22	0.02	14344	0.0973102	1.8	0.80813	2.5	0.060231	1.7	0.72	598.63	10.39	601	11	612	37	98
GJ1-23_seq b	24679	248	23	0.02	41313	0.0995327	2	0.82583	2.2	0.060176	1	0.89	611.67	11.47	611	10	610	21	100
GJ1-1_seq c	24726	244	22	0.02	41425	0.0994662	2.3	0.82478	2.5	0.06014	1	0.92	611.28	13.17	611	11	609	21	100
GJ1-2_seq c	25280	251	23	0.02	52904	0.0978825	2.4	0.81178	3.4	0.060149	2.4	0.69	601.99	13.56	603	16	609	53	99
GJ1-3_seq c	24024	240	22	0.02	134159	0.0974755	2.3	0.80954	2.6	0.060234	1.2	0.88	599.6	13.32	602	12	612	27	98
GJ1-4_seq c	24630	250	23	0.02	41221	0.0987798	2.4	0.81972	2.5	0.060186	1	0.92	607.25	13.67	608	12	610	21	100
GJ1-5_seq c	24333	238	21	0.02	40707	0.0975348	2.2	0.80904	2.5	0.06016	1	0.91	599.95	12.83	602	11	609	22	98
GJ1-6_seq c	24423	243	22	0.02	30003	0.0995409	2.1	0.82591	2.5	0.060177	1.3	0.86	611.72	12.38	611	11	610	27	100
GJ1-7_seq c	24124	240	22	0.02	40321	0.0985101	2.1	0.81811	2.4	0.060232	1.2	0.88	605.67	12.37	607	11	612	25	99
GJ1-8_seq c	24232	243	22	0.02	40583	0.0990276	2.1	0.82147	2.4	0.060164	1.2	0.88	608.71	12.44	609	11	609	25	100
GJ1-9_seq c	23814	238	22	0.02	39912	0.098341	2.2	0.81503	2.6	0.060109	1.4	0.83	604.68	12.42	605	12	608	31	100
GJ1-10_seq c	24314	242	22	0.02	40687	0.0989817	2.2	0.82143	2.4	0.060188	1	0.91	608.44	12.81	609	11	610	22	100
GJ1-11_seq c	24658	250	23	0.02	61816	0.0994537	2.2	0.82611	2.4	0.060244	0.9	0.93	611.21	13.03	611	11	612	19	100
GJ1-12_seq c	24644	251	23	0.02	41219	0.0994344	2.2	0.82576	2.4	0.06023	1	0.9	611.09	12.57	611	11	612	23	100
GJ1-13_seq c	23838	241	22	0.02	39880	0.0985638	2.2	0.81856	2.5	0.060233	1.2	0.88	605.99	12.57	607	11	612	26	99
GJ1-14_seq c	23931	245	23	0.02	40087	0.0995537	2.4	0.82521	2.7	0.060118	1.2	0.9	611.79	14.12	611	12	608	26	101
GJ1-15_seq c	24266	251	22	0.02	40654	0.0968913	2.3	0.80302	2.7	0.060109	1.5	0.83	596.16	12.93	599	12	608	33	98
GJ1-16_seq c	24390	246	22	0.02	40848	0.0981254	2.2	0.8135	2.5	0.060128	1.1	0.89	603.41	12.65	604	11	608	24	99
GJ1-17_seq c	23444	234	21	0.02	39187	0.0986377	2.2	0.81951	2.6	0.060258	1.3	0.87	606.42	12.8	608	12	613	28	99

La Sal Mountains tephra sampled in 2014

Sample and spot number	$^{207}Pb^a$	U^b	Pb^b	Th^b	$\frac{^{206}Pb^c}{^{204}Pb}$	$\frac{^{206}Pb^c}{^{238}U}$	2σ	$\frac{^{207}Pb^c}{^{235}U}$	2σ	$\frac{^{207}Pb^c}{^{206}Pb}$	2σ	ρ^d	$\frac{^{206}Pb}{^{238}U}$	2σ	$\frac{^{207}Pb}{^{235}U}$	2σ	$\frac{^{207}Pb}{^{206}Pb}$	2σ	Conc
2014-LSM-T_a02	263	1081	0	0.51	216	0.00023451	6.3	0.0034808	35.5	0.11474	34.9	0.18	1.51	0.1	-	-	-	-	-
2014-LSM-T_a04	1643	12	0	2.92	75	0.0075582	5.5	0.53679	10.2	0.51509	8.6	0.54	48.54	2.68	-	-	-	-	-
2014-LSM-T_a05	3301	23	1	3.08	35	0.0094936	4.8	0.65312	6.9	0.49896	5	0.69	60.91	2.9	-	-	-	-	-
2014-LSM-T_a06	-	-	-	-	-	-	-	-	-	-	-	-	-	-	-	-	-	-	-

Sample and spot number	$^{207}Pb^a$ (cps)	U^b (ppm)	Pb^b (ppm)	Th^b/U	$\frac{^{206}Pb^c}{^{204}Pb}$	$\frac{^{206}Pb^c}{^{238}U}$	2 σ %	$\frac{^{207}Pb^c}{^{235}U}$	2 σ %	$\frac{^{207}Pb^c}{^{206}Pb}$	2 σ %	ρ^d	$\frac{^{206}Pb}{^{238}U}$	2 σ (Ma)	$\frac{^{207}Pb}{^{235}U}$	2 σ (Ma)	$\frac{^{207}Pb}{^{206}Pb}$	2 σ (Ma)	Conc %
2014-LSM-T_a07	43	826	0	0.54	135	0.00020667	7.7	0.0010108	70.7	0.038118	70.2	0.11	1.33	0.1	-	-	-	-	-
2014-LSM-T_a08	183830	425	20	0.15	224	0.022835	31.3	0.32859	31.3	0.10436	2.3	1	145.55	45.15	288	82	1703	42	9
2014-LSM-T_a09	156	36	0	7.54	174	0.0056573	7	0.075344	36.4	0.096592	35.7	0.19	36.37	2.53	-	-	-	-	-
2014-LSM-T_a10	422	27	0	6.48	313	0.0043298	10.5	0.082527	30.5	0.13824	28.6	0.34	27.85	2.91	-	-	-	-	-
2014-LSM-T_a12	328	50	0	0.82	110	0.0015104	14.5	0.061991	34.3	0.29981	31.1	0.42	9.73	1.41	-	-	-	-	-
2014-LSM-T_a13	88	810	0	0.67	136	0.00021642	7.6	0.0016957	33.7	0.060683	32.9	0.22	1.39	0.11	-	-	-	-	-
2014-LSM-T_a14	716	15	0	5.53	117	0.0046476	8.9	0.17587	18.4	0.27445	16.1	0.49	29.89	2.67	-	-	-	-	-
2014-LSM-T_a15	18959	37	2	0.86	26	0.020385	10	1.71084	13.5	0.6087	9	0.74	130.09	12.88	1013	90	4529	131	3
2014-LSM-T_a16	652	16	1	10.41	134	0.011012	15.5	0.73189	23.8	0.48204	18	0.65	70.6	10.92	-	-	-	-	-
2014-LSM-T_a17	63	28	0	4.83	179	0.0035009	9.4	0.018333	73.2	0.03798	72.6	0.13	22.53	2.11	-	-	-	-	-
2014-LSM-T_a18	678	917	1	0.63	37	0.00038989	5.2	0.015119	19.2	0.29158	18.5	0.27	2.51	0.13	-	-	-	-	-
2014-LSM-T_a19	-	-	-	-	-	-	-	-	-	-	-	-	-	-	-	-	-	-	-
2014-LSM-T_a20	43	748	0	0.63	114	0.00018207	11.1	0.0009338	51.7	0.040283	50.5	0.21	1.17	0.13	-	-	-	-	-
2014-LSM-T_a22	66	264	0	0.59	118	0.00020058	6.2	0.0011814	89.5	0.045955	89.3	0.07	1.29	0.08	-	-	-	-	-
2014-LSM-T_a23	-	-	-	-	-	-	-	-	-	-	-	-	-	-	-	-	-	-	-
2014-LSM-T_a24	-	-	-	-	-	-	-	-	-	-	-	-	-	-	-	-	-	-	-
2014-LSM-T_a25	50	1034	0	0.62	152	0.00017413	8	0.00069362	53.4	0.031415	52.8	0.15	1.12	0.09	-	-	-	-	-
2014-LSM-T_a26	-	-	-	-	-	-	-	-	-	-	-	-	-	-	-	-	-	-	-
2014-LSM-T_a27	323	55	0	3.94	594	0.0043241	4.3	0.032737	16.1	0.054909	15.5	0.27	27.81	1.19	-	-	-	-	-
2014-LSM-T_a28	476	80	0	1.89	811	0.0035107	3.6	0.02906	10.9	0.060035	10.3	0.33	22.59	0.8	-	-	-	-	-
2014-LSM-T_a29	-	-	-	-	-	-	-	-	-	-	-	-	-	-	-	-	-	-	-
2014-LSM-T_a3	204821	118	26	0.29	3581	0.21054	3.2	3.15723	3.6	0.10876	1.7	0.88	1231.69	36.13	1447	28	1779	31	69
2014-LSM-T_a30	-	-	-	-	-	-	-	-	-	-	-	-	-	-	-	-	-	-	-
2014-LSM-T_a32	-	-	-	-	-	-	-	-	-	-	-	-	-	-	-	-	-	-	-
2014-LSM-T_a33	18866	95	25	0.23	7943	0.26542	3	3.40297	3.5	0.092987	1.8	0.85	1517.51	40.95	1505	28	1488	35	**102**
2014-LSM-T_a34	29	19	0	9.26	65	0.0029842	24.4	0.025511	88	0.062	84.5	0.28	19.21	4.68	-	-	-	-	-
2014-LSM-T_a35	-	-	-	-	-	-	-	-	-	-	-	-	-	-	-	-	-	-	-

Sample and spot number	$^{207}Pb^a$	U^b	Pb^b	Th^b	$^{206}Pb^c$	$^{206}Pb^c$	2 σ	$^{207}Pb^c$	2 σ	$^{207}Pb^c$	2 σ	ρ^d	^{206}Pb	2 σ	^{207}Pb	2 σ	^{207}Pb	2 σ	Conc
	(cps)	(ppm)	(ppm)	U	^{204}Pb	^{238}U	%	^{235}U	%	^{206}Pb	%		^{238}U	(Ma)	^{235}U	(Ma)	^{206}Pb	(Ma)	%
2014-LSM-T_a36	319	24	0	8.02	128	0.0046764	5.3	0.084356	18.7	0.13083	17.9	0.28	30.08	1.59	-	-	-	-	-
2014-LSM-T_a37	69009	59	17	0.67	756	0.24712	3.5	3.07547	4.2	0.09026	2.3	0.84	1423.62	45.39	1427	33	1431	44	99
2014-LSM-T_a38	-	-	-	-	-	-	-	-	-	-	-	-	-	-	-	-	-	-	-
2014-LSM-T_a39	-	-	-	-	-	-	-	-	-	-	-	-	-	-	-	-	-	-	-
2014-LSM-T_a40	40	37	0	3.18	126	0.0027769	9.7	0.013855	86.5	0.036186	86	0.11	17.88	1.72	-	-	-	-	-
2014-LSM-T_a42	11438	3	2	0.86	36	0.17235	23	16.11287	25.1	0.67804	10	0.92	1025.06	222.16	2884	274	4685	145	22
2014-LSM-T_a43	1551	96	1	1.62	101	0.0043982	3.9	0.095682	10.6	0.15778	9.9	0.37	28.29	1.1	-	-	-	-	-
2014-LSM-T_a44	322	28	0	4.52	504	0.0049648	3.5	0.04401	11.6	0.06429	11.1	0.3	31.93	1.11	-	-	-	-	-
2014-LSM-T_a45	-	-	-	-	-	-	-	-	-	-	-	-	-	-	-	-	-	-	-
2014-LSM-T_a46	4914	84	1	1.17	76	0.0033497	6.8	0.11002	20.6	0.2382	19.5	0.33	21.56	1.47	-	-	-	-	-
2014-LSM-T_a47	-	-	-	-	-	-	-	-	-	-	-	-	-	-	-	-	-	-	-
2014-LSM-T_a48	745	14	1	12.52	118	0.015191	9.3	1.3291	16.1	0.63456	13.2	0.57	97.19	8.96	-	-	-	-	-
2014-LSM-T_a49	118	33	0	6.27	38	0.0055057	9	0.078853	68.3	0.10387	67.7	0.13	35.39	3.16	-	-	-	-	-
2014-LSM-T_a50	567	361	2	1.9	1008	0.0044229	3.2	0.034236	15	0.05614	14.7	0.21	28.45	0.9	-	-	-	-	-
2014-LSM-T_a52	-	-	-	-	-	-	-	-	-	-	-	-	-	-	-	-	-	-	-
2014-LSM-T_a53	-	-	-	-	-	-	-	-	-	-	-	-	-	-	-	-	-	-	-
2014-LSM-T_a54	279	83	1	3.3	373	0.0045806	3.2	0.047011	18.3	0.074435	18	0.18	29.46	0.95	-	-	-	-	-
2014-LSM-T_a55	431	1133	0	0.67	180	0.000259	7	0.0085961	16.7	0.2542	15.2	0.42	1.67	0.12	-	-	-	-	-
2014-LSM-T_a56	5675	8	3	13.67	25	0.051675	19.4	5.56875	19.9	0.78159	4.4	0.98	324.8	61.73	1911	188	4889	63	7
2014-LSM-T_a57	5325	3	1	4.39	50	0.14971	10	8.92862	12.9	0.43255	8.1	0.77	899.32	84.25	2331	125	4026	121	22
2014-LSM-T_a58	-	-	-	-	-	-	-	-	-	-	-	-	-	-	-	-	-	-	-
2014-LSM-T_a59	61	1306	0	0.64	149	0.00016876	3.9	0.0009006	98.9	0.042182	98.8	0.04	1.09	0.04	-	-	-	-	-
2014-LSM-T_a60	-	-	-	-	-	-	-	-	-	-	-	-	-	-	-	-	-	-	-
2014-LSM-T_b02	2454	1404	1	0.08	46	0.00044209	8.3	0.022898	13	0.39005	10	0.64	2.85	0.02	-	-	-	-	-
2014-LSM-T_b03	40	385	0	0.68	101	0.00022136	5.7	0.0011333	94.5	0.039579	94.3	0.06	1.43	0.08	-	-	-	-	-
2014-LSM-T_b04	2518	10	1	25.85	16	0.039026	47.1	5.6756	58.2	1.054756	34.1	0.81	246.8	115.18	1928	693	5312	477	5
2014-LSM-T_b05	-	-	-	-	-	-	-	-	-	-	-	-	-	-	-	-	-	-	-

Sample and spot number	$^{207}Pb^a$ (cps)	U^b (ppm)	Pb^b (ppm)	$\underline{Th^b}$ U	$\underline{^{206}Pb^c}$ ^{204}Pb	$\underline{^{206}Pb^c}$ ^{238}U	2 σ %	$\underline{^{207}Pb^c}$ ^{235}U	2 σ %	$\underline{^{207}Pb^c}$ ^{206}Pb	2 σ %	ρ^d	$\underline{^{206}Pb}$ ^{238}U	2 σ (Ma)	$\underline{^{207}Pb}$ ^{235}U (Ma)	2 σ (Ma)	$\underline{^{207}Pb}$ ^{206}Pb (Ma)	2 σ (Ma)	Conc %
2014-LSM-T_b06	-	-	-	-	-	-	-	-	-	-	-	-	-	-	-	-	-	-	-
2014-LSM-T_b07	3835	29	3	5.03	61	0.033764	5.5	1.47568	8.3	0.31698	6.3	0.66	214.06	11.6	920	52	3555	96	6
2014-LSM-T_b08	-	-	-	-	-	-	-	-	-	-	-	-	-	-	-	-	-	-	-
2014-LSM-T_b09	-	-	-	-	-	-	-	-	-	-	-	-	-	-	-	-	-	-	-
2014-LSM-T_b10	329	746	0	0.08	321	0.00023598	4.6	0.0030476	30.9	0.1007	30.5	0.15	1.52	0.07	-	-	-	-	-
2014-LSM-T_b12	-	-	-	-	-	-	-	-	-	-	-	-	-	-	-	-	-	-	-
2014-LSM-T_b13	408	108	1	1.93	910	0.0039249	5.4	0.025175	19.1	0.046521	18.3	0.28	25.25	1.37	25	5	25	439	**102**
2014-LSM-T_b14	192120	3365	157	0.09	38	0.023055	4	1.44357	5.7	0.45412	4.1	0.7	146.94	5.83	907	35	4099	61	4
2014-LSM-T_b15	-	-	-	-	-	-	-	-	-	-	-	-	-	-	-	-	-	-	-
2014-LSM-T_b16	-	-	-	-	-	-	-	-	-	-	-	-	-	-	-	-	-	-	-
2014-LSM-T_b17	10420	247	18	1.05	5050	0.066474	2.5	0.54001	3.7	0.058919	2.6	0.7	414.88	10.24	438	13	564	57	74
2014-LSM-T_b18	-	-	-	-	-	-	-	-	-	-	-	-	-	-	-	-	-	-	-
2014-LSM-T_b19	594	17	0	6.71	20	0.0092478	5.8	0.53863	19.3	0.42243	18.4	0.3	59.34	3.44	-	-	-	-	-
2014-LSM-T_b20	797	563	0	0.64	158	0.00032614	9.3	0.022556	18	0.52384	15.4	0.52	2.1	0.2	-	-	-	-	-
2014-LSM-T_b22	62765	1017	61	0.72	104	0.039481	4.5	1.17126	5.7	0.21516	3.5	0.79	249.61	11.07	787	32	2945	56	8
2014-LSM-T_b23	-	-	-	-	-	-	-	-	-	-	-	-	-	-	-	-	-	-	-
2014-LSM-T_b24	-	-	-	-	-	-	-	-	-	-	-	-	-	-	-	-	-	-	-
2014-LSM-T_b25	508	684	0	0.68	65	0.00036346	7	0.017165	17.4	0.35585	15.9	0.41	2.34	0.01	-	-	-	-	-
2014-LSM-T_b26	-	-	-	-	-	-	-	-	-	-	-	-	-	-	-	-	-	-	-
2014-LSM-T_b27	-	-	-	-	-	-	-	-	-	-	-	-	-	-	-	-	-	-	-
2014-LSM-T_b28	3308	120	3	1.26	25	0.0075516	5.5	0.90186	9.7	0.86617	8	0.57	48.5	2.67	-	-	-	-	-
2014-LSM-T_b29	3008	23	2	4.8	22	0.039169	6	3.90501	10.9	0.72307	9.1	0.55	247.68	14.51	1615	92	4777	131	5
2014-LSM-T_b30	-	-	-	-	-	-	-	-	-	-	-	-	-	-	-	-	-	-	-
2014-LSM-T_b32	-	-	-	-	-	-	-	-	-	-	-	-	-	-	-	-	-	-	-
2014-LSM-T_b33	31	1036	0	0.68	117	0.00019014	8.4	0.00076023	105.1	0.031254	104.8	0.08	1.23	0.1	-	-	-	-	-
2014-LSM-T_b34	38887	252	61	0.34	1457	0.23958	2.8	3.4763	2.9	0.10524	0.9	0.96	1384.52	34.5	1522	23	1718	16	81
2014-LSM-T_b35	60	478	0	0.77	65	0.0002523	9.6	0.0026879	86.3	0.081555	85.7	0.11	1.63	0.002	-	-	-	-	-

Sample and spot number	$^{207}Pb^a$ (cps)	U^b (ppm)	Pb^b (ppm)	$\underline{Th^b}$ U	$\underline{^{206}Pb^c}$ ^{204}Pb	$\underline{^{206}Pb^c}$ ^{238}U	2σ %	$\underline{^{207}Pb^c}$ ^{235}U	2σ %	$\underline{^{207}Pb^c}$ ^{206}Pb	2σ %	$ρ^d$	$\underline{^{206}Pb}$ ^{238}U (Ma)	2σ (Ma)	$\underline{^{207}Pb}$ ^{235}U (Ma)	2σ (Ma)	$\underline{^{207}Pb}$ ^{206}Pb (Ma)	2σ (Ma)	Conc %
2014-LSM-T_b36	1306	423	1	0.67	177	0.0010115	8.5	0.086781	10.6	0.63021	6.3	0.81	6.52	0.05	-	-	-	-	-
2014-LSM-T_b37	-	-	-	-	-	-	-	-	-	-	-	-	-	-	-	-	-	-	-
2014-LSM-T_b38	766	892	0	0.05	588	0.0002698	3.3	0.0037824	22.6	0.10834	22.4	0.15	1.74	0.06	-	-	-	-	-
2014-LSM-T_b39	221	937	0	0.06	348	0.00022171	5.9	0.001811	25.7	0.06403	25	0.23	1.43	0.08	-	-	-	-	-
2014-LSM-T_b40	30997	206	14	0.78	20	0.021541	4.8	2.52985	5.7	0.85179	3	0.85	137.39	6.58	1281	42	5011	42	3

Standard: GJ1

Sample and spot number	$^{207}Pb^a$ (cps)	U^b (ppm)	Pb^b (ppm)	$\underline{Th^b}$ U	$\underline{^{206}Pb^c}$ ^{204}Pb	$\underline{^{206}Pb^c}$ ^{238}U	2σ %	$\underline{^{207}Pb^c}$ ^{235}U	2σ %	$\underline{^{207}Pb^c}$ ^{206}Pb	2σ %	$ρ^d$	$\underline{^{206}Pb}$ ^{238}U (Ma)	2σ (Ma)	$\underline{^{207}Pb}$ ^{235}U (Ma)	2σ (Ma)	$\underline{^{207}Pb}$ ^{206}Pb (Ma)	2σ (Ma)	Conc %
GJ1-1_seq a	28101	218	20	0.02	47948	0.098134	2.8	0.81214	3.3	0.060022	1.8	0.85	603.46	16.29	604	15	604	38	100
GJ1-2_seq a	29377	227	21	0.02	49156	0.098737	2.8	0.81868	3.3	0.060136	1.8	0.84	607	16.2	607	15	608	39	100
GJ1-3_seq a	29030	232	22	0.02	41778	0.10052	2.8	0.83758	3.2	0.06043	1.6	0.87	617.48	16.24	618	15	619	34	100
GJ1-4_seq a	28073	229	21	0.02	128516	0.10038	2.7	0.82671	3.1	0.059731	1.5	0.87	616.64	15.75	612	14	594	33	104
GJ1-5_seq a	28338	229	21	0.02	25568	0.099482	2.8	0.8275	3.4	0.060328	1.9	0.83	611.38	16.5	612	16	615	40	99
GJ1-6_seq a	28558	248	22	0.02	105495	0.096094	2.8	0.79662	3.3	0.060125	1.7	0.86	591.48	16.06	595	15	608	37	97
GJ1-7_seq a	27995	240	22	0.02	34775	0.09852	2.8	0.8148	3.2	0.059983	1.6	0.87	605.73	16.02	605	15	603	34	100
GJ1-8_seq a	28675	251	23	0.02	12020	0.097458	2.8	0.81179	3.4	0.060412	1.9	0.83	599.5	16.04	603	15	618	40	97
GJ1-9_seq a	28688	266	24	0.02	37428	0.098889	2.9	0.82202	3.4	0.060288	1.8	0.85	607.9	16.86	609	16	614	39	99
GJ1-10_seq a	28525	262	24	0.02	37246	0.097693	2.9	0.81765	3.3	0.060702	1.6	0.87	600.87	16.43	607	15	629	34	96
GJ1-11_seq a	27800	267	24	0.02	40368	0.099216	2.9	0.81447	3.2	0.059537	1.5	0.89	609.81	16.63	605	15	587	32	104
GJ1-12_seq a	27742	273	25	0.02	34319	0.098928	2.9	0.82512	3.3	0.060492	1.6	0.88	608.12	16.6	611	15	621	34	98
GJ1-13_seq a	28252	285	25	0.02	47626	0.096578	2.7	0.79419	3.1	0.059641	1.5	0.87	594.32	15.49	594	14	591	34	101
GJ1-14_seq a	27273	273	25	0.02	88869	0.098269	2.7	0.81804	3.1	0.060375	1.5	0.88	604.26	15.69	607	14	617	31	98
GJ1-15_seq a	27409	272	25	0.02	16941	0.098721	2.6	0.82428	3.1	0.060557	1.6	0.85	606.91	15.34	610	14	624	35	97
GJ1-16_seq a	26889	252	23	0.02	58057	0.098024	2.9	0.81234	3.3	0.060105	1.5	0.89	602.82	16.93	604	15	607	33	99
GJ1-17_seq a	26741	237	22	0.02	124073486	0.098339	2.7	0.81745	3.2	0.060288	1.7	0.85	604.67	15.57	607	15	614	37	98
GJ1-18_seq a	27215	247	23	0.02	46023	0.099341	3	0.81549	3.6	0.059538	2	0.83	610.54	17.33	606	16	587	43	104
GJ1-19_seq a	26854	225	20	0.02	22684	0.098312	2.8	0.82671	3.1	0.060988	1.4	0.9	604.51	16.31	612	15	639	29	95
GJ1-20_seq a	26945	216	20	0.02	12204	0.098365	2.7	0.81322	3.2	0.059961	1.7	0.85	604.82	15.53	604	14	602	36	100

Sample and spot number	$^{207}Pb^a$ (cps)	U^b (ppm)	Pb^b (ppm)	$\underline{Th^b}$ U	$\underline{^{206}Pb^c}$ ^{204}Pb	$\underline{^{206}Pb^c}$ ^{238}U	2σ %	$\underline{^{207}Pb^c}$ ^{235}U	2σ %	$\underline{^{207}Pb^c}$ ^{206}Pb	2σ %	$ρ^d$	$\underline{^{206}Pb}$ ^{238}U	2σ (Ma)	$\underline{^{207}Pb}$ ^{235}U	2σ (Ma)	$\underline{^{207}Pb}$ ^{206}Pb	2σ (Ma)	Conc %
GJ1-21_seq a	26812	219	20	0.02	44910	0.099538	2.6	0.82426	3.1	0.060058	1.7	0.85	611.7	15.48	610	14	606	36	101
GJ1-23_seq a	26652	194	18	0.02	44470	0.09966	2.7	0.82808	3.2	0.060263	1.7	0.85	612.42	15.99	613	15	613	36	100
GJ1-1_seq b	23251	242	22	0.02	38965	0.099202	3	0.8227	3.3	0.060148	1E+13	0.92	609.73	17.5	610	15	609	27	100
GJ1-2_seq b	23981	242	22	0.02	40173	0.098918	3	0.82007	3.3	0.060128	1.3	0.91	608.06	17.37	608	15	608	29	100
GJ1-3_seq b	23571	240	21	0.02	58492	0.096718	2.9	0.80578	3.3	0.060424	1.4	0.9	595.15	16.7	600	15	619	31	96
GJ1-4_seq b	23451	242	22	0.02	11451	0.099272	3	0.81845	3.2	0.059795	1.1	0.94	610.14	17.63	607	15	596	24	102
GJ1-5_seq b	23669	240	22	0.02	14188	0.098002	2.9	0.81277	3.3	0.060149	1.5	0.89	602.69	16.82	604	15	609	33	99
GJ1-7_seq b	22899	243	22	0.02	57161	0.099335	2.9	0.81892	3.2	0.059791	1.4	0.9	610.51	16.98	607	15	596	31	102
GJ1-8_seq b	22922	243	22	0.02	12576	0.097915	3	0.81367	3.4	0.06027	1.5	0.9	602.18	17.36	605	15	613	32	98
GJ1-9_seq b	22811	248	23	0.02	43048	0.098843	3	0.82509	3.3	0.060542	1.5	0.89	607.62	17.16	611	15	623	32	98
GJ1-10_seq b	22451	243	22	0.02	37407	0.098243	3	0.81927	3.2	0.060482	1.2	0.93	604.1	17.07	608	15	621	26	97
GJ1-11_seq b	23111	262	24	0.02	39022	0.099826	2.9	0.82131	3.4	0.059671	1.8	0.85	613.39	16.96	609	16	592	38	104
GJ1-12_seq b	22128	251	23	0.02	19865	0.09871	3	0.81951	3.2	0.060213	1.2	0.93	606.84	17.2	608	15	611	25	99
GJ1-13_seq b	22514	249	23	0.02	147159	0.10068	2.9	0.83556	3.1	0.060191	1.3	0.91	618.4	16.86	617	15	610	28	101
GJ1-14_seq b	22510	260	24	0.02	37703	0.097785	2.9	0.8109	3.2	0.060144	1.3	0.92	601.41	16.77	603	15	609	28	99
GJ1-15_seq b	22820	251	22	0.02	10721	0.096245	3	0.80206	3.4	0.06044	1.6	0.88	592.37	17.16	598	16	619	35	96
GJ1-16_seq b	22124	228	21	0.02	37061	0.097897	3	0.81201	3.2	0.060157	1.3	0.92	602.07	17.14	604	15	609	28	99
GJ1-17_seq b	22726	220	20	0.02	57041	0.100009	3	0.83198	3.3	0.060336	1.5	0.89	614.46	17.42	615	16	616	33	100
Tsankawi tephra sampled in 2014																			
2014-NM-Ts-seq1-a01	2674	1119	1	0.11	52	0.00052	2.8	0.02749	9.2	0.39439	8.8	0.31	3.36	0.1	-	-	-	-	-
2014-NM-Ts-seq1-a02	4428	1086	1	0.2	92	0.00025	5.9	0.01699	9.2	0.53416	7	0.65	1.59	0.09	-	-	-	-	-
2014-NM-Ts-seq1-a03	1876	1373	1	0.08	102	0.00026	8.6	0.01036	15.8	0.30805	13.3	0.54	1.68	0.14	-	-	-	-	-
2014-NM-Ts-seq1-a04	1207	711	0	0.08	78	0.00021	8.8	0.00574	21.7	0.21289	19.8	0.41	1.37	0.12	-	-	-	-	-
2014-NM-Ts-seq1-a05	46	1087	0	0.64	100	0.00022	6.7	0.00148	70.6	0.0509	70.3	0.09	1.44	0.1	-	-	-	-	-
2014-NM-Ts-seq1-a06	198	943	0	0.49	152	0.00024	8.7	0.00345	47.7	0.11246	46.9	0.18	1.52	0.13	-	-	-	-	-
2014-NM-Ts-seq1-a07	670	1002	1	0.68	45	0.00022	19.4	0.01021	27.2	0.36407	19.1	0.71	1.4	0.27	-	-	-	-	-

Sample and spot number	$^{207}Pb^a$	U^b	Pb^b	$\frac{Th^b}{U}$	$\frac{^{206}Pb^c}{^{204}Pb}$	$\frac{^{206}Pb^c}{^{238}U}$	2σ	$\frac{^{207}Pb^c}{^{235}U}$	2σ	$\frac{^{207}Pb^c}{^{206}Pb}$	2σ	ρ^d	$\frac{^{206}Pb}{^{238}U}$	2σ	$\frac{^{207}Pb}{^{235}U}$	2σ	$\frac{^{207}Pb}{^{206}Pb}$	2σ	Conc
	(cps)	(ppm)	(ppm)				%		%		%			(Ma)		(Ma)		(Ma)	%
2014-NM-Ts-seq1-a08	73	1210	0	0.71	137	0.0002	6.5	0.00146	29.8	0.05513	29.1	0.22	1.32	0.09	-	-	-	-	-
2014-NM-Ts-seq1-a09	77	1251	0	0.75	146	0.00021	6.9	0.00139	62.2	0.05082	61.8	0.11	1.36	0.09	-	-	-	-	-
2014-NM-Ts-seq1-a10	158	1268	0	0.53	102	0.00019	9.1	0.00476	44.8	0.19387	43.9	0.2	1.24	0.11	-	-	-	-	-
2014-NM-Ts-seq1-a11	8064	749	5	0.07	4	0.0042	2.5	0.29055	27.3	0.50146	27.2	0.09	27.03	0.67	-	-	-	-	-
2014-NM-Ts-seq1-a13	736	807	0	0.33	116	0.00029	4.8	0.00886	11.2	0.23528	10.1	0.43	1.86	0.09	-	-	-	-	-
2014-NM-Ts-seq1-a14	116	2019	0	0.44	255	0.00025	4	0.00149	41.4	0.04673	41.2	0.1	1.59	0.06	-	-	-	-	-
2014-NM-Ts-seq1-a15	114	991	0	0.57	163	0.00022	4	0.00218	42.1	0.07519	41.9	0.1	1.44	0.06	-	-	-	-	-
2014-NM-Ts-seq1-a16	945	1902	1	0.05	232	0.00025	4.4	0.00466	23.1	0.14775	22.7	0.19	1.58	0.07	-	-	-	-	-
2014-NM-Ts-seq1-a17	1784	1192	1	0.07	56	0.00029	7.7	0.01235	14.1	0.32642	11.8	0.55	1.87	0.14	-	-	-	-	-
2014-NM-Ts-seq1-a18	3085	1003	1	0.12	58	0.00024	6	0.01491	11.6	0.47667	9.9	0.52	1.57	0.09	-	-	-	-	-
2014-NM-Ts-seq1-a19	248	1496	0	0.06	449	0.00021	3.6	0.00145	12.8	0.05547	12.3	0.28	1.33	0.05	-	-	-	-	-
2014-NM-Ts-seq1-a20	76	1099	0	0.51	174	0.00021	4.8	0.00123	65.4	0.04626	65.3	0.07	1.33	0.06	-	-	-	-	-
2014-NM-Ts-seq1-a21	123	1111	0	0.5	176	0.00019	7.5	0.00117	125.9	0.04818	125.7	0.06	1.23	0.09	-	-	-	-	-
2014-NM-Ts-seq1-a22	383	939	0	0.15	290	0.00021	3.3	0.00357	19.2	0.13272	18.9	0.17	1.36	0.04	-	-	-	-	-
2014-NM-Ts-seq1-a23	281	1278	0	0.48	221	0.00021	5.5	0.00317	28.5	0.12017	28	0.19	1.33	0.07	-	-	-	-	-
2014-NM-Ts-seq1-a24	595	1286	0	0.46	244	0.00024	10.6	0.00697	24.1	0.22667	21.6	0.44	1.53	0.16	-	-	-	-	-
2014-NM-Ts-seq1-a25	204	1126	0	0.65	115	0.0002	6.8	0.00464	24.6	0.1802	23.7	0.28	1.29	0.09	-	-	-	-	-
2014-NM-Ts-seq1-a26	4357	1152	1	0.1	54	0.00034	11.9	0.02414	18.9	0.53548	14.7	0.63	2.21	0.26	-	-	-	-	-
2014-NM-Ts-seq1-a27	7471	1040	4	0.23	24	0.00134	5	0.12178	6.3	0.66801	3.8	0.8	8.61	0.43	-	-	-	-	-
2014-NM-Ts-seq1-a28	510	474	1	0.96	148	0.00072	5.2	0.03134	19.9	0.32001	19.2	0.26	4.65	0.24	-	-	-	-	-
2014-NM-Ts-seq1-a29	116	1298	0	0.63	135	0.00021	8.9	0.00233	68.5	0.08599	68	0.13	1.35	0.12	-	-	-	-	-
2014-NM-Ts-seq1-a30	54	185	0	0.64	28	0.00023	18.1	0.00682	61.4	0.23208	58.7	0.29	1.46	0.26	-	-	-	-	-
2014-NM-Ts-seq1-a31	98	1664	0	0.06	211	0.00009	4.5	0.0005	37.4	0.04985	37.2	0.12	0.58	0.03	-	-	-	-	-
2014-NM-Ts-seq1-a32	3757	991	1	0.15	44	0.00052	10.7	0.03791	14.3	0.54402	9.5	0.75	3.36	0.36	-	-	-	-	-
2014-NM-Ts-seq1-a33	1308	939	0	0.21	300	0.00023	7.7	0.00971	14.2	0.33215	12	0.54	1.47	0.11	-	-	-	-	-
2014-NM-Ts-seq1-a34	189	1396	0	0.07	417	0.0002	2.7	0.00116	18.9	0.04638	18.7	0.14	1.28	0.03	-	-	-	-	-
2014-NM-Ts-seq1-a35	817	785	0	0.71	198	0.00026	10.3	0.01397	14.7	0.41191	10.5	0.7	1.67	0.17	-	-	-	-	-

Sample and spot number	$^{207}Pb^a$ (cps)	U^b (ppm)	Pb^b (ppm)	$\frac{Th^b}{U}$	$\frac{^{206}Pb^c}{^{204}Pb}$	$\frac{^{206}Pb^c}{^{238}U}$	2σ %	$\frac{^{207}Pb^c}{^{235}U}$	2σ %	$\frac{^{207}Pb^c}{^{206}Pb}$	2σ %	ρ^d	$\frac{^{206}Pb}{^{238}U}$	2σ (Ma)	$\frac{^{207}Pb}{^{235}U}$	2σ (Ma)	$\frac{^{207}Pb}{^{206}Pb}$	2σ (Ma)	Conc %
2014-NM-Ts-seq1-a36	743	1226	1	0.43	252	0.00025	7	0.00979	13.1	0.30014	11.1	0.53	1.62	0.11	-	-	-	-	-
2014-NM-Ts-seq1-a37	471	1091	0	0.55	195	0.00019	10.4	0.00484	48.1	0.19393	47	0.22	1.26	0.13	-	-	-	-	-
2014-NM-Ts-seq1-a38	140	386	0	0.22	295	0.00021	5.7	0.00124	20.6	0.04663	19.8	0.27	1.35	0.08	-	-	-	-	-
2014-NM-Ts-seq1-a39	82	1251	0	0.48	165	0.0002	12.2	0.00132	36	0.05224	33.9	0.34	1.27	0.15	-	-	-	-	-
2014-NM-Ts-seq1-a40	5184	451	3	0.78	17	0.00252	9.9	0.25269	10.4	0.72684	3.1	0.95	16.23	1.61	-	-	-	-	-
2014-NM-Ts-seq1-a41	160	1005	0	0.22	314	0.0002	2.3	0.00126	43.5	0.04879	43.4	0.05	1.3	0.03	-	-	-	-	-
2014-NM-Ts-seq1-a42	774	1007	0	0.15	99	0.00026	4.4	0.00777	16.2	0.22897	15.6	0.27	1.69	0.07	-	-	-	-	-
2014-NM-Ts-seq1-a43	73	499	0	0.49	131	0.0002	4.7	0.00151	40.8	0.05782	40.6	0.11	1.31	0.06	-	-	-	-	-
2014-NM-Ts-seq1-a44	812	1118	0	0.1	412	0.00021	5.5	0.00545	17.4	0.19983	16.5	0.32	1.38	0.08	-	-	-	-	-
2014-NM-Ts-seq1-a45	2491	777	2	0.69	23	0.0014	4.8	0.11833	9.4	0.61618	8.1	0.51	9.04	0.43	-	-	-	-	-
2014-NM-Ts-seq1-a46	6988	935	2	0.3	25	0.00068	8.2	0.05927	9.7	0.64444	5.2	0.84	4.39	0.36	-	-	-	-	-
2014-NM-Ts-seq1-a47	152	1223	0	0.11	330	0.0002	4.4	0.00122	24.7	0.04715	24.3	0.18	1.31	0.06	-	-	-	-	-
2014-NM-Ts-seq1-a48	4046	350	1	0.31	27	0.00079	6.1	0.06418	8.6	0.60073	6.1	0.7	5.09	0.31	-	-	-	-	-
2014-NM-Ts-seq1-a49	96	896	0	0.52	122	0.0002	6.3	0.00183	57.4	0.07237	57	0.11	1.27	0.08	-	-	-	-	-
2014-NM-Ts-seq1-a50	51619	520	26	0.48	17	0.02311	2.2	2.64556	3.5	0.83039	2.8	0.61	147.26	3.14	1313	26	4975	40	3
2014-NM-Ts-seq1-a51	151	904	0	0.76	131	0.0002	11.5	0.00162	201	0.06162	200.7	0.06	1.31	0.15	-	-	-	-	-
2014-NM-Ts-seq1-a52	38	78	0	0.48	10	0.0003	21.6	0.00761	51.2	0.19218	46.4	0.42	1.94	0.42	-	-	-	-	-
2014-NM-Ts-seq1-a53	210	1102	0	0.55	161	0.00021	3.1	0.00349	29.8	0.13053	29.7	0.11	1.34	0.04	-	-	-	-	-
2014-NM-Ts-seq1-a54	123	993	0	0.17	266	0.0002	5.2	0.00123	28.2	0.04818	27.7	0.18	1.3	0.07	-	-	-	-	-
2014-NM-Ts-seq1-a55	36	951	0	0.66	96	0.00021	5.1	0.00132	77.1	0.0485	77	0.07	1.36	0.07	-	-	-	-	-
2014-NM-Ts-seq1-a56	220	1241	0	0.07	340	0.0002	3.4	0.00171	11.9	0.06617	11.5	0.28	1.32	0.04	-	-	-	-	-
2014-NM-Ts-seq1-a57	63	735	0	0.49	98	0.00022	5.8	0.00185	79.9	0.0651	79.7	0.07	1.42	0.08	-	-	-	-	-
2014-NM-Ts-seq1-a58	30	61	0	1.32	21	0.00022	30.3	0.00343	64.5	0.11688	56.9	0.47	1.44	0.44	-	-	-	-	-
2014-NM-Ts-seq1-a59	114	127	0	1.38	40	0.0005	32.1	0.02086	56.3	0.30739	46.3	0.57	3.23	1.04	-	-	-	-	-
2014-NM-Ts-seq1-a60	390	1135	0	0.53	195	0.00009	23.7	0.00144	58.1	0.14484	53	0.41	0.56	0.13	-	-	-	-	-
2014-NM-Ts-seq2-b02	65	947	0	0.55	123	0.0002	6.7	0.00121	67.7	0.04701	67.4	0.1	1.29	0.09	-	-	-	-	-
2014-NM-Ts-seq2-b03	824	125	0	0.71	134	0.00036	42.7	0.02964	44.7	0.61734	13.4	0.95	2.33	0.99	-	-	-	-	-

Sample and spot number	$^{207}Pb^a$ (cps)	U^b (ppm)	Pb^b (ppm)	$\frac{Th^b}{U}$	$\frac{^{206}Pb^c}{^{204}Pb}$	$\frac{^{206}Pb^c}{^{238}U}$	2σ %	$\frac{^{207}Pb^c}{^{235}U}$	2σ %	$\frac{^{207}Pb^c}{^{206}Pb}$	2σ %	$ρ^d$	$\frac{^{206}Pb}{^{238}U}$ (Ma)	2σ (Ma)	$\frac{^{207}Pb}{^{235}U}$ (Ma)	2σ (Ma)	$\frac{^{207}Pb}{^{206}Pb}$ (Ma)	2σ (Ma)	Conc %
2014-NM-Ts-seq2-b04	21	180	0	0.58	13	0.0002	24.6	0.01309	136.7	0.49756	134.5	0.18	1.32	0.32	-	-	-	-	-
2014-NM-Ts-seq2-b05	4310	420	13	1.05	8800	0.02539	2.1	0.173	3.5	0.04941	2.8	0.6	161.65	3.33	162	5	167	66	97
2014-NM-Ts-seq2-b06	205	357	0	1.43	64	0.00034	19.7	0.01448	27.6	0.31473	19.3	0.71	2.21	0.43	-	-	-	-	-
2014-NM-Ts-seq2-b07	7415	733	8	0.46	22	0.00412	2.7	0.43259	4	0.76182	2.9	0.68	26.49	0.71	-	-	-	-	-
2014-NM-Ts-seq2-b08	39	966	0	0.65	82	0.0002	8.9	0.00126	96.8	0.04939	96.4	0.09	1.28	0.11	-	-	-	-	-
2014-NM-Ts-seq2-b09	52	1425	0	0.64	123	0.0002	9.4	0.00127	67.9	0.04862	67.2	0.14	1.31	0.12	-	-	-	-	-
2014-NM-Ts-seq2-b10	171	1052	0	0.56	146	0.0002	9.2	0.00327	44.8	0.13022	43.8	0.2	1.27	0.12	-	-	-	-	-
2014-NM-Ts-seq2-b12	19170	657	21	0.45	120	0.02116	2.4	0.14221	11.5	0.04875	11.3	0.21	134.97	3.2	135	15	136	265	99
2014-NM-Ts-seq2-b13	810	1052	1	0.61	234	0.00045	4.6	0.0185	21.9	0.30418	21.5	0.21	2.93	0.13	-	-	-	-	-
2014-NM-Ts-seq2-b14	82	1062	0	0.5	148	0.0002	4.8	0.00146	32.3	0.05574	31.9	0.15	1.31	0.06	-	-	-	-	-
2014-NM-Ts-seq2-b15	85	1228	0	0.55	174	0.0002	5.8	0.00129	27.6	0.04953	27	0.21	1.3	0.08	-	-	-	-	-
2014-NM-Ts-seq2-b16	78	300	0	0.69	34	0.00006	74.5	0.00211	101.7	0.33822	69.3	0.73	0.38	0.28	-	-	-	-	-
2014-NM-Ts-seq2-b17	1540	1168	1	0.57	71	0.00021	7.4	0.01348	10.7	0.5	7.7	0.7	1.35	0.1	-	-	-	-	-
2014-NM-Ts-seq2-b18	461	1157	0	0.54	211	0.0002	6.3	0.00588	17.2	0.22641	16	0.37	1.3	0.08	-	-	-	-	-
2014-NM-Ts-seq2-b19	130	1461	0	0.42	182	0.0002	10.6	0.00178	34	0.06814	32.3	0.31	1.32	0.14	-	-	-	-	-
2014-NM-Ts-seq2-b20	82	1076	0	0.65	91	0.0002	7.7	0.00238	43.7	0.09024	43	0.18	1.32	0.1	-	-	-	-	-
2014-NM-Ts-seq2-b22	28	1545	0	0.26	195	0.00021	5.5	0.00044	141.4	0.01608	141.3	0.04	1.37	0.08	-	-	-	-	-
2014-NM-Ts-seq2-b23	86	1526	0	0.4	161	0.00017	3	0.00113	81.5	0.05344	81.5	0.04	1.08	0.03	-	-	-	-	-
2014-NM-Ts-seq2-b24	76	1230	0	0.63	111	0.00027	4.6	0.00267	70.8	0.07458	70.7	0.07	1.76	0.08	-	-	-	-	-
2014-NM-Ts-seq2-b25	53	1238	0	0.68	102	0.00021	9.8	0.00144	78.3	0.05291	77.7	0.13	1.36	0.13	-	-	-	-	-
2014-NM-Ts-seq2-b27	109	852	0	0.63	83	0.0002	12.7	0.00298	66.6	0.11421	65.4	0.19	1.31	0.17	-	-	-	-	-
2014-NM-Ts-seq2-b28	36	578	0	0.76	21	0.00027	9.5	0.00177	113.2	0.05031	112.8	0.08	1.73	0.16	-	-	-	-	-
2014-NM-Ts-seq2-b29	91	1299	0	0.69	96	0.00021	3.8	0.0025	122.9	0.09231	122.8	0.03	1.35	0.05	-	-	-	-	-
2014-NM-Ts-seq2-b30	117	552	0	0.91	64	0.00026	13.3	0.00584	61	0.16935	59.6	0.22	1.69	0.22	-	-	-	-	-
2014-NM-Ts-seq2-b32	72	1751	0	0.56	154	0.0002	5.9	0.00121	31.3	0.04725	30.7	0.19	1.29	0.08	-	-	-	-	-
2014-NM-Ts-seq2-b33	983	1218	2	0.52	39	0.00077	4.7	0.04629	18	0.44245	17.4	0.26	4.98	0.23	-	-	-	-	-
2014-NM-Ts-seq2-b34	53	891	0	0.6	65	0.00021	7.9	0.00234	49.3	0.08774	48.7	0.16	1.33	0.11	-	-	-	-	-

Sample and spot number	$^{207}Pb^a$	U^b	Pb^b	Th^b	$^{206}Pb^c$	$^{206}Pb^c$	2 σ	$^{207}Pb^c$	2 σ	$^{207}Pb^c$	2 σ	ρ^d	^{206}Pb	2 σ	^{207}Pb	2 σ	^{207}Pb	2 σ	Conc
	(cps)	(ppm)	(ppm)	U	^{204}Pb	^{238}U	%	^{235}U	%	^{206}Pb	%		^{238}U	(Ma)	^{235}U	(Ma)	^{206}Pb	(Ma)	%
2014-NM-Ts-seq2-b35	80	1461	0	0.63	125	0.0002	6.1	0.00169	46.2	0.06436	45.8	0.13	1.31	0.08	-	-	-	-	-
2014-NM-Ts-seq2-b36	95	1130	0	0.65	24	0.00021	9.4	0.00257	35.1	0.09584	33.8	0.27	1.34	0.13	-	-	-	-	-
2014-NM-Ts-seq2-b37	51	781	0	0.61	66	0.00021	11.6	0.00221	54.5	0.08173	53.3	0.21	1.35	0.16	-	-	-	-	-
2014-NM-Ts-seq2-b38	248	666	0	0.5	65	0.00021	23	0.00766	54.4	0.28646	49.3	0.42	1.34	0.31	-	-	-	-	-
2014-NM-Ts-seq2-b39	74	1869	0	0.48	142	0.00019	2.9	0.00129	91.9	0.05388	91.8	0.03	1.21	0.04	-	-	-	-	-
2014-NM-Ts-seq2-b40	51	1311	0	0.54	108	0.00015	7.4	0.00088	53.3	0.04674	52.8	0.14	0.97	0.07	-	-	-	-	-
2014-NM-Ts-seq2-b42	223	1203	0	0.53	163	0.00024	3.9	0.00396	25.2	0.12745	24.9	0.15	1.54	0.06	-	-	-	-	-
2014-NM-Ts-seq2-b43	60	90	0	1.29	112	0.00133	10.5	0.00908	46.3	0.04991	45	0.23	8.55	0.9	-	-	-	-	-
2014-NM-Ts-seq2-b44	136	950	0	0.61	85	0.0002	10.3	0.00408	39.8	0.15851	38.4	0.26	1.29	0.13	-	-	-	-	-
2014-NM-Ts-seq2-b45	54	145	0	1.58	30	0.00103	9.2	0.00833	51	0.05931	50.2	0.18	6.61	0.61	-	-	-	-	-
2014-NM-Ts-seq2-b46	64	964	0	0.53	116	0.0002	5.8	0.00149	69.8	0.05725	69.6	0.08	1.31	0.08	-	-	-	-	-
2014-NM-Ts-seq2-b47	102	1690	0	0.35	221	0.0002	3.6	0.0012	26.4	0.0468	26.1	0.14	1.3	0.05	-	-	-	-	-
2014-NM-Ts-seq2-b48	131	1246	0	0.46	152	0.00017	4.4	0.00195	25.5	0.08808	25.1	0.17	1.13	0.05	-	-	-	-	-
2014-NM-Ts-seq2-b49	191	1507	0	0.26	184	0.00017	6.6	0.00209	32.1	0.09919	31.4	0.21	1.08	0.07	-	-	-	-	-
2014-NM-Ts-seq2-b50	61	1011	0	0.59	92	0.00025	7.9	0.00208	97.2	0.06475	96.8	0.08	1.59	0.13	-	-	-	-	-
2014-NM-Ts-seq2-b52	58	1580	0	0.56	118	0.00021	4.7	0.00133	42.3	0.04955	42	0.11	1.34	0.06	-	-	-	-	-
2014-NM-Ts-seq2-b53	90	1204	0	0.51	138	0.00014	6.2	0.00124	39.7	0.0697	39.3	0.16	0.92	0.06	-	-	-	-	-
2014-NM-Ts-seq2-b54	49	1413	0	0.63	113	0.00021	4.8	0.00131	63.4	0.04775	63.2	0.07	1.37	0.07	-	-	-	-	-
2014-NM-Ts-seq2-b55	53	1241	0	0.62	113	0.0002	9.2	0.00135	76.9	0.05233	76.3	0.12	1.29	0.12	-	-	-	-	-
2014-NM-Ts-seq2-b56	65	981	0	0.51	133	0.0002	5.1	0.00127	43.7	0.04885	43.4	0.12	1.31	0.07	-	-	-	-	-
2014-NM-Ts-seq2-b57	455	1021	0	0.64	24	0.00021	13.6	0.00902	20.1	0.33746	14.8	0.68	1.34	0.18	-	-	-	-	-
2014-NM-Ts-seq2-b58	47	1230	0	0.72	100	0.00021	6.7	0.00131	68.8	0.04885	68.5	0.1	1.34	0.09	-	-	-	-	-
2014-NM-Ts-seq2-b59	29	966	0	0.55	58	0.0002	6.4	0.00139	121.6	0.05288	121.5	0.05	1.32	0.08	-	-	-	-	-
2014-NM-Ts-seq2-b60	5406	976	4	0.65	24	0.001	7.8	0.10752	8.5	0.79286	3.5	0.91	6.42	0.5	-	-	-	-	-

Standard: GJ1

| GJ1-1_seq1 | 25373 | 249.6 | 22.66 | 0 | 42621 | 0.0983 | 2 | 0.8151 | 2.35 | 0.0601 | 1.24 | 0.8 | 604.61 | 11.51 | 605 | 10.8 | 608 | 26.9 | 99 |

Sample and spot number	$^{207}Pb^{a}$	U^{b}	Pb^{b}	Th^{b}	$\frac{^{206}Pb^{c}}{^{204}Pb}$	$\frac{^{206}Pb^{c}}{^{238}U}$	2σ	$\frac{^{207}Pb^{c}}{^{235}U}$	2σ	$\frac{^{207}Pb^{c}}{^{206}Pb}$	2σ	ρ^{d}	$\frac{^{206}Pb}{^{238}U}$	2σ	$\frac{^{207}Pb}{^{235}U}$	2σ	$\frac{^{207}Pb}{^{206}Pb}$	2σ	Conc
	(cps)	(ppm)	(ppm)	U			%		%		%			(Ma)		(Ma)		(Ma)	%
GJ1-2_seq1	25012	249.6	22.83	0	41985	0.0989	2	0.8215	2.46	0.0602	1.42	0.8	608.14	11.66	609	11.3	612	30.7	99
GJ1-3_seq1	24627	246.4	22.22	0	18702	0.0976	1.8	0.8093	2.21	0.0601	1.28	0.8	600.44	10.33	602	10.1	608	27.6	99
GJ1-5_seq1	24959	246.6	22.53	0	9669	0.0988	1.8	0.8191	2.16	0.0601	1.14	0.9	607.65	10.66	608	9.93	607	24.6	100
GJ1-6_seq1	25130	247.5	22.68	0	65829	0.0992	1.7	0.8233	2.06	0.0602	1.21	0.8	609.54	9.73	610	9.5	611	26.1	100
GJ1-7_seq1	22983	234.3	21.33	0	38514	0.0986	1.6	0.8197	2.04	0.0603	1.22	0.8	606.16	9.48	608	9.39	614	26.4	99
GJ1-8_seq1	24744	246.9	22.44	0	43011	0.0983	1.8	0.8192	2.04	0.0604	0.9	0.9	604.44	10.53	608	9.35	620	19.5	98
GJ1-9_seq1	25154	248.7	22.96	0	42085	0.0999	2	0.832	2.61	0.0604	1.64	0.8	613.72	11.92	615	12.1	618	35.4	99
GJ1-10_seq1	23984	240.2	22.05	0	40298	0.0994	1.8	0.8236	2.39	0.0601	1.54	0.8	610.61	10.62	610	11	608	33.3	100
GJ1-11_seq1	24426	243.9	22.04	0	41033	0.0977	2	0.8104	2.32	0.0601	1.15	0.9	601.12	11.56	603	10.6	608	24.8	99
GJ1-12_seq1	23880	237.9	21.8	0	40081	0.0991	1.7	0.8222	2.08	0.0602	1.14	0.8	609.19	10.12	609	9.57	610	24.6	100
GJ1-13_seq1	21255	217.6	19.85	0	35871	0.0987	2	0.815	2.18	0.0599	0.98	0.9	606.88	11.31	605	10	599	21.3	101
GJ1-14_seq1	24128	242.7	22.14	0	16336	0.0986	1.9	0.8224	2.22	0.0605	1.14	0.9	606.32	11.02	609	10.2	621	24.5	98
GJ1-15_seq1	24054	245.8	22.26	0	16277	0.098	2	0.812	2.34	0.0601	1.24	0.8	602.46	11.41	604	10.7	608	26.8	99
GJ1-16_seq1	24041	240.1	21.93	0	40316	0.0988	2	0.8212	2.24	0.0603	1.01	0.9	607.64	11.6	609	10.3	613	21.9	99
GJ1-17_seq1	23640	243.1	21.96	0	26117	0.0978	1.8	0.8112	2.18	0.0602	1.22	0.8	601.39	10.36	603	9.94	610	26.3	99
GJ1-18_seq1	24467	246.8	22.49	0	18351	0.0986	1.7	0.8182	2.23	0.0602	1.47	0.8	606.26	9.71	607	10.3	610	31.9	99
GJ1-19_seq1	25348	258.7	23.3	0	54171	0.0975	1.8	0.8038	2.01	0.0598	0.84	0.9	599.96	10.45	599	9.13	595	18.3	101
GJ1-20_seq1	24189	246.4	22.32	0	40624	0.098	1.8	0.8128	2.31	0.0602	1.4	0.8	602.46	10.61	604	10.6	610	30.2	99
GJ1-21_seq1	24321	247.9	22.71	0	24073	0.0991	1.8	0.8232	2.26	0.0602	1.37	0.8	609.34	10.45	610	10.4	612	29.7	100
GJ1-22_seq1	23784	244.3	22.32	0	37484	0.0989	1.8	0.8209	2.23	0.0602	1.25	0.8	607.78	10.73	609	10.3	611	27	99
GJ1-23_seq1	23654	243	22.29	0	36947	0.0993	1.8	0.8233	2.14	0.0601	1.19	0.8	610.25	10.34	610	9.84	609	25.7	100
GJ1-1_seq2	22758	248.4	22.59	0	40397	0.0985	1.8	0.8162	2.1	0.0601	1.14	0.8	605.47	10.17	606	9.61	608	24.6	100
GJ1-2_seq2	23217	253.3	23.26	0	38898	0.0994	1.6	0.8249	1.92	0.0602	1.03	0.8	611.02	9.49	611	8.87	610	22.2	100
GJ1-3_seq2	22405	250.7	22.93	0	108398	0.099	1.7	0.8238	2.11	0.0604	1.22	0.8	608.53	10.04	610	9.74	616	26.3	99
GJ1-4_seq2	22012	245.3	22.48	0	36796	0.0992	1.6	0.8249	1.94	0.0603	1.11	0.8	609.72	9.21	611	8.92	615	24.1	99
GJ1-5_seq2	21899	235.7	21.48	0	36672	0.0986	1.6	0.8187	1.96	0.0602	1.07	0.8	606.15	9.52	607	9	612	23.1	99
GJ1-6_seq2	21756	248.8	22.32	0	36493	0.0971	1.6	0.8044	2.08	0.0601	1.28	0.8	597.32	9.35	599	9.44	607	27.6	98

Sample and spot number	$^{207}Pb^a$	U^b	Pb^b	Th^b	$^{206}Pb^c$	$^{206}Pb^c$	2σ	$^{207}Pb^c$	2σ	$^{207}Pb^c$	2σ	ρ^d	^{206}Pb	2σ	^{207}Pb	2σ	^{207}Pb	2σ	Conc
	(cps)	(ppm)	(ppm)	U	^{204}Pb	^{238}U	%	^{235}U	%	^{206}Pb	%		^{238}U	(Ma)	^{235}U	(Ma)	^{206}Pb	(Ma)	%
GJ1-7_seq2	22258	251	22.86	0	37303	0.0985	1.7	0.8174	2.13	0.0602	1.3	0.8	605.61	9.76	607	9.76	610	28	99
GJ1-8_seq2	21901	235	21.55	0	36598	0.0991	1.8	0.8247	2.18	0.0603	1.27	0.8	609.4	10.32	611	10	615	27.3	99
GJ1-9_seq2	21151	239.5	21.84	0	35467	0.0987	1.7	0.8181	2.12	0.0601	1.31	0.8	606.7	9.67	607	9.75	608	28.4	100
GJ1-10_seq2	21315	234.3	21.15	0	35899	0.0977	1.9	0.8078	2.37	0.06	1.47	0.8	601.06	10.65	601	10.8	602	31.8	100
GJ1-11_seq2	21436	245.5	22.27	0	8946.8	0.0981	2.1	0.8141	2.53	0.0602	1.42	0.8	603.22	12.07	605	11.6	611	30.7	99
GJ1-12_seq2	20895	243.6	22.15	0	11840	0.0984	1.9	0.8165	2.33	0.0602	1.3	0.8	605.16	11.17	606	10.7	610	28.2	99
GJ1-13_seq2	20782	231.3	21.19	0	64482	0.0991	1.6	0.8233	2.19	0.0602	1.51	0.7	609.33	9.25	610	10.1	612	32.6	100
GJ1-14_seq2	21043	241	21.99	0	70540	0.0987	1.5	0.821	2.19	0.0603	1.61	0.7	606.92	8.56	609	10.1	615	34.8	99
GJ1-15_seq2	21433	244.3	22.16	0	59485	0.0981	1.7	0.8145	2.17	0.0602	1.41	0.8	603.36	9.54	605	9.96	611	30.5	99
GJ1-16_seq2	20810	240.5	21.9	0	34314	0.0986	1.7	0.8188	2.16	0.0602	1.28	0.8	606	10.05	607	9.91	613	27.7	99
GJ1-17_seq2	21293	245.7	22.36	0	35669	0.0985	1.7	0.8173	2.37	0.0602	1.64	0.7	605.64	9.87	607	10.9	610	35.5	99
GJ1-18_seq2	21439	243.5	22.2	0	67854	0.0987	1.6	0.8187	1.97	0.0602	1.21	0.8	606.61	9	607	9.04	610	26.1	99
GJ1-19_seq2	20582	244.7	22.28	0	34432	0.0985	1.9	0.8192	2.16	0.0603	1.06	0.9	605.79	10.86	608	9.91	614	23	99
GJ1-20_seq2	21142	244.3	22.33	0	35533	0.0989	1.7	0.8191	2.16	0.0601	1.3	0.8	607.97	9.98	608	9.9	606	28.1	100
GJ1-21_seq2	21247	256.3	23.19	0	35726	0.0979	1.8	0.81	2.19	0.06	1.3	0.8	602.27	10.18	602	10	603	28.1	100
GJ1-22_seq2	20650	241.8	22.3	0	34546	0.0998	1.6	0.8298	1.88	0.0603	1.04	0.8	613.48	9.19	614	8.71	614	22.5	100
GJ1-23_seq2	21190	247.6	22.54	0	105872	0.0985	1.7	0.8141	1.87	0.0599	0.77	0.9	605.83	9.87	605	8.57	601	16.8	101
Guaje tephra sampled in 2014																			
2014 NM-GU_a02	168	775	0	0.64	36	0.00020645	7.4	0.0038206	30.1	0.14311	29.2	0.25	1.33	0.1	-	-	-	-	-
2014 NM-GU_a03	1877	225	3	0.51	3953	0.013138	2.1	0.08651	4.8	0.047756	4.3	0.44	84.14	1.76	84	4	87	101	97
2014 NM-GU_a04	22434	87	32	1.23	21931	0.30159	1.9	4.28391	2.3	0.10302	1.2	0.84	1699.2	29.17	1690	19	1679	23	101
2014 NM-GU_a05	50023	204	66	0.8	17567	0.29798	1.3	4.2725	1.8	0.10399	1.2	0.76	1681.28	19.82	1688	15	1697	21	99
2014 NM-GU_a06	736	2	0	0.59	32	0.02597	9.1	2.45112	22.9	0.68453	21.1	0.4	165.28	14.88	1258	181	4699	303	4
2014 NM-GU_a07	151	1997	0	0.09	371	0.00025955	4.2	0.0013452	37.6	0.040103	37.3	0.11	1.67	0.07	-	-	-	-	-
2014 NM-GU_a08	217	103	0	0.72	548	0.0011077	3.3	0.0060769	17.8	0.040193	17.5	0.19	7.14	0.24	-	-	-	-	-
2014 NM-GU_a09	1677	2100	1	0.06	121	0.00037229	3	0.010635	18.8	0.21673	18.6	0.16	2.4	0.07	-	-	-	-	-

Sample and spot number	$^{207}Pb^a$ (cps)	U^b (ppm)	Pb^b (ppm)	$\underline{Th^b}$ U	$\underline{^{206}Pb^c}$ ^{204}Pb	$\underline{^{206}Pb^c}$ ^{238}U	2σ %	$\underline{^{207}Pb^c}$ ^{235}U	2σ %	$\underline{^{207}Pb^c}$ ^{206}Pb	2σ %	ρ^d	$\underline{^{206}Pb}$ ^{238}U	2σ (Ma)	$\underline{^{207}Pb}$ ^{235}U	2σ (Ma)	$\underline{^{207}Pb}$ ^{206}Pb	2σ (Ma)	Conc %
2014 NM-GU_a10	312	2438	1	0.05	665	0.00022272	3.2	0.0013506	16.6	0.047508	16.3	0.19	1.44	0.05	-	-	-	-	-
2014 NM-GU_a12	384	2067	1	0.05	572	0.00023847	4.3	0.0020777	20.1	0.06789	19.6	0.21	1.54	0.07	-	-	-	-	-
2014 NM-GU_a13	-					-	-	-	-	-	-		-		-	-	-	-	-
2014 NM-GU_a14	599	2603	1	0.04	1327	0.00023692	2.5	0.0013907	10.9	0.045776	10.6	0.23	1.53	0.04	-	-	-	-	-
2014 NM-GU_a15	247	2072	1	0.05	573	0.00026482	3.6	0.001501	19.1	0.043841	18.7	0.19	1.71	0.06	-	-	-	-	97
2014 NM-GU_a16	4356	275	4	0.5	9163	0.014419	1.8	0.095245	3.5	0.047908	3.1	0.51	92.29	1.66	92	3	95	72	-
2014 NM-GU_a17	619	2082	0	0.04	551	0.00023754	1.8	0.001871	18.3	0.061415	18.2	0.1	1.53	0.03	-	-	-	-	-
2014 NM-GU_a18	3088	2676	2	0.08	57	0.00037941	5.8	0.020543	9.2	0.41024	7.1	0.63	2.45	0.14	-	-	-	-	102
2014 NM-GU_a19	2237	18	4	1	3021	0.1856	3.2	1.92236	5.2	0.075121	4.2	0.6	1097.49	31.97	1089	36	1072	84	45
2014 NM-GU_a20	42304	171	25	0.36	39827	0.13032	4.4	1.92974	4.8	0.10739	1.8	0.93	789.7	33.14	1091	33	1756	33	-
2014 NM-GU_a22	27228	406	17	1.19	19	0.010813	2.2	1.27469	3.1	0.855	2.1	0.72	69.33	1.51	-	-	-	-	-
2014 NM-GU_a23	74	2536	1	0.07	424	0.00025263	3.4	0.00054027	67.9	0.016588	67.8	0.05	1.63	0.06	-	-	-	-	-
2014 NM-GU_a24	396	3756	1	0.05	935	0.0002297	3.3	0.001256	14.7	0.042732	14.3	0.23	1.48	0.05	-	-	-	-	-
2014 NM-GU_a25	411	109	2	1.01	1044	0.013765	2.7	0.075595	13.9	0.039829	13.6	0.19	88.13	2.33	-	-	-	-	-
2014 NM-GU_a26	553	1960	1	0.16	286	0.00028613	5.2	0.0060129	15.1	0.16131	14.2	0.34	1.84	0.1	-	-	-	-	-
	-					-	-	-	-	-	-		-			-	-	-	-
2014 NM-GU_a28	237	2604	1	0.05	615	0.0002234	3.5	0.0011095	22.8	0.038892	22.6	0.15	1.44	0.05	-	-	-	-	-
2014 NM-GU_a29	118	2207	0	0.05	526	0.00023103	3.6	0.00065198	36.9	0.022041	36.7	0.1	1.49	0.05	-	-	-	-	-
2014 NM-GU_a30	631	357	0	0.41	766	0.0010057	2.7	0.010965	19.4	0.080136	19.2	0.14	6.48	0.18	-	-	-	-	-
2014 NM-GU_a32	3724	734	1	0.1	40	0.00041818	12.2	0.028296	16.1	0.51043	10.4	0.76	2.7	0.33	-	-	-	-	102
2014 NM-GU_a33	828	8422	2	0.04	363	0.00024392	2.5	0.0020342	17.7	0.064893	17.5	0.14	1.57	0.04	-	-	-	-	-
2014 NM-GU_a34	22898	36	12	0.79	22175	0.30842	1.9	4.41996	2.2	0.10394	1.2	0.84	1732.94	28.64	1716	19	1696	22	-
2014 NM-GU_a35	342	3666	1	0.06	536	0.00024355	4.1	0.0020623	27.5	0.065862	27.2	0.15	1.57	0.06	-	-	-	-	-
2014 NM-GU_a36	276	2704	1	0.05	601	0.0002539	6.4	0.0015024	32	0.045905	31.3	0.2	1.64	0.1	-	-	-	-	-
2014 NM-GU_a37	812	1221	1	0.77	252	0.00022676	24.7	0.009975	32.4	0.33694	21	0.76	1.46	0.36	-	-	-	-	87
2014 NM-GU_a38	1106	471	1	0.57	19	0.0011074	6.8	0.08978	13.2	0.59449	11.3	0.52	7.13	0.49	-	-	-	-	-
2014 NM-GU_a39	75773	159	44	0.41	7709	0.25921	1.8	3.75547	2	0.10508	1	0.86	1485.82	23.47	1583	17	1716	19	-

139

Sample and spot number	$^{207}Pb^a$	U^b	Pb^b	Th^b	$\frac{^{206}Pb^c}{^{204}Pb}$	$\frac{^{206}Pb^c}{^{238}U}$	2σ	$\frac{^{207}Pb^c}{^{235}U}$	2σ	$\frac{^{207}Pb^c}{^{206}Pb}$	2σ	ρ^d	$\frac{^{206}Pb}{^{238}U}$	2σ	$\frac{^{207}Pb}{^{235}U}$	2σ	$\frac{^{207}Pb}{^{206}Pb}$	2σ	Conc
	(cps)	(ppm)	(ppm)	U	^{204}Pb	^{238}U	%	^{235}U	%	^{206}Pb	%		^{238}U	(Ma)	^{235}U	(Ma)	^{206}Pb	(Ma)	%
2014 NM-GU_a40	472	2746	1	0.04	1137	0.00025219	2.6	0.0013616	9.5	0.041912	9.1	0.27	1.63	0.04	-	-	-	-	101
2014 NM-GU_a42	391	2260	1	0.04	885	0.00023147	3.7	0.0013237	12.9	0.044675	12.4	0.29	1.49	0.06	-	-	-	-	99
2014 NM-GU_a43	11080	61	21	0.96	10768	0.30481	2.5	4.36248	3.4	0.1038	2.3	0.74	1715.1	37.42	1705	28	1693	42	-
2014 NM-GU_a44	129850	227	71	0.44	68662	0.29827	1.5	4.29983	1.6	0.10455	0.7	0.91	1682.74	21.52	1693	13	1706	12	-
2014 NM-GU_a45	275	2546	1	0.05	530	0.00026811	3.3	0.001792	24.2	0.051654	23.9	0.14	1.73	0.06	-	-	-	-	-
2014 NM-GU_a46	246	2845	1	0.05	572	0.00024333	3.7	0.0013229	22.2	0.042303	21.9	0.17	1.57	0.06	-	-	-	-	-
2014 NM-GU_a47	35	179	0	1	173	0.0010355	8.3	0.0024993	137.3	0.017669	137.1	0.06	6.67	0.55	-	-	-	-	-
2014 NM-GU_a48	126	2768	1	0.07	343	0.00023055	4.4	0.0010992	38.9	0.037227	38.7	0.11	1.49	0.07	-	-	-	-	-
2014 NM-GU_a49	2404	1456	1	0.06	60	0.000040823	119.9	0.0010025	121.2	0.2992	17.7	0.99	0.26	0.32	-	-	-	-	-
2014 NM-GU_a50	941	3558	1	0.04	339	0.00025108	2.4	0.0023236	8.9	0.071869	8.6	0.27	1.62	0.04	-	-	-	-	-
2014 NM-GU_a52	1250	2293	1	0.04	163	0.00026119	2.7	0.0047646	10.1	0.14125	9.8	0.26	1.68	0.04	-	-	-	-	-
2014 NM-GU_a53	242	3319	1	0.07	399	0.0002142	5.5	0.0016874	32.8	0.061876	32.3	0.17	1.38	0.08	-	-	-	-	-
2014 NM-GU_a54	257	3862	1	0.06	440	0.00023855	2.2	0.0017289	28.7	0.056457	28.6	0.08	1.54	0.03	-	-	-	-	-
2014 NM-GU_a55	123	2013	1	0.09	348	0.00029213	4.1	0.0013921	30.1	0.036596	29.8	0.14	1.88	0.08	-	-	-	-	-
2014 NM-GU_a56	279	3317	1	0.05	568	0.00025934	2.9	0.0015997	24.8	0.047778	24.6	0.12	1.67	0.05	-	-	-	-	-
2014 NM-GU_a57	-	-	-	-	-	-	-	-	-	-	-	-	-	-	-	-	-	-	-
2014 NM-GU_a58	-	-	-	-	-	-	-	-	-	-	-	-	-	-	-	-	-	-	-
2014 NM-GU_a59	638	3934	1	0.04	689	0.00022961	2.1	0.001515	8.4	0.051578	8.2	0.25	1.48	0.03	-	-	-	-	101
2014 NM-GU_a60	254	737	1	0.21	584	0.0010795	4.3	0.0065166	18.6	0.044375	18.1	0.23	6.95	0.3	-	-	-	-	92
2014 NM-GU_a61	-	-	-	-	-	-	-	-	-	-	-	-	-	-	-	-	-	-	-
2014 NM-GU_b03	132	2151	1	0.05	523	0.00024223	4.3	0.00077765	24.8	0.024986	24.4	0.17	1.56	0.07	-	-	-	-	-
2014 NM-GU_b04	567	118	0	0.86	604	0.001593	4.1	0.020354	27.2	0.092672	26.9	0.15	10.26	0.42	-	-	-	-	-
2014 NM-GU_b05	147484	221	72	0.56	14614	0.30148	1.5	4.2898	1.9	0.1032	1.1	0.79	1698.66	22.13	1691	15	1682	21	-
2014 NM-GU_b06	1524	102	3	0.82	3061	0.030148	1.9	0.20898	7.5	0.050274	7.2	0.26	191.48	3.61	193	13	208	168	-
2014 NM-GU_b07	-	-	-	-	-	-	-	-	-	-	-	-	-	-	-	-	-	-	-
2014 NM-GU_b08	761	2777	1	0.04	514	0.00026092	3.2	0.0019955	15.1	0.059232	14.7	0.21	1.68	0.05	-	-	-	-	-
2014 NM-GU_b09	28	50	0	1.3	55	0.0035843	5.8	0.011055	912.8	0.022369	912.8	0.01	23.06	1.33	-	-	-	-	98

156

Sample and spot number	$^{207}Pb^a$	U^b	Pb^b	Th^b	$\frac{^{206}Pb^c}{^{204}Pb}$	$\frac{^{206}Pb^c}{^{238}U}$	2 σ	$\frac{^{207}Pb^c}{^{235}U}$	2 σ	$\frac{^{207}Pb^c}{^{206}Pb}$	2 σ	ρ^d	$\frac{^{206}Pb}{^{238}U}$	2 σ	$\frac{^{207}Pb}{^{235}U}$	2 σ	$\frac{^{207}Pb}{^{206}Pb}$	2 σ	Conc
	(cps)	(ppm)	(ppm)	U	^{204}Pb	^{238}U	%	^{235}U	%	^{206}Pb	%		^{238}U	(Ma)	^{235}U	(Ma)	^{206}Pb	(Ma)	%
2014 NM-GU_b10	-	-	-	-	-	-	-	-	-	-	-	-	-	-	-	-	-	-	15
2014 NM-GU_b12	1104	1916	1	0.04	258	0.00028444	2.8	0.0043576	10.1	0.11797	9.7	0.28	1.83	0.05	-	-	-	-	-
2014 NM-GU_b13	579	1433	0	0.03	1043	0.00027509	2.5	0.001503	14.4	0.042174	14.2	0.18	1.77	0.05	-	-	-	-	-
2014 NM-GU_b14	723	1882	1	0.05	470	0.00026357	2.8	0.0029498	12.7	0.086596	12.4	0.22	1.7	0.05	-	-	-	-	**102**
2014 NM-GU_b15	1574	2465	1	0.03	2160	0.00023112	2.3	0.0022289	11.2	0.075366	10.9	0.21	1.49	0.03	-	-	-	-	-
2014 NM-GU_b16	1434	2703	1	0.03	384	0.00021931	2.9	0.0016452	10.3	0.058881	9.8	0.29	1.41	0.04	-	-	-	-	-
2014 NM-GU_b17	63858	51	17	0.64	61830	0.29635	1.7	4.27257	1.9	0.10456	0.8	0.9	1673.19	25	1688	16	1707	15	-
2014 NM-GU_b18	-	-	-	-	-	-	-	-	-	-	-	-	-	-	-	-	-	-	-
2014 NM-GU_b19	2197	4	2	4.82	17	0.12002	4	14.24862	6.1	0.86106	4.6	0.66	730.65	27.8	2766	59	5026	65	-
2014 NM-GU_b20	242	2190	1	0.06	498	0.00027129	3	0.0017406	23.6	0.049538	23.4	0.13	1.75	0.05	-	-	-	-	-
2014 NM-GU_b22	63	2937	1	0.06	398	0.0002561	5.6	0.0005005	135.3	0.015146	135.2	0.04	1.65	0.09	-	-	-	-	-
2014 NM-GU_b23	-	-	-	-	-	-	-	-	-	-	-	-	-	-	-	-	-	-	-
2014 NM-GU_b24	3356	264	4	0.82	1774	0.012464	1.9	0.081769	14.9	0.047582	14.8	0.13	79.85	1.48	80	12	79	352	-
2014 NM-GU_b25	914	2919	1	0.05	203	0.000238	3.9	0.0040419	14.7	0.13235	14.2	0.26	1.53	0.06	-	-	-	-	-
2014 NM-GU_b26	291	1620	0	0.05	681	0.00023579	2.8	0.0013262	14.6	0.043869	14.3	0.19	1.52	0.04	-	-	-	-	-
2014 NM-GU_b27	24	2528	1	0.1	340	0.00023318	1.9	0.00022272	232.2	0.0074453	232.2	0.01	1.5	0.03	-	-	-	-	-
2014 NM-GU_b28	40	74	0	0.79	50	0.00144875	7.7	0.011483	101.7	0.057878	101.4	0.08	9.33	0.72	-	-	-	-	-
2014 NM-GU_b29	888	2030	1	0.06	461	0.00024073	6.7	0.00609	14.8	0.19695	13.2	0.45	1.55	0.1	-	-	-	-	-
2014 NM-GU_b30	270	2091	1	0.06	484	0.00024798	3.7	0.0018529	14	0.058039	13.5	0.26	1.6	0.06	-	-	-	-	-
2014 NM-GU_b32	104	2878	1	0.07	367	0.00024034	2.3	0.00088853	48.4	0.028778	48.3	0.05	1.55	0.04	-	-	-	-	-
2014 NM-GU_b33	430	1950	0	0.04	794	0.00025458	2.6	0.0017911	9.9	0.054573	9.5	0.27	1.64	0.04	-	-	-	-	-
2014 NM-GU_b34	-	-	-	-	-	-	-	-	-	-	-	-	-	-	-	-	-	-	-
2014 NM-GU_b35	151	2376	1	0.06	497	0.00024048	3	0.000951	22.5	0.030792	22.3	0.13	1.55	0.05	-	-	-	-	-
2014 NM-GU_b36	248	2394	1	0.06	535	0.00027122	7	0.0016547	23.5	0.047104	22.5	0.3	1.75	0.12	-	-	-	-	59
2014 NM-GU_b37	338	1767	1	0.07	375	0.00033104	3.2	0.0037206	25.7	0.085745	25.5	0.12	2.13	0.07	-	-	-	-	-
2014 NM-GU_b38	369	1903	0	0.04	695	0.00024592	3.4	0.0015473	14.5	0.048927	14.1	0.23	1.59	0.05	-	-	-	-	-
2014 NM-GU_b39	159	1919	1	0.07	464	0.00027293	4.2	0.0012667	27.3	0.035811	26.9	0.16	1.76	0.07	-	-	-	-	-

Sample and spot number	$^{207}Pb^a$ (cps)	U^b (ppm)	Pb^b (ppm)	Th^b/U	$\frac{^{206}Pb^c}{^{204}Pb}$	$\frac{^{206}Pb^c}{^{238}U}$	2 σ %	$\frac{^{207}Pb^c}{^{235}U}$	2 σ %	$\frac{^{207}Pb^c}{^{206}Pb}$	2 σ %	ρ^d	$\frac{^{206}Pb}{^{238}U}$ (Ma)	2 σ (Ma)	$\frac{^{207}Pb}{^{235}U}$ (Ma)	2 σ (Ma)	$\frac{^{207}Pb}{^{206}Pb}$ (Ma)	2 σ (Ma)	Conc %
2014 NM-GU_b40	123	2854	1	0.07	367	0.00023417	4.8	0.0010107	21.7	0.033659	21.2	0.22	1.51	0.07	-	-	-	-	-
2014 NM-GU_b42	234	337	0	1.36	100	0.0002286	18.6	0.0070525	31.2	0.23226	25.1	0.59	1.47	0.27	-	-	-	-	-
2014 NM-GU_b43	155	3290	1	0.06	413	0.00023884	4	0.0011654	35.9	0.038003	35.7	0.11	1.54	0.06	-	-	-	-	-
2014 NM-GU_b44	233	3195	1	0.06	626	0.00024414	3.3	0.0011886	23.5	0.037867	23.3	0.14	1.57	0.05	-	-	-	-	-
2014 NM-GU_b45	56	112	0	1.49	221	0.00126668	4.3	0.0043104	92.1	0.024798	92	0.05	8.16	0.35	-	-	-	-	-
2014 NM-GU_b46	139	2816	1	0.05	497	0.00024217	4.2	0.00086541	55.2	0.027817	55	0.08	1.56	0.07	-	-	-	-	-
2014 NM-GU_b47	5046	254	7	0.8	9864	0.025353	2.7	0.18099	6.2	0.051776	5.5	0.44	161.4	4.35	169	10	276	127	59
2014 NM-GU_b48	91	141	0	0.83	238	0.0011027	4.9	0.00552011	48.1	0.036656	47.9	0.1	7.1	0.35	-	-	-	-	-
2014 NM-GU_b49	451	2899	1	0.04	1034	0.00025492	2.9	0.00145899	10.3	0.044394	9.9	0.28	1.64	0.05	-	-	-	-	-
2014 NM-GU_b50	-	-	-	-	-	-	-	-	-	-	-	-	-	-	-	-	-	-	-
Standard: GJ1																			
GJ1-1_seq a	25859	234	21	0.02	4948093	0.09874	1.7	0.81878	2.1	0.06014	1.3	0.8	607.04	9.93	607	10	609	28	**100**
GJ1-2_seq a	25497	229	21	0.02	42659	0.09817	1.5	0.81461	1.8	0.06018	1	0.83	603.68	8.57	605	8	610	21	**99**
GJ1-3_seq a	25656	237	22	0.02	42958	0.0988	1.6	0.81936	1.9	0.06014	1	0.84	607.4	9.16	608	9	609	22	**100**
GJ1-4_seq a	24779	228	21	0.02	41453	0.09894	1.6	0.82098	2	0.06018	1.3	0.78	608.21	9.25	609	9	610	27	**100**
GJ1-5_seq a	25226	234	22	0.02	42210	0.10003	1.7	0.82987	2.2	0.06017	1.3	0.79	614.56	10.09	614	10	610	29	**101**
GJ1-6_seq a	25397	259	23	0.02	27890	0.09814	1.4	0.81266	1.9	0.06006	1.3	0.72	603.5	8.03	604	9	606	29	**100**
GJ1-7_seq a	23959	240	22	0.02	40137	0.09848	1.6	0.81626	2.2	0.06012	1.4	0.75	605.48	9.43	606	10	608	31	**100**
GJ1-8_seq a	24540	245	22	0.02	35591	0.09782	1.6	0.81116	2.1	0.06014	1.4	0.75	601.63	9.09	603	10	609	30	**99**
GJ1-9_seq a	23143	249	23	0.02	38715	0.09774	1.6	0.81077	2	0.06017	1.2	0.81	601.12	9.42	603	9	610	26	**99**
GJ1-10_seq a	23112	256	23	0.02	56544	0.0984	1.9	0.81778	2.3	0.06027	1.2	0.85	605.05	11.22	607	11	613	26	**99**
GJ1-11_seq a	23588	260	23	0.02	39370	0.09754	1.8	0.81118	2	0.06032	0.8	0.91	599.96	10.59	603	9	615	18	**98**
GJ1-12_seq a	23144	271	25	0.02	38769	0.09892	1.8	0.82036	2.4	0.06015	1.6	0.75	608.07	10.39	608	11	609	34	**100**
GJ1-13_seq a	21040	259	24	0.02	35388	0.09901	1.4	0.81721	2	0.05986	1.4	0.72	608.61	8.38	606	9	599	30	**102**
GJ1-14_seq a	22723	271	25	0.02	28262	0.09945	1.7	0.82655	2	0.06028	1.1	0.85	611.18	9.93	612	9	614	23	**100**
GJ1-15_seq a	21433	261	24	0.02	35871	0.09887	1.5	0.82021	1.9	0.06016	1.1	0.8	607.81	8.94	608	9	609	25	**100**

Sample and spot number	$^{207}Pb^a$ (cps)	U^b (ppm)	Pb^b (ppm)	Th^b U	$\frac{^{206}Pb^c}{^{204}Pb}$	$\frac{^{206}Pb^c}{^{238}U}$	2σ %	$\frac{^{207}Pb^c}{^{235}U}$	2σ %	$\frac{^{207}Pb^c}{^{206}Pb}$	2σ %	ρ^d	$\frac{^{206}Pb}{^{238}U}$	2σ (Ma)	$\frac{^{207}Pb}{^{235}U}$	2σ (Ma)	$\frac{^{207}Pb}{^{206}Pb}$	2σ (Ma)	Conc %
GJ1-16_seq a	22445	258	24	0.02	11766	0.09881	1.7	0.81885	2	0.06011	1.1	0.82	607.41	9.69	607	9	607	25	100
GJ1-17_seq a	21829	242	22	0.02	13679	0.09816	2	0.8152	2.7	0.06023	1.9	0.73	603.64	11.42	605	12	612	40	99
GJ1-18_seq a	21331	240	22	0.02	63553	0.09852	1.6	0.81779	2	0.0602	1.2	0.81	605.75	9.28	607	9	611	25	99
GJ1-19_seq a	21762	234	21	0.02	115118	0.0987	1.6	0.82272	2.1	0.06046	1.2	0.8	606.77	9.49	610	9	620	27	98
GJ1-20_seq a	21346	221	20	0.02	12064	0.09905	1.8	0.82256	2.1	0.06023	1.1	0.86	608.82	10.61	609	10	612	23	99
GJ1-21_seq a	21182	234	21	0.02	35333	0.09821	1.8	0.81423	2.1	0.06013	1.1	0.86	603.9	10.28	605	9	608	23	99
GJ1-22_seq a	21443	224	20	0.02	34714	0.09864	1.7	0.82097	2	0.06036	1.1	0.83	606.46	9.76	609	9	617	25	98
GJ1-23_seq a	21579	215	20	0.02	36205	0.09865	1.6	0.81726	2.5	0.06008	1.9	0.64	606.52	9.15	607	11	607	41	100
GJ1-1_seq b	21827	236	21	0.02	174925	0.09866	1.8	0.81607	2.2	0.05999	1.2	0.84	606.57	10.48	606	10	603	25	101
GJ1-2_seq b	22439	245	22	0.02	37655	0.09898	1.9	0.82342	2.3	0.06033	1.4	0.81	608.46	11.05	610	11	616	30	99
GJ1-3_seq b	21728	234	21	0.02	36541	0.099	1.8	0.82275	2.3	0.06027	1.4	0.79	608.56	10.45	610	11	613	31	99
GJ1-4_seq b	20745	226	21	0.02	34975	0.09865	1.7	0.81676	2	0.06005	1.2	0.82	606.51	9.73	606	9	605	25	100
GJ1-6_seq b	20338	233	21	0.02	34301	0.09862	1.8	0.81626	2.3	0.06003	1.4	0.79	606.29	10.51	606	11	605	31	100
GJ1-7_seq b	21208	247	23	0.02	35670	0.09929	2.1	0.82441	2.6	0.06022	1.6	0.78	610.25	12.05	611	12	611	36	100
GJ1-8_seq b	21112	250	23	0.02	35508	0.0975	1.7	0.80923	2	0.06019	1	0.86	599.75	9.76	602	9	611	21	98
GJ1-10_seq b	21083	256	23	0.02	35332	0.09802	1.6	0.8158	2	0.06036	1.2	0.8	602.78	9.17	606	9	617	26	98
GJ1-11_seq b	20438	252	23	0.02	34359	0.09847	1.7	0.81729	2.1	0.06019	1.2	0.81	605.46	9.95	607	10	611	27	99
GJ1-12_seq b	20517	272	25	0.02	34570	0.09869	1.6	0.81776	2.3	0.0601	1.7	0.69	606.74	9.35	607	11	607	37	100
GJ1-13_seq b	19921	266	24	0.02	33581	0.09802	1.5	0.81148	2	0.06004	1.3	0.76	602.78	8.89	603	9	605	28	100
GJ1-14_seq b	19889	266	24	0.02	33381	0.09766	1.6	0.81228	2	0.06032	1.1	0.83	600.7	9.37	604	9	615	24	98
GJ1-15_seq b	20436	268	25	0.02	34324	0.09915	1.7	0.824	2.1	0.06028	1.2	0.81	609.41	9.64	610	9	613	26	99
GJ1-16_seq b	19709	247	23	0.02	26150	0.09937	1.5	0.82486	1.8	0.0602	1	0.84	610.72	8.93	611	8	611	21	100
GJ1-17_seq b	19941	228	21	0.02	33613	0.09896	1.5	0.81981	2	0.06008	1.3	0.75	608.3	8.65	608	9	607	28	100
GJ1-18_seq b	19614	231	21	0.02	33022	0.09925	1.7	0.82313	2.1	0.06015	1.2	0.83	610.02	10.06	610	10	609	25	100
GJ1-19_seq b	19683	223	20	0.02	27783	0.098	1.6	0.8128	1.9	0.06015	1	0.84	602.7	9.36	604	9	609	23	99
GJ1-20_seq b	20078	209	19	0.02	33727	0.09849	1.9	0.81849	2.3	0.06027	1.4	0.79	605.56	10.77	607	11	613	31	99

6.2 Supplementary material chapter 3 'Cover beds older than the mid-Pleistocene revolution and the provenance of their eolian components, La Sal Mountains, Utah, USA'

Table 3.3 U-Pb ratios and calculated ages for all samples.
[a] within-run background-corrected mean ^{207}Pb signal in counts per second.
[b] U and Pb content and Th/U ratio were calculated relative to GJ-1 and are accurate to approximately 10%.
[c] corrected for background, mass bias, laser-induced U-Pb fractionation and common Pb (if detectable, see Materials and Methods section) using STACEY & KRAMERS' (1975) model-Pb composition. ^{207}Pb/^{235}U calculated using ^{207}Pb/^{206}Pb/(^{238}U/^{206}Pb × 1/137.88). Errors are propagated by quadratic addition of within-run errors (2SE) and the reproducibility of GJ-1 (2SD).
[d] error correlation defined as err^{206}Pb/^{238}U/err^{207}Pb/^{235}U.
[e] Conc % printed in bold indicate grains assumed to be concordant.
Measurements used to calculate the concordia age for the La Sal Mountains tephra layer (T13 in Fig. 3.1; lab denomination LSM-T in this table) are printed in bold.

sample and spot number	^{207}Pb[a] (cps)	U[b] (ppm)	Pb[b] (ppm)	Th[b] U	^{206}Pb[c] ^{204}Pb	^{206}Pb[c] ^{238}U	2 σ %	^{207}Pb[c] ^{235}U	2 σ %	^{207}Pb[c] ^{206}Pb	2 σ %	ρ[d]	^{206}Pb ^{238}U	2 σ (Ma)	^{207}Pb ^{235}U	2 σ (Ma)	^{207}Pb ^{206}Pb	2 σ (Ma)	Conc %[e]
LSM-T-seq1-a01	52620	401	73	0.51	1676	0.17114	2.5	1.90093	3.1	0.080559	1.8	0.82	1018.39	23.79	1081	21	1211	35	84
LSM-T-seq1-a02	**204**	**1267**	**0**	**0.07**	**417**	**0.00020472**	**2.6**	**0.0012507**	**20.3**	**0.048198**	**20.2**	**0.13**	**1.32**	**0.03**					
LSM-T-seq1-a03	878	133	0	0.80	158	0.00029149	17.3	0.021515	20.3	0.56028	10.6	0.85	1.88	0.33					
LSM-T-seq1-a04	346	1751	0	0.06	219	0.00021089	6.8	0.0018057	18.8	0.067393	17.5	0.36	1.36	0.09					
LSM-T-seq1-a05	-	-	-	-	-	-	-	-	-	-	-	-	-	-	-	-	-	-	-
LSM-T-seq1-a06	56051	63	20	0.28	20431	0.31569	2.1	4.66362	2.4	0.10714	1.2	0.88	1768.66	33.34	1761	21	1751	21	**101**
LSM-T-seq1-a07	4560	2018	1	0.07	46	0.00015481	10.6	0.0072290	20.5	0.37919	17.6	0.52	1.00	0.11					
LSM-T-seq1-a08	10969	10	6	5.07	102	0.23579	3.7	7.47014	7.0	0.22977	6.0	0.52	1364.79	45.16	2169	65	3050	96	45
LSM-T-seq1-a09	874	432	2	0.91	1665	0.0041189	2.7	0.030137	8.5	0.053066	8.1	0.32	26.50	0.72					
LSM-T-seq1-a10	-	-	-	-	-	-	-	-	-	-	-	-	-	-	-	-	-	-	-
LSM-T-seq1-a11	1717	13	0	3.62	64	0.011926	3.3	0.72799	6.1	0.44271	5.1	0.53	76.43	2.47					
LSM-T-seq1-a12	16971	569	28	0.61	669	0.044690	3.3	0.32839	6.4	0.053295	5.5	0.52	281.84	9.15	288	16	341	124	83
LSM-T-seq1-a13	1353	754	1	0.27	84	0.00035114	6.8	0.014940	12.5	0.32282	10.5	0.54	2.26	0.15					
LSM-T-seq1-a14	24183	164	31	1.25	1223	0.16316	3.0	1.82276	4.0	0.081024	2.7	0.74	974.32	26.93	1054	27	1222	53	80

sample and spot number	$^{207}Pb^a$ (cps)	U^b (ppm)	Pb^b (ppm)	Th^b / U	$^{206}Pb^c$ / ^{204}Pb	$^{206}Pb^c$ / ^{238}U	2 σ %	$^{207}Pb^c$ / ^{235}U	2 σ %	$^{207}Pb^c$ / ^{206}Pb	2 σ %	$ρ^d$ -	^{206}Pb / ^{238}U	2 σ (Ma)	^{207}Pb / ^{235}U	2 σ (Ma)	^{207}Pb / ^{206}Pb	2 σ (Ma)	Conc %^{e} -
LSM-T-seq1-a17	**112**	**1439**	**0**	**0.44**	**238**	**0.00020832**	**3.9**	**0.0012805**	**28.3**	**0.04798**	**28.0**	**0.14**	**1.34**	**0.05**					
LSM-T-seq1-a18	8152	70	1	0.72	23	0.0036939	6.0	0.17118	59.4	0.33611	59.1	0.10	23.77	1.43					
LSM-T-seq1-a19	32148	398	39	0.91	3724	0.084139	2.4	0.67625	3.3	0.058292	2.3	0.72	520.78	11.79	525	14	541	50	**96**
LSM-T-seq1-a20	246990	979	125	0.12	252	0.10392	2.1	2.12475	2.4	0.14828	1.1	0.89	637.36	12.91	1157	17	2326	19	27
LSM-T-seq2-b01	1066	94	0	0.66	20	0.0010504	19.1	0.086790	22.1	0.60656	11.0	0.87	6.77	1.29					
LSM-T-seq2-b02	4440	735	1	0.44	23	0.00046805	6.2	0.036310	8.3	0.58074	5.4	0.75	3.02	0.19					
LSM-T-seq2-b03	3939	2	1	5.02	23	0.16840	2.6	10.34267	30.5	0.44544	30.4	0.09	1003.29	24.54	2466	331	4070	452	25
LSM-T-seq2-b04	6354	68	5	1.75	11707	0.052314	2.4	0.39322	3.7	0.054515	2.8	0.64	328.71	7.58	337	11	392	63	84
LSM-T-seq2-b05	37958	305	24	0.49	639	0.076676	1.9	0.57990	3.0	0.054852	2.4	0.61	476.25	8.55	464	11	406	54	117
LSM-T-seq2-b06	2521	181	0	1.07	30	0.00043677	16.8	0.025658	38.1	0.43768	34.2	0.44	2.81	0.47					
LSM-T-seq2-b07	-	-	-	-	-	-	-	-	-	-	-	-	-	-	-	-	-	-	-
LSM-T-seq2-b08	104021	608	122	0.69	14714	0.17996	2.9	2.52307	3.9	0.10168	2.5	0.76	1066.77	28.75	1279	28	1655	47	64
LSM-T-seq2-b09	-	-	-	-	-	-	-	-	-	-	-	-	-	-	-	-	-	-	-
LSM-T-seq2-b10	5661	456	6	0.67	1199	0.011071	1.8	0.099480	8.2	0.065172	8.0	0.23	70.97	1.30					
LSM-T-seq2-b11	-	-	-	-	-	-	-	-	-	-	-	-	-	-	-	-	-	-	-
LSM-T-seq2-b12	-	-	-	-	-	-	-	-	-	-	-	-	-	-	-	-	-	-	-
LSM-T-seq2-b13	-	-	-	-	-	-	-	-	-	-	-	-	-	-	-	-	-	-	-
LSM-T-seq2-b14	-	-	-	-	-	-	-	-	-	-	-	-	-	-	-	-	-	-	-
LSM-T-seq2-b15	31426	499	25	0.61	654	0.045584	3.3	0.33860	4.2	0.053874	2.6	0.78	287.35	9.20	296	11	366	59	79
LSM-T-seq2-b16	34642	647	51	0.21	973	0.075984	2.3	0.62528	3.3	0.059683	2.4	0.69	472.11	10.56	493	13	592	52	80
LSM-T-seq2-b17	4852	140	14	0.99	1960	0.093302	2.2	0.83302	3.9	0.064753	3.2	0.56	575.04	11.98	615	18	766	68	75
LSM-T-seq2-b18	-	-	-	-	-	-	-	-	-	-	-	-	-	-	-	-	-	-	-
LSM-T-seq2-b19	-	-	-	-	-	-	-	-	-	-	-	-	-	-	-	-	-	-	-
LSM-T-seq2-b20	271	94	0	0.81	14	0.00085473	12.2	0.064946	30.7	0.55915	28.1	0.40	5.51	0.67					
LSM-T-seq2-b21	-	-	-	-	-	-	-	-	-	-	-	-	-	-	-	-	-	-	-
LSM-T-seq2-b22	-	-	-	-	-	-	-	-	-	-	-	-	-	-	-	-	-	-	-
LSM-T-seq2-b23	8158	5	5	5.56	19	0.30033	4.8	17.79081	36.8	0.42964	36.5	0.13	1692.93	71.30	2978	435	4016	545	42
LSM-T-seq2-b24	-	-	-	-	-	-	-	-	-	-	-	-	-	-	-	-	-	-	-
LSM-T-seq2-b25	45952	6	5	2.00	21	0.26175	3.3	10.38629	44.8	0.28778	44.7	0.07	1498.81	44.00	2470	533	3406	695	44
LSM-T-seq2-b26	3236	663	2	0.70	38	0.00079126	6.3	0.052200	26.7	0.48641	26.0	0.24	5.10	0.32					
LSM-T-seq2-b27	82	89	0	0.87	47	0.00019897	7.3	0.0049005	44.0	0.19082	43.4	0.17	1.28	0.09					
LSM-T-seq2-b29	51	58	0	0.59	41	0.00020031	14.5	0.0036449	60.4	0.14193	58.6	0.24	1.29	0.19					

sample and spot number	$^{207}Pb^a$ (cps)	U^b (ppm)	Pb^b (ppm)	$\frac{Th^b}{U}$	$\frac{^{206}Pb^c}{^{204}Pb}$	$\frac{^{206}Pb^c}{^{238}U}$	2 σ %	$\frac{^{207}Pb^c}{^{235}U}$	2 σ %	$\frac{^{207}Pb^c}{^{206}Pb}$	2 σ %	ρ^d 0.11	$\frac{^{206}Pb}{^{238}U}$	2 σ (Ma)	$\frac{^{207}Pb}{^{235}U}$	2 σ (Ma)	$\frac{^{207}Pb}{^{206}Pb}$	2 σ (Ma)	Conc %^{e}
LSM-T-seq2-b30	-	-	-	-	-	-	-	-	-	-	-	-	-	-	-	-	-	-	-
LSM-T-seq2-b31	875	1569	0	0.06	531	0.00025130	3.1	0.0052868	12.9	0.16333	12.5	0.24	1.62	0.05					
LSM-T-seq2-b32	-	-	-	-	-	-	-	-	-	-	-	-	-	-	-	-	-	-	-
LSM-T-seq2-b33	133	362	0	0.46	221	0.00020535	4.8	0.0016824	25.5	0.063994	25.1	0.19	1.32	0.06					
LSM-T-seq2-b34	10410	1	2	4.37	18	0.48413	8.7	31.24638	115.0	0.46810	114.7	0.08	2545.24	185.01	3527	-	4144	1700	61
LSM-T-seq2-b35	4969	3	1	3.94	20	0.15742	3.8	15.98160	7.2	0.73632	6.1	0.52	942.41	33.16	2876	71	4803	88	20
LSM-T-seq2-b36	-	-	-	-	-	-	-	-	-	-	-	-	-	-	-	-	-	-	-
LSM-T-seq2-b37	-	-	-	-	-	-	-	-	-	-	-	-	-	-	-	-	-	-	-
LSM-T-seq2-b38	-	-	-	-	-	-	-	-	-	-	-	-	-	-	-	-	-	-	-
LSM-T-seq2-b39	-	-	-	-	-	-	-	-	-	-	-	-	-	-	-	-	-	-	-
LSM-T-seq2-b40	229	239	0	0.67	201	0.00020512	6.8	0.0031077	18.5	0.11772	17.2	0.37	1.32	0.09					
LSM-T-seq2-b41	-	-	-	-	-	-	-	-	-	-	-	-	-	-	-	-	-	-	-
LSM-T-seq2-b42	-	-	-	-	-	-	-	-	-	-	-	-	-	-	-	-	-	-	-
LSM-T-seq2-b43	5449	147	1	1.14	17	0.0027636	3.3	0.29177	5.0	0.76570	3.8	0.65	17.79	0.58					
LSM-T-seq2-b44	-	-	-	-	-	-	-	-	-	-	-	-	-	-	-	-	-	-	-
LSM-T-seq2-b45	100	58	0	0.61	24	0.00032865	17.9	0.015705	96.8	0.36183	95.2	0.18	2.12	0.38					
LSM-T-seq2-b46	322	252	0	0.65	196	0.00020144	8.4	0.0039960	24.2	0.15443	22.7	0.35	1.30	0.11					
LSM-T-seq2-b47	93478	389	49	0.28	803	0.11838	2.1	1.40703	2.3	0.086203	1.0	0.90	721.23	14.27	892	14	1343	19	54
LSM-T-seq2-b48	89	71	0	0.60	59	0.00021237	12.7	0.0044232	34.4	0.16172	32.0	0.37	1.37	0.17					
LSM-T-seq2-b49	93708	91	29	0.58	9441	0.29360	1.9	4.20655	2.2	0.10391	1.0	0.90	1659.49	28.27	1675	18	1695	18	**98**
LSM-T-seq2-b50	619	1363	0	0.04	539	0.00022415	2.3	0.0017162	7.1	0.059985	6.8	0.32	1.44	0.03					
LSM-T-seq2-b51	29663	2	3	4.17	18	0.90867	2.9	30.49485	42.0	0.24340	41.9	0.07	4167.01	91.08	3503	530	3142	666	133
LSM-T-seq2-b52	-	-	-	-	-	-	-	-	-	-	-	-	-	-	-	-	-	-	-
LSM-T-seq2-b53	9118	2	2	2.86	20	0.39215	3.8	15.38208	84.5	0.28449	84.4	0.04	2132.78	69.35	2839	1601	3388	1316	63
LSM-T-seq2-b54	4914	39	3	0.62	6267	0.063204	2.5	0.48691	4.2	0.055873	3.4	0.60	395.08	9.77	403	14	447	75	88
LSM-T-seq2-b55	2083	3	2	6.41	44	0.14684	5.8	12.42169	6.7	0.61352	3.4	0.86	883.24	47.92	2637	65	4540	49	19
LSM-T-seq2-b56	-	-	-	-	-	-	-	-	-	-	-	-	-	-	-	-	-	-	-
LSM-T-seq2-b57	**83**	**1207**	**0**	**0.50**	**171**	**0.00020100**	**7.2**	**0.0011891**	**36.7**	**0.046231**	**36.0**	**0.20**	**1.30**	**0.09**					
LSM-T-seq2-b58	1601	3	1	3.20	64	0.14884	6.6	11.68284	11.5	0.56927	9.4	0.58	894.48	55.52	2579	113	4431	137	20
LSM-T-seq2-b59	5424	152	3	0.59	10380	0.016542	2.3	0.11996	5.8	0.052596	5.3	0.40	105.76	2.45	115	6	311	122	34
LSM-T-seq2-b60	1522	3	1	4.41	33	0.15259	4.2	11.09837	14.9	0.52751	14.3	0.28	915.47	35.62	2531	149	4320	210	21

sample and spot number	$^{207}Pb^a$ (cps)	U^b (ppm)	Pb^b (ppm)	$\underline{Th^b}$ U	$\underline{^{206}Pb^c}$ ^{204}Pb	$\underline{^{206}Pb^c}$ ^{238}U	2 σ %	$\underline{^{207}Pb^c}$ ^{235}U	2 σ %	$\underline{^{207}Pb^c}$ ^{206}Pb	2 σ %	ρ^d -	$\underline{^{206}Pb}$ ^{238}U	2 σ (Ma)	$\underline{^{207}Pb}$ ^{235}U	2 σ (Ma)	$\underline{^{207}Pb}$ ^{206}Pb	2 σ (Ma)	Conc %^{e} -
LSM-T-seq3-c03	3057	13	2	5.45	19	0.056121	3.2	4.24662	21.6	0.54880	21.4	0.15	351.99	11.08	1683	196	4378	313	8
LSM-T-seq3-c04	3040	2	1	3.13	21	0.19677	4.5	4.58115	75.3	0.16885	75.1	0.06	1157.95	48.10	1746	977	2546	1259	45
LSM-T-seq3-c05	2313	12	0	3.80	120	0.0097485	5.3	0.58259	15.8	0.43343	14.9	0.33	62.54	3.27					
LSM-T-seq3-c06	797	647	0	0.51	159	0.00015340	39.2	0.0097472	40.6	0.50879	10.6	0.97	0.99	0.39					
LSM-T-seq3-c07	**216**	**1480**	**0**	**0.05**	**447**	**0.00020238**	**3.0**	**0.0012600**	**19.2**	**0.049187**	**19.0**	**0.15**	**1.30**	**0.04**					
LSM-T-seq3-c08	806	498	0	0.70	251	0.00028857	10.2	0.012255	15.2	0.32309	11.2	0.68	1.86	0.19					
LSM-T-seq3-c09	26976	154	52	0.63	22847	0.30538	2.3	4.42134	3.1	0.10501	2.0	0.76	1717.94	35.31	1716	26	1714	37	**100**
LSM-T-seq3-c10	-	-	-	-	-	-	-	-	-	-	-	-	-	-	-	-	-	-	-
LSM-T-seq3-c11	2888	98	11	1.15	4923	0.093282	3.4	0.76285	6.4	0.059312	5.4	0.53	574.92	18.73	576	29	579	118	**99**
LSM-T-seq3-c12	31	253	0	0.60	67	0.00020529	5.0	0.0012344	35.5	0.046798	35.1	0.14	1.32	0.07					
LSM-T-seq3-c13	**115**	**1026**	**0**	**0.33**	**261**	**0.00020434**	**3.4**	**0.0012124**	**32.3**	**0.046504**	**32.2**	**0.10**	**1.32**	**0.04**					
LSM-T-seq3-c14	-	-	-	-	-	-	-	-	-	-	-	-	-	-	-	-	-	-	-
LSM-T-seq3-c15	405	102	0	0.46	42	0.00029355	18.3	0.010666	32.1	0.27731	26.3	0.57	1.89	0.35					
LSM-T-seq3-c16	2733	671	0	0.08	73	0.00020266	6.5	0.0099202	15.6	0.38639	14.2	0.42	1.31	0.09					
LSM-T-seq3-c17	39974	187	51	0.57	42687	0.24934	2.2	3.24237	3.0	0.094313	2.1	0.73	1435.07	28.81	1467	24	1514	39	**95**
LSM-T-seq3-c18	30760	81	13	0.70	22971	0.15198	1.9	1.55758	2.1	0.074332	1.0	0.88	912.03	15.89	954	13	1050	20	87
LSM-T-seq3-c19	-	-	-	-	-	-	-	-	-	-	-	-	-	-	-	-	-	-	-
LSM-T-seq3-c20	8233	167	17	0.65	13615	0.095652	2.5	0.79180	4.3	0.060037	3.5	0.59	588.88	14.33	592	19	605	75	**97**
LSM-T-seq3-c21	3281	164	5	1.14	1029	0.028311	3.0	0.23356	5.5	0.059834	4.6	0.55	179.97	5.35	213	11	598	99	30
LSM-T-seq3-c22	-	-	-	-	-	-	-	-	-	-	-	-	-	-	-	-	-	-	-
LSM-T-seq3-c23	**98**	**1472**	**0**	**0.36**	**210**	**0.00020734**	**4.9**	**0.0012735**	**26.3**	**0.048055**	**25.9**	**0.19**	**1.34**	**0.07**					
LSM-T-seq3-c24	7305	222	28	1.39	11878	0.10239	2.3	0.86682	4.1	0.061400	3.4	0.56	628.40	13.81	634	20	653	73	**96**
LSM-T-seq3-c25	202	806	0	0.36	229	0.00024580	6.5	0.0027869	24.7	0.087621	23.9	0.26	1.58	0.10					
LSM-T-seq3-c26	32988	184	31	0.53	6314	0.16351	2.1	1.66738	3.3	0.073960	2.6	0.63	976.24	18.97	996	21	1040	52	**94**
LSM-T-seq3-c27	1110	117	1	0.22	1066	0.0044985	2.0	0.039248	8.3	0.063278	8.1	0.23	28.93	0.56					
LSM-T-seq3-c28	1502	17	0	2.34	65	0.0041971	15.9	0.23344	18.1	0.40338	8.8	0.88	27.00	4.28					
LSM-T-seq3-c29	-	-	-	-	-	-	-	-	-	-	-	-	-	-	-	-	-	-	-
LSM-T-seq3-c30	8644	17	1	6.07	26	0.021087	6.4	0.58077	72.9	0.19975	72.6	0.09	134.52	8.58	465	316	2824	1185	5
LSM-T-seq3-c31	1365	291	1	0.55	22	0.0010998	13.7	0.083597	14.6	0.55791	5.0	0.94	7.09	0.97					
LSM-T-seq3-c32	-	-	-	-	-	-	-	-	-	-	-	-	-	-	-	-	-	-	-
LSM-T-seq3-c33	277	1059	0	0.05	603	0.00018653	3.3	0.0010855	13.3	0.046337	12.9	0.25	1.20	0.04					
LSM-T-seq3-c35	34214	284	43	0.75	546	0.12092	3.6	1.91581	5.3	0.11491	3.9	0.68	735.84	25.21	1087	36	1879	71	39

sample and spot number	$^{207}Pb^a$ (cps)	U^b (ppm)	Pb^b (ppm)	$\underline{Th^b}$ U	$\underline{^{206}Pb^c}$ ^{204}Pb	$\underline{^{206}Pb^c}$ ^{238}U	2 σ %	$\underline{^{207}Pb^c}$ ^{235}U	2 σ %	$\underline{^{207}Pb^c}$ ^{206}Pb	2 σ %	ρ^d	$\underline{^{206}Pb}$ ^{238}U	2 σ (Ma)	$\underline{^{207}Pb}$ ^{235}U	2 σ (Ma)	$\underline{^{207}Pb}$ ^{206}Pb	2 σ (Ma)	Conc %[e]
LSM-T-seq3-c36	5983	320	14	0.66	10369	0.040899	2.3	0.29024	3.5	0.051467	2.6	0.66	258.41	5.90	259	8	262	60	**99**
LSM-T-seq3-c37	-	-	-	-	-	-	-	-	-	-	-	-	-	-	-	-	-	-	-
LSM-T-seq3-c38	1007	112	0	0.95	16	0.00043538	6.8	0.033153	11.1	0.56816	8.9	0.61	2.81	0.19					
LSM-T-seq3-c39	1233	1028	5	0.16	2651	0.0047637	1.9	0.030683	6.5	0.046714	6.2	0.30	30.64	0.59					
LSM-T-seq3-c40	1677	39	1	1.38	3418	0.026535	2.2	0.18126	4.7	0.049542	4.2	0.46	168.82	3.59	169	7	174	98	**97**

sample and spot number	$^{207}Pb^a$ (cps)	U^b (ppm)	Pb^b (ppm)	Th^b / U	$\frac{^{206}Pb^c}{^{204}Pb}$	$\frac{^{206}Pb^c}{^{238}U}$	2 σ %	$\frac{^{207}Pb^c}{^{235}U}$	2 σ %	$\frac{^{207}Pb^c}{^{206}Pb}$	2 σ %	ρ^d -	$\frac{^{206}Pb}{^{238}U}$	2 σ (Ma)	$\frac{^{207}Pb}{^{235}U}$	2 σ (Ma)	$\frac{^{207}Pb}{^{206}Pb}$	2 σ (Ma)	Conc %^{e} -
LSM-3a-seq1-a01	3664	328	78	0.43	4214	0.21819	3.7	2.75646	4.4	0.09162	2.3	0.85	1272	43	1344	33	1460	44	87
LSM-3a-seq1-a02	742	308	12	0.28	704	0.03466	4.0	0.36161	10.0	0.07567	9.2	0.40	220	9	313	27	1086	184	20
LSM-3a-seq1-a03	451	34	8	0.48	588	0.20430	2.9	2.30403	7.6	0.08179	7.0	0.38	1198	31	1214	55	1241	137	97
LSM-3a-seq1-a04	119	40	3	0.43	228	0.06523	3.5	0.49920	16.1	0.05550	15.7	0.22	407	14	411	56	432	349	94
LSM-3a-seq1-a05	5779	1639	46	1.34	249	0.01998	6.6	0.47269	8.1	0.17158	4.8	0.81	128	8	393	27	2573	80	5
LSM-3a-seq1-a06	330	148	8	0.24	640	0.05217	2.8	0.39193	7.9	0.05449	7.4	0.35	328	9	336	23	391	166	84
LSM-3a-seq1-a07	2462	190	42	0.23	3151	0.21495	2.6	2.44247	4.6	0.08241	3.8	0.56	1255	29	1255	34	1255	75	100
LSM-3a-seq1-a08	273	25	5	0.33	374	0.18725	3.0	1.98651	6.8	0.07694	6.1	0.45	1106	31	1111	47	1120	121	99
LSM-3a-seq1-a09	6633	804	133	0.06	10296	0.17496	3.2	1.63929	4.1	0.06795	2.6	0.77	1039	31	985	27	867	54	120
LSM-3a-seq1-a10	1464	69	25	0.72	1474	0.30034	2.2	4.30251	6.2	0.10390	5.8	0.35	1693	32	1694	53	1695	108	100
LSM-3a-seq1-a11	55	323	2	0.44	121	0.00430	6.7	0.02850	20.8	0.04802	19.7	0.32	28	2	29	6	100	465	28
LSM-3a-seq1-a12	106	108	3	0.21	220	0.02563	4.0	0.17998	20.0	0.05093	19.6	0.20	163	6	168	31	238	451	69
LSM-3a-seq1-a13	340	135	11	1.06	651	0.06372	3.4	0.48638	8.5	0.05536	7.8	0.40	398	13	402	29	427	174	93
LSM-3a-seq1-a14	52	7	1	0.33	94	0.13456	9.5	1.01383	37.4	0.05465	36.1	0.25	814	73	711	212	398	810	205
LSM-3a-seq1-a15	866	73	16	0.45	1036	0.21825	3.0	2.54597	6.5	0.08460	5.8	0.46	1273	35	1285	49	1306	112	97
LSM-3a-seq1-a16	298	117	9	0.87	565	0.06691	3.0	0.50926	11.0	0.05520	10.6	0.27	418	12	418	39	420	237	99
LSM-3a-seq1-a17	1585	59	20	0.33	1500	0.31481	2.4	4.82917	4.4	0.11125	3.7	0.54	1764	37	1790	38	1820	67	97
LSM-3a-seq1-a18	406	26	6	0.43	505	0.22885	3.3	2.68066	6.6	0.08495	5.8	0.49	1328	39	1323	50	1314	112	101
LSM-3a-seq1-a19	205	67	6	0.46	436	0.07790	3.6	0.53871	10.7	0.05015	10.1	0.33	484	17	438	39	202	234	239
LSM-3a-seq1-a20	1285	122	28	0.94	1775	0.17930	2.9	1.88029	4.9	0.07606	4.0	0.58	1063	28	1074	33	1097	81	97
LSM-3a-seq1-a21	1284	266	26	0.45	2054	0.08973	3.3	0.81768	5.9	0.06609	5.0	0.55	554	18	607	28	809	104	68
LSM-3a-seq1-a22	2113	95	27	0.22	2269	0.27684	3.6	3.73686	4.8	0.09790	3.0	0.77	1575	51	1579	39	1584	57	99
LSM-3a-seq1-a23	5205	79	47	0.64	2877	0.48532	2.6	12.84277	4.0	0.19193	3.1	0.64	2550	55	2668	39	2759	51	92
LSM-3a-seq1-a24	2853	165	45	0.45	3089	0.25289	5.3	3.37093	7.6	0.09667	5.4	0.69	1453	69	1498	61	1561	102	93
LSM-3a-seq1-a25	104	51	3	0.41	203	0.05388	3.0	0.41452	15.6	0.05580	15.3	0.19	338	10	352	48	444	341	76
LSM-3a-seq1-a26	436	43	10	0.91	585	0.18390	2.5	1.99686	6.6	0.07875	6.1	0.37	1088	25	1114	46	1166	121	93
LSM-3a-seq1-a28	929	168	23	0.52	1416	0.12665	3.3	1.22641	7.1	0.07023	6.3	0.47	769	24	813	40	935	128	82
LSM-3a-seq1-a29	435	116	11	0.60	790	0.08291	3.3	0.66699	8.7	0.05835	8.0	0.38	513	16	519	36	543	176	95
LSM-3a-seq1-a30	136	253	4	0.42	295	0.01492	3.5	0.09973	12.3	0.04849	11.8	0.29	95	3	97	11	123	277	77
LSM-3a-seq1-a31	857	39	15	1.03	839	0.29892	3.1	4.44578	5.6	0.10787	4.6	0.56	1686	46	1721	47	1764	85	96
LSM-3a-seq1-a32	361	194	15	2.12	700	0.04830	2.8	0.36440	8.5	0.05472	8.0	0.34	304	8	315	23	401	179	76
LSM-3a-seq1-a33	178	22	4	0.41	286	0.16295	5.3	1.48250	12.0	0.06599	10.7	0.44	973	48	923	75	806	224	121

| sample and | $^{207}Pb^a$ | U^b | Pb^b | Th^b | $^{206}Pb^c$ | $^{206}Pb^c$ | 2 σ | $^{207}Pb^c$ | 2 σ | $^{207}Pb^c$ | 2 σ | $ρ^d$ | ^{206}Pb | 2 σ | ^{207}Pb | 2 σ | ^{207}Pb | 2 σ | Conc %[e] |
spot number	(cps)	(ppm)	(ppm)	U	^{204}Pb	^{238}U	%	^{235}U	%	^{206}Pb	%	-	^{238}U	(Ma)	^{235}U	(Ma)	^{206}Pb	(Ma)	-
LSM-3a-seq1-a34	2462	1157	23	1.36	120	0.01099	2.3	0.32202	8.5	0.21256	8.2	0.27	70	2	283	21	2925	133	2
LSM-3a-seq1-a35	312	116	5	1.19	420	0.02415	12.6	0.25945	19.2	0.07793	14.5	0.65	154	19	234	41	1145	288	13
LSM-3a-seq1-a36	60	56	1	1.29	65	0.01430	7.4	0.22444	38.7	0.11382	38.0	0.19	92	7	206	75	1861	686	5
LSM-3a-seq1-a37	170	51	4	0.62	251	0.06142	3.4	0.61899	18.5	0.07309	18.2	0.18	384	13	489	75	1016	369	38
LSM-3a-seq1-a38	293	132	9	0.50	564	0.06064	3.2	0.46315	9.4	0.05539	8.8	0.34	380	12	386	31	428	196	89
LSM-3a-seq1-a39	983	128	24	0.77	1495	0.15510	2.4	1.48066	5.6	0.06924	5.0	0.43	929	21	922	34	906	104	**103**
LSM-3a-seq1-a40	5896	291	103	0.64	6259	0.29536	3.1	4.05788	4.0	0.09964	2.5	0.78	1668	46	1646	33	1617	47	**103**
LSM-3a-seq1-a41	2491	144	36	0.27	1427	0.23811	4.3	3.28753	6.6	0.10014	5.0	0.65	1377	53	1478	53	1627	93	85
LSM-3a-seq1-a42	332	33	7	0.70	456	0.18481	3.7	1.94176	9.7	0.07620	9.0	0.38	1093	37	1096	67	1100	180	**99**
LSM-3a-seq1-a43	15324	226	129	0.32	8467	0.50929	2.6	13.40237	3.2	0.19086	1.7	0.83	2654	57	2708	30	2750	29	**97**
LSM-3a-seq1-a44	173	97	4	0.61	259	0.03457	3.7	0.33662	13.8	0.07062	13.3	0.27	219	8	295	36	946	273	23
LSM-3a-seq1-a45	733	72	14	0.30	954	0.18758	2.8	2.09549	5.3	0.08102	4.5	0.53	1108	28	1147	37	1222	88	**91**
LSM-3a-seq1-a46	670	137	13	0.16	978	0.09680	3.9	0.90672	7.0	0.06793	5.8	0.56	596	22	655	34	866	120	69
LSM-3a-seq1-a47	7641	83	63	0.81	3720	0.57461	2.7	17.13087	3.1	0.21623	1.6	0.86	2927	63	2942	30	2953	25	**99**
LSM-3a-seq1-a48	1459	62	20	0.33	1425	0.30802	2.8	4.59066	5.8	0.10809	5.1	0.49	1731	43	1748	50	1768	93	**98**
LSM-3a-seq1-a49	1388	123	27	0.34	1804	0.20776	2.4	2.32259	4.8	0.08108	4.1	0.51	1217	27	1219	35	1223	81	**99**
LSM-3a-seq1-a50	1097	283	22	0.33	417	0.07275	3.4	0.93795	7.8	0.09350	7.1	0.43	453	15	672	39	1498	134	30
LSM-3a-seq2-b01	364	250	13	1.20	667	0.03936	2.8	0.31503	8.5	0.05805	8.0	0.33	249	7	278	21	532	176	47
LSM-3a-seq2-b02	126	25	4	1.24	210	0.11045	5.8	0.96642	16.2	0.06346	15.2	0.36	675	37	687	84	724	321	**93**
LSM-3a-seq2-b03	52	49	1	0.52	84	0.02201	8.6	0.19871	23.4	0.06548	21.8	0.37	140	12	184	40	790	457	18
LSM-3a-seq2-b04	128	32	3	0.64	164	0.07403	5.2	0.83267	18.5	0.08157	17.8	0.28	460	23	615	89	1235	349	37
LSM-3a-seq2-b05	145	20	4	0.38	213	0.18124	4.0	1.56991	13.0	0.06282	12.4	0.31	1074	40	958	84	702	264	153
LSM-3a-seq2-b06	493	44	10	0.49	629	0.21543	3.1	2.46080	6.4	0.08285	5.6	0.49	1258	36	1261	47	1266	110	**99**
LSM-3a-seq2-b07	2953	400	62	0.10	4312	0.16231	2.4	1.62148	4.0	0.07245	3.2	0.61	970	22	979	25	999	64	**97**
LSM-3a-seq2-b08	991	288	29	0.52	1766	0.09032	2.8	0.73694	5.4	0.05917	4.7	0.51	557	15	561	24	573	102	**97**
LSM-3a-seq2-b09	203	86	5	0.13	386	0.06023	4.9	0.46154	18.9	0.05558	18.3	0.26	377	18	385	63	436	407	87
LSM-3a-seq2-b11	381	224	10	0.22	705	0.04455	2.7	0.34867	9.5	0.05676	9.1	0.29	281	8	304	25	482	201	58
LSM-3a-seq2-b12	4360	187	57	0.09	4240	0.30768	2.4	4.59160	3.4	0.10823	2.5	0.69	1729	36	1748	29	1770	46	**98**
LSM-3a-seq2-b13	546	53	12	0.28	659	0.21614	1.7	2.61294	9.8	0.08768	9.7	0.18	1261	20	1304	75	1375	186	**92**
LSM-3a-seq2-b14	1350	95	25	0.39	1549	0.24104	2.4	3.03141	4.9	0.09121	4.3	0.49	1392	31	1416	38	1451	81	**96**
LSM-3a-seq2-b15	669	179	18	0.44	1184	0.09397	2.6	0.77165	6.2	0.05956	5.6	0.42	579	14	581	28	588	122	**99**
LSM-3a-seq2-b16	232	203	8	0.77	484	0.03432	3.7	0.23910	9.2	0.05053	8.3	0.41	218	8	218	18	220	193	**99**

sample and spot number	$^{207}Pb^a$ (cps)	U^b (ppm)	Pb^b (ppm)	Th^b U	$\frac{^{206}Pb^c}{^{204}Pb}$	$\frac{^{206}Pb^c}{^{238}U}$	2σ %	$\frac{^{207}Pb^c}{^{235}U}$	2σ %	$\frac{^{207}Pb^c}{^{206}Pb}$	2σ %	ρ^d -	$\frac{^{206}Pb}{^{238}U}$	2σ (Ma)	$\frac{^{207}Pb}{^{235}U}$	2σ (Ma)	$\frac{^{207}Pb}{^{206}Pb}$	2σ (Ma)	Conc %^{e} -
LSM-3a-seq2-b17	1969	160	37	0.29	2432	0.22478	2.7	2.65308	4.9	0.08560	4.0	0.57	1307	33	1315	36	1329	77	**98**
LSM-3a-seq2-b18	1406	42	17	0.28	1129	0.38956	2.0	7.08749	5.5	0.13195	5.1	0.37	2121	37	2122	50	2124	90	**100**
LSM-3a-seq2-b19	2690	193	48	0.56	3050	0.21546	3.0	2.76734	4.0	0.09315	2.7	0.74	1258	34	1347	30	1491	52	84
LSM-3a-seq2-b20	748	106	17	0.32	1063	0.16907	2.7	1.72608	8.0	0.07405	7.5	0.34	1007	25	1018	53	1043	151	**97**
LSM-3a-seq2-b21	35	34	1	1.10	54	0.02398	8.5	0.23973	43.9	0.07251	43.1	0.19	153	13	218	90	1000	874	15
LSM-3a-seq2-b22	99	48	4	1.57	182	0.05422	5.0	0.44488	18.4	0.05951	17.7	0.27	340	17	374	59	586	385	58
LSM-3a-seq2-b23	358	39	8	0.42	493	0.18425	2.9	1.95629	8.3	0.07701	7.8	0.35	1090	30	1101	57	1121	155	**97**
LSM-3a-seq2-b24	6289	233	50	0.78	470	0.13256	3.0	3.77359	4.0	0.20646	2.6	0.76	802	23	1587	33	2878	43	28
LSM-3a-seq2-b25	3944	275	85	0.90	4539	0.24642	2.4	3.06101	3.5	0.09009	2.6	0.67	1420	30	1423	27	1427	49	**99**
LSM-3a-seq2-b26	138	53	4	0.81	249	0.06932	3.8	0.55662	15.8	0.05824	15.3	0.24	432	16	449	59	539	334	80
LSM-3a-seq2-b27	4635	99	51	0.27	2940	0.46918	2.2	10.78446	3.5	0.16671	2.8	0.62	2480	45	2505	33	2525	46	**98**
LSM-3a-seq2-b28	200	56	5	0.33	343	0.08439	2.3	0.71851	12.5	0.06175	12.3	0.18	522	11	550	54	665	262	78
LSM-3a-seq2-b29	406	133	11	0.43	724	0.07230	4.2	0.58222	8.2	0.05840	7.0	0.52	450	18	466	31	545	154	83
LSM-3a-seq2-b30	2787	95	45	1.02	2324	0.36262	2.1	6.22263	3.3	0.12446	2.6	0.64	1995	37	2008	30	2021	46	**99**
LSM-3a-seq2-b31	2049	211	29	0.46	558	0.11372	3.5	1.89590	5.4	0.12091	4.1	0.65	694	23	1080	36	1970	72	35
LSM-3a-seq2-b32	2170	254	36	0.16	2680	0.13904	3.0	1.63019	5.5	0.08503	4.6	0.55	839	24	982	35	1316	89	64
LSM-3a-seq2-b33	180	374	6	0.41	391	0.01437	2.8	0.09592	9.9	0.04840	9.5	0.28	92	3	93	9	119	224	78
LSM-3a-seq2-b34	289	41	7	0.52	460	0.16027	2.7	1.46875	9.2	0.06647	8.8	0.30	958	24	918	57	821	183	117
LSM-3a-seq2-b35	453	208	14	0.43	827	0.06417	2.5	0.51109	6.2	0.05777	5.7	0.40	401	10	419	21	521	124	77
LSM-3a-seq2-b36	153	66	5	1.16	291	0.06041	5.2	0.47092	12.6	0.05653	11.5	0.41	378	19	392	42	473	254	80
LSM-3a-seq2-b37	474	119	14	0.48	814	0.10474	2.9	0.88374	7.4	0.06119	6.8	0.39	642	18	643	36	646	147	**99**
LSM-3a-seq2-b38	315	125	11	0.58	578	0.08143	2.5	0.65219	10.1	0.05809	9.8	0.25	505	12	510	41	533	214	**95**
LSM-3a-seq2-b39	555	55	12	0.33	692	0.20922	2.4	2.42065	8.6	0.08391	8.3	0.28	1225	27	1249	64	1290	161	**95**
LSM-3a-seq2-b40	5215	234	63	0.50	1146	0.22759	3.4	5.54486	4.4	0.17670	2.8	0.77	1322	41	1908	38	2622	46	50
LSM-3a-seq2-b41	199	26	5	0.31	283	0.18214	3.4	1.86610	10.0	0.07431	9.5	0.34	1079	34	1069	69	1050	191	**103**
LSM-3a-seq2-b42	135	152	6	1.25	261	0.02839	4.5	0.21417	13.6	0.05470	12.9	0.33	180	8	197	25	400	288	45
LSM-3a-seq2-b44	327	147	12	0.66	620	0.06967	3.3	0.54082	8.3	0.05630	7.6	0.39	434	14	439	30	464	169	**94**
LSM-3a-seq2-b45	2708	163	47	0.26	2861	0.28044	2.2	3.87080	4.1	0.10011	3.5	0.54	1594	31	1608	34	1626	64	**98**
LSM-3a-seq2-b46	698	317	16	0.25	744	0.04917	3.4	0.43413	6.3	0.06404	5.3	0.54	309	10	366	20	743	113	42
LSM-3a-seq2-b47	142	31	4	0.55	242	0.11423	4.8	0.99512	10.5	0.06318	9.3	0.46	697	32	701	54	714	197	**98**
LSM-3a-seq2-b48	1697	64	26	0.33	1402	0.37892	2.2	6.65101	5.0	0.12730	4.5	0.43	2071	38	2066	45	2061	80	**100**
LSM-3a-seq2-b49	24	90	1	0.96	30	0.00566	11.4	0.07131	32.7	0.09145	30.6	0.35	36	4	70	22	1456	583	2

sample and spot number	$^{207}Pb^a$ (cps)	U^b (ppm)	Pb^b (ppm)	$\underline{Th^b}$ U	$\underline{^{206}Pb^c}$ ^{204}Pb	$\underline{^{206}Pb^c}$ ^{238}U	2 σ %	$\underline{^{207}Pb^c}$ ^{235}U	2 σ %	$\underline{^{207}Pb^c}$ ^{206}Pb	2 σ %	ρ^d -	$\underline{^{206}Pb}$ ^{238}U	2 σ (Ma)	$\underline{^{207}Pb}$ ^{235}U	2 σ (Ma)	$\underline{^{207}Pb}$ ^{206}Pb	2 σ (Ma)	Conc %^{e} -
LSM-3a-seq2-b50	248	135	8	0.29	485	0.05523	3.3	0.41106	9.5	0.05398	9.0	0.34	347	11	350	29	370	202	**94**
LSM-3a-seq2-b51	2535	266	57	0.38	3300	0.19896	2.4	2.22117	3.7	0.08097	2.9	0.64	1170	25	1188	26	1221	57	**96**
LSM-3a-seq2-b52	370	32	11	1.35	434	0.24036	2.6	2.94444	8.4	0.08884	8.0	0.31	1389	32	1393	65	1401	152	**99**
LSM-3a-seq2-b53	78	87	2	0.10	146	0.02950	3.6	0.23544	17.5	0.05789	17.1	0.20	187	7	215	34	526	375	36
LSM-3a-seq2-b54	77	89	3	0.77	160	0.02766	4.7	0.19322	18.9	0.05066	18.3	0.25	176	8	179	31	225	422	78
LSM-3a-seq2-b55	1286	181	12	0.16	1100	0.05932	4.9	1.00859	8.2	0.12331	6.5	0.60	372	18	708	43	2005	116	19
LSM-3a-seq2-b56	182	60	6	0.66	331	0.08487	3.4	0.68009	10.6	0.05812	10.0	0.32	525	17	527	44	534	219	**98**
LSM-3a-seq2-b57	266	36	7	0.53	377	0.17851	4.6	1.83551	10.4	0.07457	9.4	0.44	1059	45	1058	71	1057	189	**100**
LSM-3a-seq2-b58	2849	132	50	0.46	2690	0.33911	2.1	5.25670	3.4	0.11243	2.8	0.60	1882	34	1862	30	1839	50	**102**
LSM-3a-seq2-b59	65	109	2	0.42	127	0.01776	4.3	0.12845	33.6	0.05245	33.3	0.13	113	5	123	40	305	759	37
LSM-3a-seq2-b60	140	189	5	0.48	293	0.02541	2.1	0.17582	11.9	0.05018	11.8	0.17	162	3	164	18	203	273	80
LSM-3a-seq3-c01	2669	353	42	0.17	3291	0.11556	5.1	1.34658	6.4	0.08451	3.8	0.80	705	34	866	38	1304	74	54
LSM-3a-seq3-c02	385	233	15	1.02	653	0.05320	2.2	0.45471	10.7	0.06199	10.5	0.21	334	7	381	35	674	224	50
LSM-3a-seq3-c03	442	581	14	1.01	801	0.01819	4.9	0.14320	8.6	0.05711	7.1	0.56	116	6	136	11	496	156	23
LSM-3a-seq3-c04	206	114	7	0.61	382	0.05830	2.7	0.45160	11.9	0.05618	11.6	0.23	365	10	378	38	459	257	80
LSM-3a-seq3-c05	145	80	5	0.79	288	0.05396	4.1	0.40190	11.5	0.05401	10.7	0.36	339	14	343	34	372	241	**91**
LSM-3a-seq3-c06	3660	137	49	0.13	2966	0.34663	2.7	6.14264	4.1	0.12853	3.1	0.66	1918	45	1996	36	2078	54	**92**
LSM-3a-seq3-c07	426	52	9	0.23	589	0.17033	3.8	1.77065	7.1	0.07539	6.0	0.53	1014	36	1035	47	1079	121	**94**
LSM-3a-seq3-c08	203	106	6	0.21	319	0.05927	2.9	0.49914	13.0	0.06108	12.6	0.23	371	11	411	45	642	271	58
LSM-3a-seq3-c09	92	126	3	0.34	190	0.02339	4.6	0.16264	15.7	0.05043	15.1	0.29	149	7	153	23	215	349	69
LSM-3a-seq3-c10	7113	644	87	0.11	2614	0.13273	2.9	1.85183	3.9	0.10119	2.6	0.73	803	22	1064	26	1646	49	49
LSM-3b-seq1-a01	588	59	11	0.29	725	0.17823	3.1	2.06498	7.1	0.08403	6.4	0.43	1057	30	1137	50	1293	125	82
LSM-3b-seq1-a02	638	69	14	0.43	866	0.18551	2.6	1.98000	6.4	0.07741	5.8	0.41	1097	26	1109	44	1132	116	**97**
LSM-3b-seq1-a03	95	105	4	1.34	176	0.02372	4.1	0.18324	26.5	0.05602	26.2	0.15	151	6	171	43	453	582	33
LSM-3b-seq1-a04	180	75	5	0.41	358	0.06917	4.0	0.50551	12.4	0.05300	11.7	0.32	431	17	415	43	329	266	131
LSM-3b-seq1-a06	74	27	2	0.26	127	0.06519	4.3	0.55613	20.5	0.06187	20.1	0.21	407	17	449	77	670	429	61
LSM-3b-seq1-a07	188	80	5	0.20	361	0.05836	4.8	0.43928	11.1	0.05459	10.0	0.43	366	17	370	35	396	225	**92**
LSM-3b-seq1-a08	2321	438	63	0.21	3508	0.14690	1.8	1.40736	3.5	0.06949	3.0	0.51	884	15	892	21	913	63	**97**
LSM-3b-seq1-a09	2398	120	36	0.23	2487	0.29220	2.7	4.07647	3.6	0.10118	2.4	0.76	1652	40	1650	30	1646	44	**100**
LSM-3b-seq1-a10	1093	457	32	0.30	1730	0.06786	3.9	0.62040	9.5	0.06630	8.7	0.41	423	16	490	38	816	182	52
LSM-3b-seq1-a11	88	568	3	0.61	155	0.00390	5.4	0.03207	21.5	0.05971	20.9	0.25	25	1	32	7	593	452	4

sample and spot number	$^{207}Pb^a$ (cps)	U^b (ppm)	Pb^b (ppm)	$\frac{Th^b}{U}$	$\frac{^{206}Pb^c}{^{204}Pb}$	$\frac{^{206}Pb^c}{^{238}U}$	2 σ %	$\frac{^{207}Pb^c}{^{235}U}$	2 σ %	$\frac{^{207}Pb^c}{^{206}Pb}$	2 σ %	ρ^d -	$\frac{^{206}Pb}{^{238}U}$	2 σ (Ma)	$\frac{^{207}Pb}{^{235}U}$	2 σ (Ma)	$\frac{^{207}Pb}{^{206}Pb}$	2 σ (Ma)	Conc %^{e} -
LSM-3b-seq1-a14	1449	104	29	0.45	1619	0.25564	2.2	3.32002	5.0	0.09419	4.5	0.44	1467	29	1486	40	1512	84	**97**
LSM-3b-seq1-a15	424	107	12	0.37	720	0.10479	1.9	0.88749	9.0	0.06142	8.8	0.21	642	12	645	44	654	189	**98**
LSM-3b-seq1-a16	481	26	10	1.10	452	0.30325	3.4	4.69368	8.2	0.11226	7.5	0.42	1707	52	1766	71	1836	135	**93**
LSM-3b-seq1-a17	647	160	17	0.35	1070	0.10116	2.3	0.87929	5.9	0.06304	5.5	0.39	621	14	641	29	710	116	88
LSM-3b-seq1-a18	2525	180	53	0.69	2837	0.25102	1.9	3.24289	3.5	0.09370	3.0	0.52	1444	24	1467	28	1502	57	**96**
LSM-3b-seq1-a19	286	75	8	0.46	472	0.09824	3.6	0.86615	9.7	0.06394	9.0	0.37	604	21	633	47	740	191	82
LSM-3b-seq1-a20	76	71	2	0.42	127	0.03172	4.1	0.28282	17.5	0.06466	17.0	0.24	201	8	253	40	763	358	26
LSM-3b-seq1-a21	2198	118	37	0.38	2166	0.29009	2.6	4.17416	3.9	0.10436	2.9	0.67	1642	38	1669	32	1703	53	**96**
LSM-3b-seq1-a22	1477	188	35	0.17	1903	0.19442	1.9	2.19390	5.1	0.08184	4.7	0.37	1145	20	1179	36	1242	92	**92**
LSM-3b-seq1-a23	4201	264	69	0.34	917	0.23812	2.8	3.60217	4.5	0.10972	3.5	0.62	1377	34	1550	36	1795	64	77
LSM-3b-seq1-a24	144	320	5	0.55	301	0.01498	3.3	0.10359	13.0	0.05015	12.5	0.26	96	3	100	12	202	291	47
LSM-3b-seq1-a25	377	56	10	0.42	521	0.17387	2.5	1.81697	6.4	0.07579	5.9	0.38	1033	23	1052	43	1090	118	**95**
LSM-3b-seq1-a26	2625	128	42	0.34	2506	0.30483	1.9	4.58939	3.3	0.10919	2.8	0.56	1715	28	1747	28	1786	51	**96**
LSM-3b-seq1-a27	1223	114	27	0.35	1583	0.22449	2.3	2.51474	5.1	0.08124	4.6	0.44	1306	27	1276	38	1227	90	**106**
LSM-3b-seq1-a28	31	77	1	0.80	61	0.01163	7.3	0.08823	36.4	0.05502	35.6	0.20	75	5	86	30	413	797	18
LSM-3b-seq1-a29	2060	73	31	0.54	1710	0.36278	1.8	6.33504	4.5	0.12665	4.1	0.40	1995	31	2023	40	2052	72	**97**
LSM-3b-seq1-a30	182	213	6	0.72	352	0.02591	3.8	0.19241	11.4	0.05386	10.7	0.33	165	6	179	19	365	241	45
LSM-3b-seq1-a31	1767	475	44	0.03	3066	0.10047	1.9	0.83863	4.5	0.06054	4.1	0.42	617	11	618	21	623	89	**99**
LSM-3b-seq1-a32	6927	569	47	0.09	7552	0.07946	3.2	1.05718	4.0	0.09650	2.4	0.80	493	15	732	21	1557	46	32
LSM-3b-seq1-a33	1306	68	22	0.32	1288	0.30936	2.0	4.56418	5.4	0.10700	5.0	0.38	1738	31	1743	46	1749	91	**99**
LSM-3b-seq1-a34	1637	59	31	1.21	1329	0.37332	2.1	6.66195	4.3	0.12942	3.7	0.50	2045	38	2068	39	2090	66	**98**
LSM-3b-seq1-a35	80	111	3	0.86	167	0.02382	3.2	0.16596	22.9	0.05052	22.7	0.14	152	5	156	34	219	524	69
LSM-3b-seq1-a36	1355	460	26	1.08	419	0.03851	2.8	0.59655	7.2	0.11236	6.6	0.38	244	7	475	28	1838	120	13
LSM-3b-seq1-a37	332	34	8	0.49	437	0.20619	3.0	2.28279	10.4	0.08030	10.0	0.29	1209	33	1207	76	1204	196	**100**
LSM-3b-seq1-a38	321	143	9	0.29	598	0.06240	2.9	0.48429	10.9	0.05628	10.5	0.27	390	11	401	37	464	233	84
LSM-3b-seq1-a40	1078	59	21	0.49	1119	0.31537	1.5	4.40970	7.7	0.10141	7.5	0.20	1767	23	1714	65	1650	139	**107**
LSM-3b-seq1-a41	158	38	5	0.91	242	0.09735	5.1	0.94536	18.8	0.07043	18.0	0.27	599	29	676	97	941	370	64
LSM-3b-seq1-a42	329	158	13	0.87	630	0.06457	2.5	0.49184	9.9	0.05525	9.5	0.25	403	10	406	34	422	213	**96**
LSM-3b-seq1-a43	723	59	17	0.33	877	0.27841	1.4	3.31995	4.4	0.08648	4.2	0.33	1583	20	1486	35	1349	80	117
LSM-3b-seq1-a44	3004	161	56	0.56	3002	0.30459	2.7	4.41717	3.9	0.10518	2.8	0.69	1714	41	1716	33	1717	52	**100**
LSM-3b-seq1-a45	705	112	15	0.10	1089	0.14165	2.6	1.32366	5.4	0.06777	4.8	0.47	854	21	856	32	862	100	**99**
LSM-3b-seq1-a46	936	106	20	0.43	1193	0.18967	2.5	2.15372	6.9	0.08236	6.4	0.36	1120	25	1166	49	1254	125	89

sample and spot number	$^{207}Pb^a$ (cps)	U^b (ppm)	Pb^b (ppm)	Th^b U	$\frac{^{206}Pb^c}{^{204}Pb}$	$\frac{^{206}Pb^c}{^{238}U}$	2 σ %	$\frac{^{207}Pb^c}{^{235}U}$	2 σ %	$\frac{^{207}Pb^c}{^{206}Pb}$	2 σ %	ρ^d -	$\frac{^{206}Pb}{^{238}U}$	2 σ (Ma)	$\frac{^{207}Pb}{^{235}U}$	2 σ (Ma)	$\frac{^{207}Pb}{^{206}Pb}$	2 σ (Ma)	Conc $\%^e$ -
LSM-3b-seq1-a47	143	69	4	0.42	279	0.06052	3.7	0.44883	12.0	0.05378	11.4	0.31	379	14	376	38	362	258	**105**
LSM-3b-seq1-a48	1597	91	34	0.99	1560	0.27968	2.3	4.13095	4.3	0.10712	3.6	0.54	1590	33	1660	36	1751	66	**91**
LSM-3b-seq1-a49	145	139	5	0.44	320	0.03586	4.0	0.23327	12.1	0.04718	11.4	0.33	227	9	213	23	58	272	390
LSM-3b-seq1-a50	74	188	3	0.36	161	0.01392	5.2	0.09388	21.2	0.04892	20.5	0.25	89	5	91	19	144	481	62
LSM-3b-seq1-a51	14	30	0	0.45	21	0.00550	30.2	0.05582	46.7	0.07359	35.5	0.65	35	11	55	25	1030	719	3
LSM-3b-seq1-a52	1846	99	32	0.35	1784	0.29829	3.0	4.47390	4.6	0.10878	3.5	0.64	1683	44	1726	39	1779	65	**95**
LSM-3b-seq1-a53	167	78	6	0.52	304	0.07212	3.1	0.57429	10.9	0.05775	10.4	0.29	449	14	461	41	520	228	86
LSM-3b-seq1-a54	1885	117	36	0.34	1872	0.28246	2.3	4.09102	4.3	0.10505	3.6	0.53	1604	32	1653	36	1715	67	**94**
LSM-3b-seq1-a55	83	39	3	0.27	158	0.06475	3.8	0.49066	17.5	0.05496	17.1	0.22	404	15	405	60	411	383	**98**
LSM-3b-seq1-a56	262	114	9	0.67	467	0.06364	4.0	0.53275	16.3	0.06072	15.8	0.24	398	15	434	59	629	340	63
LSM-3b-seq1-a57	357	37	8	0.21	452	0.21021	3.1	2.40951	7.8	0.08313	7.2	0.39	1230	35	1245	58	1272	140	**97**
LSM-3b-seq1-a58	730	138	20	0.56	1034	0.13324	3.0	1.32798	7.2	0.07228	6.5	0.42	806	23	858	43	994	133	81
LSM-3b-seq1-a59	294	21	5	0.33	319	0.22185	2.9	2.92955	9.4	0.09577	8.9	0.31	1292	34	1390	74	1543	168	84
LSM-3b-seq1-a60	3314	196	61	0.18	3333	0.30339	2.3	4.37762	5.0	0.10465	4.4	0.46	1708	34	1708	42	1708	82	**100**
LSM-3b-seq2-b01	265	27	6	0.97	313	0.18534	4.9	2.25807	9.3	0.08836	7.9	0.53	1096	50	1199	68	1390	152	79
LSM-3b-seq2-b02	360	156	13	0.69	654	0.06960	3.0	0.55587	8.8	0.05793	8.3	0.33	434	12	449	33	527	183	82
LSM-3b-seq2-b03	109	285	4	0.64	219	0.01280	5.4	0.09049	17.2	0.05127	16.3	0.31	82	4	88	15	253	375	32
LSM-3b-seq2-b04	1417	106	30	0.85	1643	0.23614	3.2	2.92474	6.2	0.08983	5.3	0.52	1367	40	1388	48	1422	102	**96**
LSM-3b-seq2-b05	1307	211	32	0.13	1862	0.16040	2.9	1.62156	6.1	0.07332	5.4	0.47	959	26	979	39	1023	109	**94**
LSM-3b-seq2-b06	2635	331	68	0.16	2584	0.20309	2.8	2.37501	5.1	0.08482	4.2	0.54	1192	30	1235	37	1311	82	**91**
LSM-3b-seq2-b07	493	133	14	0.77	734	0.08763	3.6	0.84482	7.1	0.06992	6.2	0.50	542	19	622	34	926	127	58
LSM-3b-seq2-b08	4728	89	50	0.21	2758	0.51310	3.0	12.65204	3.9	0.17884	2.4	0.78	2670	67	2654	37	2642	40	**101**
LSM-3b-seq2-b09	273	339	13	1.54	535	0.02714	3.3	0.20160	10.4	0.05387	9.8	0.32	173	6	186	18	366	222	47
LSM-3b-seq2-b10	3832	224	65	0.43	3854	0.28533	3.0	4.03702	4.0	0.10261	2.7	0.75	1618	43	1642	33	1672	49	**97**
LSM-3b-seq2-b11	651	83	16	0.33	884	0.18362	3.6	1.93556	6.5	0.07645	5.4	0.55	1087	36	1093	45	1107	109	**98**
LSM-3b-seq2-b13	13541	987	134	0.02	54	0.07768	3.2	3.66274	8.4	0.34197	7.8	0.38	482	15	1563	69	3672	119	13
LSM-3b-seq2-b14	744	106	18	0.35	1046	0.16213	2.8	1.65444	5.7	0.07401	5.0	0.49	969	25	991	37	1042	101	**93**
LSM-3b-seq2-b15	514	23	9	0.64	473	0.32095	9.0	4.99361	14.7	0.11285	11.6	0.61	1794	142	1818	132	1846	210	**97**
LSM-3b-seq2-b16	755	48	16	0.78	813	0.28555	3.3	3.80493	5.8	0.09664	4.8	0.56	1619	47	1594	48	1560	91	**104**
LSM-3b-seq2-b17	56	226	1	0.67	67	0.00487	10.1	0.05877	34.9	0.08748	33.4	0.29	31	3	58	20	1371	643	2
LSM-3b-seq2-b18	1842	342	34	0.15	2286	0.09569	3.2	1.10075	5.6	0.08343	4.6	0.57	589	18	754	30	1279	90	46
LSM-3b-seq2-b20	499	231	16	0.29	945	0.06850	3.3	0.52366	7.4	0.05544	6.7	0.44	427	14	428	26	430	149	**99**

sample and spot number	$^{207}Pb^a$ (cps)	U^b (ppm)	Pb^b (ppm)	Th^b U	$\frac{^{206}Pb^c}{^{204}Pb}$	$\frac{^{206}Pb^c}{^{238}U}$	2σ %	$\frac{^{207}Pb^c}{^{235}U}$	2σ %	$\frac{^{207}Pb^c}{^{206}Pb}$	2σ %	$ρ^d$ -	$\frac{^{206}Pb}{^{238}U}$	2σ (Ma)	$\frac{^{207}Pb}{^{235}U}$	2σ (Ma)	$\frac{^{207}Pb}{^{206}Pb}$	2σ (Ma)	Conc %^{e} -
LSM-3b-seq2-b21	4525	628	110	0.09	6133	0.18122	3.4	1.86524	4.2	0.07465	2.4	0.82	1074	34	1069	28	1059	48	101
LSM-3b-seq2-b22	231	107	8	0.46	443	0.06487	4.6	0.49035	8.5	0.05482	7.1	0.54	405	18	405	29	405	160	100
LSM-3b-seq2-b23	636	573	22	1.08	1088	0.02857	3.5	0.24507	8.0	0.06220	7.1	0.44	182	6	223	16	681	152	27
LSM-3b-seq2-b24	1198	334	40	0.66	2089	0.10007	3.2	0.83286	6.3	0.06036	5.4	0.51	615	19	615	29	617	116	100
LSM-3b-seq2-b25	220	48	6	1.04	248	0.08941	4.0	1.15536	18.2	0.09372	17.8	0.22	552	21	780	104	1502	336	37
LSM-3b-seq2-b26	222	38	7	0.70	335	0.14376	3.6	1.34719	10.3	0.06797	9.6	0.35	866	30	866	62	868	199	100
LSM-3b-seq2-b27	130	163	5	0.75	272	0.02645	4.3	0.18336	12.9	0.05027	12.2	0.33	168	7	171	20	207	282	81
LSM-3b-seq2-b28	2123	329	50	0.20	2974	0.15481	2.7	1.59100	4.2	0.07453	3.2	0.65	928	24	967	26	1056	64	88
LSM-3b-seq2-b29	329	153	11	0.31	619	0.06735	3.2	0.51304	7.0	0.05524	6.2	0.45	420	13	420	24	422	139	100
LSM-3b-seq2-b30	1512	840	29	0.34	508	0.02963	4.5	0.38939	5.9	0.09532	3.9	0.76	188	8	334	17	1535	73	12
LSM-3b-seq2-b31	68	105	3	1.24	96	0.01704	5.6	0.17046	19.1	0.07255	18.3	0.29	109	6	160	29	1001	371	11
LSM-3b-seq2-b32	1407	71	25	0.39	1383	0.31579	3.3	4.63322	5.2	0.10641	4.0	0.64	1769	51	1755	44	1739	74	102
LSM-3b-seq2-b33	164	1361	7	0.63	354	0.00437	3.7	0.02919	12.0	0.04844	11.4	0.31	28	1	29	3	121	270	23
LSM-3b-seq2-b34	1034	144	26	0.38	1415	0.16903	2.8	1.76844	5.5	0.07588	4.7	0.52	1007	26	1034	36	1092	94	92
LSM-3b-seq2-b35	34	179	1	0.47	45	0.00542	9.9	0.06594	57.2	0.08817	56.3	0.17	35	3	65	37	1386	1081	3
LSM-3b-seq2-b36	873	128	20	0.53	1170	0.13953	4.1	1.49136	7.8	0.07752	6.6	0.52	842	32	927	48	1135	132	74
LSM-3b-seq2-b37	807	253	17	0.29	360	0.05985	4.5	0.77023	7.2	0.09333	5.7	0.62	375	16	580	32	1495	107	25
LSM-3b-seq2-b38	432	132	13	0.32	761	0.09233	3.7	0.75370	8.5	0.05921	7.6	0.44	569	20	570	38	575	166	99
LSM-3b-seq2-b39	2105	105	32	0.17	2187	0.30466	2.7	4.28914	5.7	0.10211	5.0	0.47	1714	41	1691	48	1663	93	103
LSM-3b-seq2-b40	1966	110	40	0.78	1927	0.29903	2.7	4.28006	5.0	0.10381	4.2	0.55	1687	41	1690	42	1693	77	100
LSM-3b-seq2-b41	185	56	7	0.70	339	0.09896	5.3	0.78045	18.4	0.05720	17.7	0.29	608	31	586	86	499	389	122
LSM-3b-seq2-b42	799	133	23	1.80	1209	0.13992	3.2	1.32493	8.9	0.06868	8.3	0.36	844	25	857	53	889	172	95
LSM-3b-seq2-b43	207	68	6	0.42	370	0.08228	5.0	0.65407	13.4	0.05766	12.4	0.37	510	24	511	55	517	272	99
LSM-3b-seq2-b44	473	214	18	0.87	890	0.06840	4.1	0.52544	7.6	0.05571	6.3	0.54	427	17	429	27	441	141	97
LSM-3b-seq2-b45	280	241	9	0.56	424	0.03084	3.4	0.30244	14.3	0.07113	13.9	0.24	196	7	268	34	961	285	20
LSM-3b-seq2-b47	48	359	2	0.41	90	0.00438	7.4	0.03480	39.7	0.05765	39.0	0.19	28	2	35	14	517	856	5
LSM-3b-seq2-b48	717	111	20	0.57	987	0.15967	3.5	1.67243	7.4	0.07597	6.5	0.47	955	31	998	48	1094	131	87
LSM-3b-seq2-b49	2378	145	43	0.15	2438	0.29443	2.8	4.14562	4.5	0.10212	3.5	0.62	1664	41	1663	37	1663	65	100
LSM-3b-seq2-b50	3251	1251	50	0.22	171	0.03093	2.8	0.67581	3.9	0.15848	2.7	0.71	196	5	524	16	2439	47	8
LSM-3b-seq2-b51	11163	206	112	0.04	6130	0.52463	3.5	13.75480	4.4	0.19015	2.7	0.79	2719	78	2733	43	2743	44	99
LSM-3b-seq2-b52	157	85	5	0.50	282	0.05721	3.5	0.45652	14.9	0.05788	14.5	0.23	359	12	382	49	525	318	68
LSM-3b-seq2-b53	877	272	26	0.19	1100	0.09607	3.2	0.82201	5.4	0.06205	4.4	0.59	591	18	609	25	676	94	87

sample and spot number	$^{207}Pb^{a}$ (cps)	U^{b} (ppm)	Pb^{b} (ppm)	Th^{b} / U	$\frac{^{206}Pb^{c}}{^{204}Pb}$	$\frac{^{206}Pb^{c}}{^{238}U}$	2σ %	$\frac{^{207}Pb^{c}}{^{235}U}$	2σ %	$\frac{^{207}Pb^{c}}{^{206}Pb}$	2σ %	$ρ^{d}$	$\frac{^{206}Pb}{^{238}U}$	2σ (Ma)	$\frac{^{207}Pb}{^{235}U}$	2σ (Ma)	$\frac{^{207}Pb}{^{206}Pb}$	2σ (Ma)	Conc %^{e}
LSM-3b-seq2-b54	242	666	9	0.29	519	0.01330	2.9	0.08967	10.6	0.04892	10.2	0.27	85	2	87	9	144	240	59
LSM-3b-seq2-b55	100	129	3	0.49	215	0.02401	5.1	0.16538	18.3	0.04996	17.6	0.28	153	8	155	27	193	409	79
LSM-3b-seq2-b56	103	149	4	0.32	212	0.02568	4.9	0.17727	16.1	0.05006	15.4	0.30	163	8	166	25	198	357	83
LSM-3b-seq2-b57	72	129	2	0.45	110	0.01703	5.5	0.16026	17.1	0.06825	16.2	0.32	109	6	151	24	876	336	12
LSM-3b-seq2-b58	406	25	9	0.55	424	0.29256	4.6	4.12304	10.2	0.10221	9.1	0.45	1654	67	1659	87	1665	168	**99**
LSM-3b-seq2-b59	1252	65	26	0.66	993	0.33981	3.3	5.38375	6.6	0.11491	5.7	0.50	1886	54	1882	58	1878	103	**100**
LSM-3b-seq2-b60	462	239	15	0.12	843	0.06750	2.8	0.53422	8.0	0.05740	7.5	0.35	421	11	435	29	507	164	83
2014-LSM-4T-a02	25745	81	9	0.31	6191	0.10856	2.0	0.92864	2.7	0.06204	1.9	0.72	664.37	12.45	667	13	676	41	**98**
2014-LSM-4T-a03	122184	24	17	1.31	32213	0.53785	1.8	14.12939	2.3	0.19053	1.3	0.81	2774.43	41.39	2758	22	2747	22	**101**
2014-LSM-4T-a04	1839	282	1	0.63	3072	0.00331	3.1	0.02843	12.4	0.06229	12.1	0.25	21.30	0.66	-	-	-	-	-
2014-LSM-4T-a06	172477	496	90	0.42	11017	0.17412	2.0	1.84343	2.2	0.07679	0.8	0.93	1034.76	19.41	1061	15	1116	16	93
2014-LSM-4T-a07	2991	21	2	0.52	2359	0.10195	1.9	0.86310	4.3	0.06140	3.9	0.45	625.81	11.62	632	21	653	83	96
2014-LSM-4T-a08	7287	50	8	0.45	10426	0.15970	1.7	1.55593	3.1	0.07066	2.6	0.53	955.10	14.84	953	20	948	54	**101**
2014-LSM-4T-a09	10876	257	9	0.32	21798	0.03410	2.4	0.23745	3.3	0.05051	2.2	0.74	216.13	5.20	216	6	218	51	**99**
2014-LSM-4T-a10	7114	232	15	0.47	12928	0.06532	1.8	0.49941	3.0	0.05545	2.4	0.61	407.88	7.23	411	10	431	53	95
2014-LSM-4T-a12	1761	24	3	1.85	630	0.07770	3.0	0.77824	5.9	0.07264	5.0	0.51	482.41	13.85	584	26	1004	102	48
2014-LSM-4T-a13	12078	94	21	0.77	13778	0.20463	2.5	2.50223	3.9	0.08869	3.0	0.63	1200.12	26.89	1273	29	1397	58	86
2014-LSM-4T-a14	1864	19	4	0.89	690	0.17568	1.8	1.83800	6.4	0.07588	6.2	0.29	1043.35	17.80	1059	43	1092	124	**96**
2014-LSM-4T-a15	48881	76	51	1.79	26987	0.48756	1.7	12.35802	2.5	0.18383	1.8	0.69	2560.10	37.00	2632	24	2688	30	**95**
2014-LSM-4T-a16	12024	234	19	0.67	4739	0.07592	2.9	0.59183	3.4	0.05654	1.8	0.86	471.72	13.40	472	13	474	39	**100**
2014-LSM-4T-a17	697	80	0	0.88	12	0.00182	7.9	0.19243	20.3	0.76719	18.7	0.39	11.72	0.92	-	-	-	-	-
2014-LSM-4T-a18	59584	607	85	0.62	6499	0.13436	2.0	1.73354	2.3	0.09357	1.1	0.88	812.71	15.60	1021	15	1500	20	54
2014-LSM-4T-a19	805	37	1	3.29	165	0.00626	8.7	0.31440	13.3	0.36408	10.1	0.65	40.25	3.50	-	-	-	-	-
2014-LSM-4T-a20	10629	41	10	0.47	12084	0.23948	2.7	2.93887	3.7	0.08900	2.6	0.71	1383.98	33.37	1392	29	1404	50	**99**
2014-LSM-4T-a23	7720	63	7	0.97	3618	0.09748	2.0	0.81545	3.7	0.06067	3.2	0.52	599.60	11.19	606	17	628	69	**96**
2014-LSM-4T-a24	25331	142	44	1.02	5324	0.28323	1.6	3.97016	3.2	0.10166	2.8	0.48	1607.60	22.21	1628	27	1655	53	**97**
2014-LSM-4T-a25	19033	258	18	0.64	9984	0.06402	1.4	0.48287	2.6	0.05470	2.2	0.55	400.03	5.56	400	9	400	48	**100**
2014-LSM-4T-a26	14000	369	12	0.63	2376	0.03060	1.7	0.23490	3.8	0.05567	3.4	0.44	194.32	3.21	214	7	439	75	44
2014-LSM-4T-a27	851	103	1	0.60	1645	0.00780	1.6	0.05670	6.6	0.05269	6.4	0.24	50.11	0.79	-	-	-	-	-
2014-LSM-4T-a28	776	98	2	0.47	1632	0.01601	3.8	0.10642	9.8	0.04820	9.1	0.39	102.40	3.87	103	10	109	214	**94**
2014-LSM-4T-a29	38897	294	64	0.21	29904	0.22104	2.0	2.61354	3.3	0.08575	2.6	0.60	1287.38	22.92	1304	24	1333	51	**97**
2014-LSM-4T-a30	23737	447	[illegible]	0.12	11658	0.15995	1.6	1.59068	3.3	0.07253	1.5	0.73	901.25	13.20	931	13	1001	30	**90**

sample and spot number	$^{207}Pb^a$ (cps)	U^b (ppm)	Pb^b (ppm)	Th^b U	$^{206}Pb^c$ ^{204}Pb	$^{206}Pb^c$ ^{238}U	2 σ %	$^{207}Pb^c$ ^{235}U	2 σ %	$^{207}Pb^c$ ^{206}Pb	2 σ %	$ρ^d$	^{206}Pb ^{238}U	2 σ (Ma)	^{207}Pb ^{235}U	2 σ (Ma)	^{207}Pb ^{206}Pb	2 σ (Ma)	Conc $\%^e$
2014-LSM-4T-a32	51867	1875	50	0.16	112	0.01421	1.6	0.38181	3.2	0.19481	2.7	0.51	90.99	1.46	-	-	-	-	-
2014-LSM-4T-a33	49254	155	51	0.65	9388	0.32336	2.3	4.84498	2.8	0.10867	1.6	0.83	1806.10	36.46	1793	24	1777	29	**102**
2014-LSM-4T-a34	63183	256	37	0.37	2828	0.13973	1.4	1.53999	2.0	0.07993	1.4	0.71	843.12	11.32	946	13	1195	28	71
2014-LSM-4T-a35	39065	533	30	0.13	1532	0.04787	2.7	0.44444	3.6	0.06733	2.4	0.74	301.45	7.86	373	11	848	50	36
2014-LSM-4T-a36	11097	111	18	0.99	2277	0.14317	1.9	1.39105	4.0	0.07047	3.6	0.47	862.54	15.25	885	24	942	73	**92**
2014-LSM-4T-a37	27046	307	23	0.74	1301	0.06548	2.6	0.50770	3.6	0.05623	2.5	0.72	408.89	10.18	417	12	462	54	89
2014-LSM-4T-a40	19916	142	9	0.57	7444	0.06289	1.5	0.49239	3.2	0.05679	2.8	0.47	393.17	5.71	407	11	483	62	81
2014-LSM-4T-a42	1151	243	1	0.91	822	0.00446	3.2	0.03561	8.2	0.05794	7.5	0.39	28.67	0.90	-	-	-	-	-
2014-LSM-4T-a43	48873	641	43	0.03	3015	0.07170	1.6	0.60371	1.9	0.06107	1.1	0.83	446.38	6.89	480	7	642	23	70
2014-LSM-4T-a44	9736	40	7	0.53	13259	0.17269	1.9	1.77227	3.0	0.07443	2.3	0.64	1026.94	18.47	1035	20	1053	47	**98**
2014-LSM-4T-a45	12010	192	18	0.35	7050	0.09253	3.0	0.81310	3.8	0.06373	2.3	0.79	570.50	16.37	604	17	733	50	78
2014-LSM-4T-a46	47989	50	36	1.60	15036	0.53502	2.7	14.26075	3.5	0.19332	2.2	0.77	2762.59	59.99	2767	33	2771	37	**100**
2014-LSM-4T-a47	138754	138	38	0.39	19060	0.26065	1.4	3.82152	1.6	0.10633	0.8	0.86	1493.17	18.64	1597	13	1738	15	86
2014-LSM-4T-a48	6579	120	15	1.34	8421	0.10215	2.5	0.82880	4.5	0.05885	3.7	0.56	626.98	15.11	613	21	561	81	112
2014-LSM-4T-a49	59189	205	63	0.37	55989	0.29567	1.9	4.29703	2.7	0.10540	2.0	0.69	1669.81	27.69	1693	23	1721	36	**97**
2014-LSM-4T-a50	14323	59	13	1.07	2677	0.17811	2.3	1.84975	3.4	0.07532	2.4	0.69	1056.66	22.44	1063	22	1077	49	**98**
2014-LSM-4T-a52	318	84	1	0.73	750	0.01667	2.9	0.09851	17.9	0.04285	17.7	0.16	106.59	3.12	-	-	-	-	-
2014-LSM-4T-a53	9881	86	30	1.99	11797	0.25652	1.8	2.99490	3.7	0.08468	3.2	0.49	1472.02	23.99	1406	29	1308	63	113
2014-LSM-4T-a54	72389	335	84	0.76	1868	0.23608	1.9	3.54436	2.3	0.10889	1.3	0.84	1366.26	23.76	1537	18	1781	23	77
2014-LSM-4T-a55	20012	275	26	0.84	1338	0.08995	2.0	0.82200	3.5	0.06628	2.9	0.56	555.23	10.39	609	16	815	60	68
2014-LSM-4T-a56	107181	263	65	0.57	4023	0.22926	3.0	3.07415	3.4	0.09725	1.7	0.87	1330.62	36.10	1426	27	1572	32	85
2014-LSM-4T-a57	10044	60	16	1.18	12231	0.22268	1.5	2.54722	2.3	0.08296	1.8	0.62	1296.02	17.15	1286	17	1268	36	**102**
2014-LSM-4T-a58	27548	123	34	0.48	29217	0.26279	1.9	3.43986	2.4	0.09494	1.6	0.77	1504.09	25.26	1514	19	1527	29	**99**
2014-LSM-4T-a59	81108	559	84	0.45	1529	0.15137	2.3	1.54144	3.0	0.07386	1.9	0.77	908.63	19.91	947	19	1038	39	88
2014-LSM-4T-b02	6236	70	5	1.14	5380	0.06338	1.6	0.47803	4.2	0.05470	3.9	0.38	396.17	6.26	397	14	400	88	**99**
2014-LSM-4T-b03	2176	18	3	0.72	3295	0.13349	2.5	1.22624	6.2	0.06662	5.7	0.40	807.73	19.14	813	35	826	119	**98**
2014-LSM-4T-b04	10785	213	10	1.09	2442	0.04022	2.0	0.32137	4.3	0.05795	3.8	0.46	254.19	4.96	283	11	528	84	48
2014-LSM-4T-b06	13408	154	9	0.19	7113	0.06415	1.7	0.48675	2.8	0.05503	2.2	0.60	400.79	6.48	403	9	414	50	**97**
2014-LSM-4T-b07	12579	277	22	1.31	5025	0.06880	1.4	0.52335	3.9	0.05517	3.6	0.36	428.94	5.81	427	14	419	81	**102**
2014-LSM-4T-b10	107703	969	93	0.22	1903	0.09679	8.2	1.27001	8.6	0.09517	2.6	0.95	595.56	46.72	832	50	1531	48	39
2014-LSM-4T-b12	63323	164	49	0.27	7392	0.29747	2.3	4.40833	2.9	0.10748	1.7	0.79	1678.74	33.82	1714	24	1757	32	**96**
2014-LSM-4T-b13	5516	37	5	1.07	9085	0.10425	2.1	0.87084	3.8	0.06058	3.1	0.56	639.27	12.94	636	18	625	67	**102**

sample and spot number	$^{207}Pb^a$ (cps)	U^b (ppm)	Pb^b (ppm)	$\frac{Th^b}{U}$	$\frac{^{206}Pb^c}{^{204}Pb}$	$\frac{^{206}Pb^c}{^{238}U}$	2 σ %	$\frac{^{207}Pb^c}{^{235}U}$	2 σ %	$\frac{^{207}Pb^c}{^{206}Pb}$	2 σ %	ρ^d -	$\frac{^{206}Pb}{^{238}U}$	2 σ (Ma)	$\frac{^{207}Pb}{^{235}U}$	2 σ (Ma)	$\frac{^{207}Pb}{^{206}Pb}$	2 σ (Ma)	Conc %^{e} -
2014-LSM-4T-b14	16801	60	10	0.67	2421	0.16863	1.3	1.76721	2.5	0.07601	2.1	0.51	1004.54	11.64	1033	16	1095	42	**92**
2014-LSM-4T-b15	24298	86	24	0.56	26499	0.26163	1.7	3.33717	2.4	0.09251	1.7	0.70	1498.16	22.96	1490	19	1478	33	**101**
2014-LSM-4T-b16	19595	102	28	0.65	1480	0.25260	2.3	3.23831	4.0	0.09298	3.3	0.57	1451.86	29.74	1466	32	1487	62	**98**
2014-LSM-4T-b17	7079	65	12	0.31	5785	0.18442	2.5	1.91041	4.0	0.07513	3.1	0.63	1091.09	25.19	1085	27	1072	63	**102**
2014-LSM-4T-b18	3439	251	10	0.77	6757	0.03717	2.0	0.26020	4.8	0.05078	4.4	0.42	235.24	4.62	235	10	231	101	**102**
2014-LSM-4T-b19	35010	51	11	0.89	26352	0.20166	1.8	2.25583	2.2	0.08113	1.4	0.78	1184.25	19.06	1199	16	1225	27	**97**
2014-LSM-4T-b20	56990	380	61	0.06	1780	0.16881	3.0	1.73830	3.5	0.07468	1.7	0.87	1005.54	28.02	1023	23	1060	34	**95**
2014-LSM-4T-b22	8070	249	13	0.12	2980	0.05677	1.7	0.41916	2.7	0.05355	2.1	0.64	355.98	6.06	355	8	352	48	**101**
2014-LSM-4T-b23	968	589	3	0.91	865	0.00475	2.0	0.03052	6.2	0.04664	5.9	0.33	30.52	0.62	31	2	31	141	**99**
2014-LSM-4T-b24	18999	508	27	0.32	6687	0.05351	2.1	0.44361	3.3	0.06012	2.6	0.63	336.05	6.86	373	10	608	56	55
2014-LSM-4T-b25	3319	105	3	0.87	6655	0.03061	1.9	0.21122	4.0	0.05005	3.5	0.48	194.35	3.66	195	7	197	82	**99**
2014-LSM-4T-b26	94456	251	50	0.42	3859	0.18883	2.6	2.50586	2.9	0.09625	1.4	0.87	1115.02	26.26	1274	22	1553	27	72
2014-LSM-4T-b27	6535	61	7	1.38	10923	0.09820	1.9	0.81577	4.7	0.06025	4.3	0.41	603.84	11.11	606	22	613	93	**99**
2014-LSM-4T-b28	6178	311	6	0.85	6118	0.01698	1.6	0.11750	3.2	0.05018	2.7	0.52	108.56	1.77	113	3	203	63	53
2014-LSM-4T-b29	30147	103	12	0.85	29348	0.08542	5.4	1.22296	5.9	0.10383	2.5	0.91	528.42	27.37	811	34	1694	46	31
2014-LSM-4T-b30	4303	40	4	1.24	7230	0.09791	2.2	0.81242	5.3	0.06018	4.8	0.42	602.16	12.71	604	24	610	104	**99**
2014-LSM-4T-b32	18018	167	17	1.01	3815	0.09637	1.1	0.79962	3.0	0.06018	2.8	0.38	593.11	6.49	597	14	610	61	**97**
2014-LSM-4T-b33	10045	159	10	1.02	7380	0.05387	2.6	0.42296	3.8	0.05695	2.8	0.68	338.22	8.58	358	12	489	62	69
2014-LSM-4T-b34	56353	240	48	1.04	5259	0.17038	1.9	1.73833	2.8	0.07400	2.1	0.68	1014.21	17.93	1023	18	1041	41	**97**
2014-LSM-4T-b35	127800	267	44	0.29	141412	0.16245	1.5	1.62804	1.8	0.07268	0.9	0.85	970.40	13.62	981	11	1005	19	**97**
2014-LSM-4T-b36	12342	138	6	0.56	3924	0.03857	2.5	0.30163	3.7	0.05672	2.7	0.69	243.97	6.10	268	9	481	59	51
2014-LSM-4T-b38	66151	231	79	0.93	65964	0.28979	2.4	4.03856	2.9	0.10107	1.6	0.83	1640.50	34.56	1642	24	1644	30	**100**
2014-LSM-4T-b39	6488	113	11	0.43	10971	0.09748	4.9	0.80197	7.2	0.05967	5.3	0.68	599.63	27.99	598	33	592	115	**101**
2014-LSM-4T-b40	275	91	1	0.72	428	0.00531	2.8	0.04750	29.9	0.06484	29.8	0.09	34.16	0.94	-	-	-	-	-
2014-LSM-4T-b43	75183	333	36	0.49	9683	0.10233	1.5	0.85810	1.9	0.06082	1.1	0.81	628.03	9.06	629	9	633	24	**99**
2014-LSM-4T-b44	26210	195	13	0.48	8047	0.06516	1.3	0.49716	2.1	0.05534	1.6	0.62	406.91	5.04	410	7	426	36	**96**
2014-LSM-4T-b45	10135	279	26	0.72	3013	0.08768	2.2	0.79125	4.5	0.06545	3.9	0.49	541.78	11.33	592	20	789	82	69
2014-LSM-4T-b46	3384	147	11	0.37	3640	0.07324	3.0	0.56671	4.4	0.05612	3.2	0.69	455.65	13.37	456	16	457	71	**100**
2014-LSM-4T-b47	3911	64	5	0.98	7261	0.06246	2.3	0.47173	4.0	0.05477	3.2	0.59	390.59	8.86	392	13	403	72	**97**
2014-LSM-4T-b48	12323	123	14	1.21	4673	0.09791	1.7	0.81701	2.5	0.06052	1.8	0.69	602.14	9.96	606	12	622	39	**97**
2014-LSM-4T-b49	16073	172	20	1.36	10263	0.09682	1.7	0.80112	2.8	0.06001	2.3	0.59	595.74	9.45	597	13	604	50	**99**

sample and spot number	$^{207}Pb^a$ (cps)	U^b (ppm)	Pb^b (ppm)	$\underline{Th^b}$ U	$\underline{^{206}Pb^c}$ ^{204}Pb	$\underline{^{206}Pb^c}$ ^{238}U	2 σ %	$\underline{^{207}Pb^c}$ ^{235}U	2 σ %	$\underline{^{207}Pb^c}$ ^{206}Pb	2 σ %	$ρ^d$ -	$\underline{^{206}Pb}$ ^{238}U	2 σ (Ma)	$\underline{^{207}Pb}$ ^{235}U	2 σ (Ma)	$\underline{^{207}Pb}$ ^{206}Pb	2 σ (Ma)	Conc %^{e} -
2014-LSM-4T-b53	27619	568	24	0.12	1068	0.03479	1.9	0.32653	3.3	0.06807	2.7	0.57	220.48	4.13	287	8	871	57	25
2014-LSM-4T-b54	20715	353	27	0.58	36997	0.07416	2.4	0.57593	2.9	0.05632	1.6	0.83	461.17	10.84	462	11	465	36	**99**
2014-LSM-4T-b55	8117	399	11	0.18	15631	0.01905	2.3	0.13843	5.5	0.05269	5.0	0.41	121.68	2.76	132	7	316	114	39
2014-LSM-4T-b56	9603	188	11	0.43	16160	0.05531	1.7	0.40949	2.5	0.05369	1.9	0.69	347.06	5.91	349	8	358	42	**97**
2014-LSM-4T-b57	24190	326	37	0.22	31441	0.11154	6.3	1.19319	7.7	0.07758	4.5	0.82	681.71	40.97	797	44	1136	89	60
2014-LSM-4T-b58	15762	222	18	0.63	28211	0.07279	1.6	0.56493	2.3	0.05629	1.6	0.71	452.92	7.15	455	8	464	36	**98**
2014-LSM-4T-b59	13223	477	16	0.44	7477	0.03318	1.6	0.24596	2.4	0.05376	1.8	0.67	210.43	3.28	223	5	361	40	58
2014-LSM-4T-b60	-	-	-	-	-	-	-	-	-	-	-	-	-	-	-	-	-	-	-

6.3 Supplementary material chapter 4

6.3.1 SI1 Raw U-Pb ratios and calculated ages

U-Pb ratios and calculated ages for all samples (except for those that have already been published in KRAUTZ ET AL. (2018a), which are part of the set Colorado Plateau).

[a] within-run background-corrected mean ^{207}Pb signal in counts per second

[b] U and Pb content and Th/U ratio were calculated relative to GJ-1 and are accurate to approximately 10%.

[c] corrected for background, mass bias, laser induced U-Pb fractionation and common Pb (if detectable, see analytical method) using Stacey & Kramers (1975) model Pb composition. $^{207}Pb/^{235}U$ calculated using $^{207}Pb/^{206}Pb/(^{238}U/^{206}Pb \times 1/137.88)$. Errors are propagated by quadratic addition oif within-run errors (2SE) and the reproducibility of GJ-1 (2SD).

[d] Rho is the error correlation defined as $err^{206}Pb/^{238}U/err^{207}Pb/^{235}U$.

[e] conc % printed in bold indicate grains assumed to be concordant.

The samples are grouped as to the sets.

Sample set "Great Basin"

DDS3

spot number	207Pb[a] (cps)	U[b] (ppm)	Pb[b] (ppm)	Th[b]/U	206Pb[c]/204Pb	206Pb[c]/238U	2 s %	207Pb[c]/235U	2 s %	207Pb[c]/206Pb	2 s %	rho[d]	206Pb/238U (Ma)	2 s (Ma)	207Pb/235U (Ma)	2 s (Ma)	207Pb/206Pb (Ma)	2 s (Ma)	conc %[e]
A-89	1349	567	3	0.47	0	0.00522	4.6	0.03379	9	0.04698	7.7	0.52	34	2	34	3	48	183	70
A-90	1007	578	3	0.84	6	0.00477	5.1	0.03156	18.3	0.04801	18	0.28	31	2	32	6	99	417	31
A-91	64962	293	75	0.49	0	0.2507	4.4	3.139	4.9	0.09082	2	0.91	1442	57	1442	38	1442	38	**100**
A-92	40791	402	31	0.63	0	0.07339	7.2	1.169	8	0.1156	3.3	0.91	457	32	786	44	1888	60	24
A-93	1713	475	3	0.57	0	0.00672	4.7	0.0483	8.7	0.05216	7.3	0.54	43	2	48	4	292	167	15
A-94	2560	553	3	1.01	3	0.00507	4.8	0.04912	11.8	0.07025	11	0.41	33	2	49	6	935	221	3
A-95	1240	907	5	0.86	0	0.00507	5.2	0.03525	9.2	0.05042	7.6	0.56	33	2	35	3	214	177	15
A-96	691	605	3	0.43	1	0.00529	5.2	0.03185	10.7	0.04366	9.4	0.49	34	2	32	3	-130	232	-26
A-101	1678	665	3	0.24	0	0.00518	4.6	0.03587	7.2	0.05024	5.5	0.65	33	2	36	3	205	128	16
A-102	1370	511	2	0.96	1	0.00463	4.8	0.03426	11.5	0.05364	10	0.42	30	1	34	4	355	235	8
A-103	5868	2068	12	3.28	0	0.00586	4.6	0.04151	6.9	0.05143	5.1	0.67	38	2	41	3	259	116	15
A-104	782	294	1	1.29	3	0.00444	5.9	0.02753	20.9	0.04497	20	0.28	29	2	28	6	-58	489	-49
A-105	370	190	1	0.7	2	0.00449	6	0.01892	23.8	0.0306	23	0.25	29	2	19	4	-1099	697	-3
A-106	1148	395	2	0.61	1	0.00518	5.1	0.03684	11.9	0.05157	11	0.43	33	2	37	4	266	245	13
A-107	1110	720	4	0.84	1	0.00569	5.3	0.04439	10.5	0.05659	9.1	0.51	37	2	44	5	475	201	8
A-108	35524	162	38	0.42	0	0.2325	4.9	2.853	5.3	0.08904	2.1	0.92	1348	59	1370	40	1404	40	**96**
A-109	775	248	1	0.55	2	0.00529	5.9	0.03536	17.1	0.04846	16	0.34	34	2	35	6	121	379	28
A-110	724	283	1	0.41	1	0.0046	4.7	0.03468	12.8	0.05467	12	0.37	30	1	35	4	398	267	7
A-111	368	340	2	1.33	1	0.00518	5.7	0.03075	20.2	0.04306	19	0.28	33	2	31	6	-165	483	-20
A-112	11422	99	14	0.43	0	0.1467	5	1.414	6.1	0.06994	3.5	0.81	882	41	895	36	926	73	**95**
A-113	1646	537	3	1.01	1	0.00609	5.3	0.04242	11.1	0.05051	9.7	0.48	39	2	42	5	218	224	18
A-114	815	309	5	0.41	1	0.01499	5.6	0.0993	17.3	0.04805	16	0.32	96	5	96	16	101	388	**95**
A-115	363	175	1	0.9	1	0.00506	6	0.03252	21.9	0.04667	21	0.27	33	2	32	7	32	504	**103**
A-116	540	270	1	1.26	1	0.00429	5.4	0.02361	21.5	0.03995	21	0.25	28	1	24	5	-355	539	-8
A-117	11192	367	10	0.33	0	0.06235	4.3	0.5062	5.2	0.05852	3.0	0.83	392	16	416	18	540	63	71

spot number	$^{207}Pb^a$ (cps)	U^b (ppm)	Pb^b (ppm)	$\frac{Th^b}{U}$	$\frac{^{206}Pb^c}{^{204}Pb}$	$\frac{^{206}Pb^c}{^{238}U}$	2 s %	$\frac{^{207}Pb^c}{^{235}U}$	2 s %	$\frac{^{207}Pb^c}{^{206}Pb}$	2 s %	rho^d	$\frac{^{206}Pb}{^{238}U}$	2 s (Ma)	$\frac{^{207}Pb}{^{235}U}$	2 s (Ma)	$\frac{^{207}Pb}{^{206}Pb}$	2 s (Ma)	conc %^{e}
A-118	192	247	1	0.92	4	0.00493	7.1	0.01338	111.7	0.01969	110	0.06	32	2	13	15	-2706	4983	-1
A-119	654	788	4	0.37	2	0.00487	4.9	0.02776	23.6	0.04134	23	0.21	31	2	28	6	-267	586	-12
A-120	33111	281	45	0	0	0.1605	4.5	1.61	4.9	0.07279	1.9	0.92	959	41	974	31	1008	38	**95**
A-125	275	118	1	0.52	2	0.00448	6.5	0.0239	34.2	0.03872	34	0.19	29	2	24	8	-437	883	-7
A-126	816	472	2	1.61	2	0.00445	7.4	0.02323	20.1	0.03788	19	0.37	29	2	23	5	-495	497	-6
A-127	1758	628	3	0.48	0	0.00529	4.9	0.03756	8	0.05155	6.4	0.61	34	2	37	3	265	146	13
A-128	498	296	1	0.79	1	0.00479	5	0.03082	23.2	0.04668	23	0.22	31	2	31	7	32	542	**96**
A-129	1687	559	3	0.83	1	0.00526	4.8	0.03693	13.8	0.05098	13	0.35	34	2	37	5	239	299	14
A-130	5666	2237	12	0.49	0	0.00541	4.6	0.03609	6.5	0.04843	4.6	0.7	35	2	36	2	120	109	29
A-131	61746	383	83	0.44	0	0.2139	4.9	2.471	5.6	0.08381	2.8	0.87	1250	56	1264	41	1288	55	**97**
A-132	886	385	2	0.78	2	0.00449	5.1	0.03077	13.5	0.0497	13	0.38	29	1	31	4	180	292	16
A-133	2396	2916	15	0.4	0	0.00535	4.8	0.03382	10.6	0.04589	9.5	0.45	34	2	34	4	-9	230	-396
A-134	1772	921	4	0.87	0	0.00403	4.5	0.02591	9	0.04664	7.8	0.5	26	1	26	2	30	188	86
A-135	1816	486	2	0.38	3	0.00509	4.9	0.03835	10.8	0.05463	9.6	0.45	33	2	38	4	396	215	8
A-136	4110	1442	9	2.39	0	0.00609	4.8	0.04501	7.9	0.05361	6.3	0.6	39	2	45	3	354	142	11
A-137	1436	517	3	0.45	0	0.00564	5.1	0.03956	10.9	0.0509	9.7	0.46	36	2	39	4	236	224	15
A-138	2478	977	5	0.53	0	0.00559	4.6	0.03581	9	0.0465	7.7	0.51	36	2	36	3	23	185	157
A-139	891	406	2	1.13	0	0.00491	5.3	0.03122	13.7	0.04617	13	0.39	32	2	31	4	6	302	538
A-140	11045	4402	23	0.22	0	0.00539	4.7	0.03509	5.7	0.04721	3.3	0.82	35	2	35	2	59	78	59
A-141	4522	1888	9	0.39	0	0.00507	4.4	0.03421	5.8	0.04895	3.8	0.76	33	1	34	2	145	89	23
A-142	1135	434	2	0.8	1	0.00558	4.9	0.03639	9.6	0.04732	8.2	0.51	36	2	36	3	65	195	56
A-143	2163	945	5	0.47	0	0.00497	4.4	0.03038	6.2	0.04436	4.4	0.71	32	1	30	2	-91	107	-35
A-144	1136	520	2	0.81	0	0.0049	4.6	0.02964	13.3	0.04392	12	0.34	31	1	30	4	-116	307	-27
A-149	18012	6767	36	1.86	0	0.00537	4.5	0.03588	5.6	0.04848	3.4	0.8	35	2	36	2	122	80	28
A-150	9662	3713	19	0.42	0	0.00529	4.4	0.03535	5.6	0.04851	3.4	0.79	34	1	35	2	123	81	28
A-151	2521	989	5	0.7	0	0.00492	4.6	0.0334	7.1	0.04929	5.4	0.65	32	1	33	2	161	126	20
A-152	21392	556	38	0.64	0	0.06872	4.9	0.5258	5.4	0.05551	2.3	0.9	428	20	429	19	432	51	**99**

spot number	$^{207}Pb^a$ (cps)	U^b (ppm)	Pb^b (ppm)	$\frac{Th^b}{U}$	$\frac{^{206}Pb^c}{^{204}Pb}$	$\frac{^{206}Pb^c}{^{238}U}$	2 s %	$\frac{^{207}Pb^c}{^{235}U}$	2 s %	$\frac{^{207}Pb^c}{^{206}Pb}$	2 s %	rho^d	$\frac{^{206}Pb}{^{238}U}$ (Ma)	2 s (Ma)	$\frac{^{207}Pb}{^{235}U}$ (Ma)	2 s (Ma)	$\frac{^{207}Pb}{^{206}Pb}$ (Ma)	2 s (Ma)	conc %^{e}
A-153	1557	491	3	0.78	0	0.00574	5.1	0.04436	8.7	0.05607	7.1	0.58	37	2	44	4	455	157	8
A-154	2157	410	2	0.99	26	0.0047	8.5	0.04493	34.5	0.06932	33	0.25	30	3	45	15	908	689	3
A-155	271	361	2	0.61	1	0.00493	8.3	0.02829	56.8	0.04166	56	0.15	32	3	28	16	-248	1421	-13
A-156	910	256	1	0.72	3	0.00422	5.1	0.02712	27.1	0.04666	27	0.19	27	1	27	7	31	639	87
A-157	1717	698	4	0.53	0	0.00533	4.5	0.03663	8	0.04986	6.6	0.56	34	2	37	3	188	155	18
A-158	1596	544	3	0.73	0	0.00576	4.7	0.03918	8.2	0.04938	6.7	0.57	37	2	39	3	165	157	22
A-159	1302	621	3	1.7	0	0.00495	4.8	0.03078	12.7	0.04511	12	0.38	32	2	31	4	-50	286	-63
A-160	3385	4148	8	1.69	0	0.00185	4.5	0.01193	6.5	0.04668	4.7	0.69	12	1	12	1	32	113	37
A-161	3551	1584	8	0.66	0	0.00501	4.4	0.03288	6.2	0.04764	4.4	0.71	32	1	33	2	81	105	40
A-162	465	202	1	2.22	2	0.00455	5.4	0.03278	17.6	0.05227	17	0.31	29	2	33	6	297	382	10
A-163	2347	1227	6	0.42	0	0.00499	4.5	0.03804	7.2	0.05533	5.6	0.63	32	1	38	3	425	124	8
A-164	1049	485	2	0.59	0	0.00478	4.8	0.0289	12.5	0.04386	11	0.39	31	1	29	4	-119	283	-26
A-165	538	260	1	0.95	1	0.00472	5.4	0.02761	17.3	0.04243	16	0.31	30	2	28	5	-202	413	-15
A-166	1952	67	5	0.72	0	0.06897	4.7	0.5255	9.1	0.05528	7.9	0.51	430	19	429	32	423	176	102
A-167	609	281	1	0.63	1	0.00499	5.5	0.02852	11.4	0.04143	10	0.49	32	2	29	3	-262	253	-12
A-168	241	215	1	0.34	2	0.00523	5.9	0.02232	37.4	0.03097	37	0.16	34	2	22	8	-1063	1112	-3
A-173	1506	448	2	0.76	2	0.00509	5.4	0.03278	13.4	0.0467	12	0.4	33	2	33	4	33	294	99
A-174	753	282	2	1.83	1	0.00582	5.1	0.03388	13.4	0.04224	12	0.38	37	2	34	4	-213	310	-18
A-175	3200	431	7	0.44	0	0.01588	4.7	0.1087	7.4	0.04968	5.7	0.63	102	5	105	7	179	134	57
A-176	474	225	1	0.58	1	0.00462	5.8	0.02845	17.2	0.04465	16	0.34	30	2	28	5	-75	397	-40
A-177	577	406	2	0.77	2	0.00464	6.2	0.02654	28.3	0.0415	28	0.22	30	2	27	7	-257	700	-12
A-178	1365	397	3	0.74	1	0.00699	5.6	0.05005	11.2	0.05192	9.7	0.5	45	3	50	5	281	223	16
A-179	2561	1157	5	0.54	0	0.00482	4.9	0.03153	8.1	0.04742	6.4	0.61	31	2	32	3	70	152	45
A-180	1668	954	5	1.81	1	0.00561	5	0.04624	10.2	0.05979	8.8	0.49	36	2	46	5	596	192	6
A-181	3866	1740	9	0.46	0	0.0052	4.4	0.03385	6.7	0.04726	5	0.66	33	1	34	2	62	119	54
A-182	525	190	1	0.83	4	0.00447	4.9	0.02333	27.4	0.03785	27	0.18	29	1	23	6	-496	718	-6

spot number	$^{207}Pb^a$ (cps)	U^b (ppm)	Pb^b (ppm)	$\frac{Th^b}{U}$	$\frac{^{206}Pb^c}{^{204}Pb}$	$\frac{^{206}Pb^c}{^{238}U}$	2 s %	$\frac{^{207}Pb^c}{^{235}U}$	2 s %	$\frac{^{207}Pb^c}{^{206}Pb}$	2 s %	rho^d	$\frac{^{206}Pb}{^{238}U}$	2 s (Ma)	$\frac{^{207}Pb}{^{235}U}$	2 s (Ma)	$\frac{^{207}Pb}{^{206}Pb}$	2 s (Ma)	conc %^{e}
A-184	1026	518	2	0.7	0	0.00456	4.7	0.02796	13.4	0.04445	13	0.35	29	1	28	4	-87	309	-34
A-185	4883	131	9	1	0	0.06837	4.6	0.5272	7.2	0.05594	5.5	0.65	426	19	430	25	449	122	**95**
A-186	1205	419	2	0.47	0	0.00607	5.5	0.03987	8.1	0.04767	6	0.68	39	2	40	3	82	141	47
A-187	1446	450	3	0.57	1	0.00615	5.1	0.0389	10.4	0.04592	9.1	0.49	39	2	39	4	-7	220	-557
A-188	14128	46	15	0.54	0	0.3111	4.5	4.463	5.3	0.1041	2.9	0.84	1746	69	1724	44	1698	53	**103**

DDS4

spot number	$^{207}Pb^a$ (cps)	U^b (ppm)	Pb^b (ppm)	$\frac{Th^b}{U}$	$\frac{^{206}Pb^c}{^{204}Pb}$	$\frac{^{206}Pb^c}{^{238}U}$	2 s %	$\frac{^{207}Pb^c}{^{235}U}$	2 s %	$\frac{^{207}Pb^c}{^{206}Pb}$	2 s %	rho^d	$\frac{^{206}Pb}{^{238}U}$	2 s (Ma)	$\frac{^{207}Pb}{^{235}U}$	2 s (Ma)	$\frac{^{207}Pb}{^{206}Pb}$	2 s (Ma)	conc %^{e}
A-5	649	265	1	0.43	1	0.00428	5.2	0.02977	12.5	0.05046	11	0.42	28	1	30	4	216	263	13
A-6	1648	567	3	0.44	0	0.00552	4.4	0.04087	7.7	0.05368	6.3	0.57	36	2	41	3	357	143	10
A-7	1016	427	2	0.4	0	0.00495	4.7	0.03106	12.5	0.0455	12	0.38	32	1	31	4	-29	281	-109
A-8	912	412	2	0.42	0	0.00507	4.8	0.03361	13.1	0.04809	12	0.37	33	2	34	4	103	289	32
A-9	2676	1094	6	0.44	0	0.0054	4.3	0.03426	7.4	0.04602	6	0.58	35	1	34	2	-2	146	-1602
A-10	931	305	2	0.51	1	0.00548	4.7	0.03555	12	0.04704	11	0.39	35	2	35	4	51	263	70
A-11	1476	623	3	0.61	0	0.00528	4.7	0.03307	9.1	0.04543	7.8	0.51	34	2	33	3	-33	190	-102
A-12	2430	875	5	0.69	1	0.00541	4.9	0.03894	8.5	0.05222	6.9	0.58	35	2	39	3	294	157	12
A-13	2626	406	2	0.74	8	0.00546	4.7	0.0511	14.1	0.06788	13	0.33	35	2	51	7	864	276	4
A-14	980	445	2	0.38	0	0.00486	4.6	0.02799	12.6	0.04179	12	0.37	31	1	28	3	-240	297	-13
A-15	896	557	3	0.49	0	0.00475	5.3	0.02785	15	0.0425	14	0.35	31	2	28	4	-198	353	-15
A-16	863	386	2	0.97	2	0.00539	4.7	0.03907	16.5	0.05265	16	0.28	35	2	39	6	313	360	11
A-17	1251	1067	5	0.66	0	0.00492	4.3	0.03003	9.4	0.04425	8.4	0.46	32	1	30	3	-97	205	-32
A-18	2681	1140	5	0.87	0	0.00475	4.7	0.03343	9.4	0.05103	8.1	0.5	31	1	33	3	241	187	13
A-19	2965	1126	6	0.38	0	0.00551	4.9	0.03615	7.4	0.0476	5.6	0.66	35	2	36	3	79	132	45
A-20	1231	709	3	0.71	0	0.00444	4.6	0.02711	11	0.04427	10	0.42	29	1	27	3	-97	245	-30
A-21	146926	569	152	0.52	0	0.262	4.7	3.369	5	0.09331	1.7	0.94	1500	63	1497	39	1494	33	**100**
A-22	692	247	1	0.45	1	0.0056	5.9	0.03555	21.2	0.04609	20	0.28	36	2	35	7	2	491	2066
A-23	1022	372	2	0.62	1	0.00509	4.8	0.03217	17.2	0.04581	17	0.28	33	2	32	5	-13	400	-255
A-24	846	379	2	0.61	1	0.00482	4.9	0.02823	13.2	0.0425	12	0.37	31	2	28	4	-197	307	-16

spot number	$^{207}Pb^a$ (cps)	U^b (ppm)	Pb^b (ppm)	$\underline{Th^b}$ U	$\underline{^{206}Pb^c}$ ^{204}Pb	$\underline{^{206}Pb^c}$ ^{238}U	2 s %	$\underline{^{207}Pb^c}$ ^{235}U	2 s %	$\underline{^{207}Pb^c}$ ^{206}Pb	2 s %	rho^d	$\underline{^{206}Pb}$ ^{238}U	2 s (Ma)	$\underline{^{207}Pb}$ ^{235}U	2 s (Ma)	$\underline{^{207}Pb}$ ^{206}Pb	2 s (Ma)	conc %^{e}
A-29	3095	540	3	1.59	6	0.0059	4.7	0.06845	12.2	0.08414	11	0.38	38	2	67	8	1295	220	3
A-30	1001	545	3	0.52	0	0.00562	4.6	0.03624	11.5	0.04679	11	0.4	36	2	36	4	38	253	**96**
A-31	1427	518	3	0.45	0	0.00522	4.9	0.03654	9.3	0.05077	7.9	0.52	34	2	36	3	230	183	15
A-32	4529	1655	8	0.28	0	0.00509	4.6	0.03124	6.6	0.04454	4.8	0.69	33	1	31	2	-81	116	-40
A-33	889	331	2	0.37	0	0.00506	5.7	0.03493	12	0.05011	11	0.47	33	2	35	4	199	245	16
A-34	339	163	1	0.25	1	0.00516	5	0.02144	32	0.03017	32	0.15	33	2	22	7	-1143	969	-3
A-35	749	323	2	0.43	1	0.00508	5	0.02858	16.9	0.04086	16	0.29	33	2	29	5	-297	412	-11
A-36	643	305	1	0.38	1	0.005	5.5	0.02411	23.8	0.03496	23	0.23	32	2	24	6	-712	645	-5
A-37	3504	22	5	0.42	0	0.2064	4.6	2.171	7	0.07629	5.3	0.65	1210	50	1172	49	1102	106	**110**
A-38	2986	1504	8	0.82	0	0.00567	4.3	0.03867	6.5	0.04946	4.9	0.66	36	2	39	2	169	115	22
A-39	1155	373	2	0.81	0	0.00559	5.2	0.04091	10.6	0.0531	9.2	0.49	36	2	41	4	332	209	11
A-40	1680	566	3	0.53	0	0.00557	4.5	0.04075	8.2	0.05313	6.8	0.55	36	2	41	3	334	154	11
A-41	1029	367	2	0.57	0	0.00548	5.1	0.03924	11.8	0.05198	11	0.44	35	2	39	5	284	243	12
A-42	850	328	2	0.58	0	0.00551	5.2	0.0326	13.2	0.04296	12	0.4	35	2	33	4	-170	302	-21
A-43	715	284	1	0.62	1	0.00514	5.5	0.03239	14.7	0.04571	14	0.37	33	2	32	5	-19	330	-178
A-44	83	100	0	0.86	10	0.00131	11.7	0.00136	958.9	0.00754	960	0.01	8	1	1	13	-21018	2E+06	0
A-45	919	467	3	0.55	0	0.00663	5.1	0.0457	12.9	0.05001	12	0.39	43	2	45	6	195	276	22
A-46	454	193	1	0.3	1	0.00491	6.4	0.02857	25.1	0.04219	24	0.26	32	2	29	7	-216	611	-15
A-47	743	322	2	0.42	1	0.00497	5.5	0.02879	19.1	0.04201	18	0.29	32	2	29	5	-227	460	-14
A-48	403	11	1	6.47	1	0.1079	5.4	0.9427	18.5	0.06336	18	0.29	661	34	674	91	720	375	**92**
A-53	1235	436	2	1.44	0	0.00576	5.1	0.03803	10.2	0.04793	8.8	0.51	37	2	38	4	95	208	39
A-54	1657	633	4	0.24	0	0.00594	4.9	0.03547	11.1	0.04334	9.9	0.45	38	2	35	4	-149	246	-26
A-55	1609	632	3	0.51	0	0.00551	4.5	0.03551	8.2	0.04678	6.9	0.55	35	2	35	3	37	164	**95**
A-56	2133	799	4	0.42	0	0.00513	4.9	0.03466	8.3	0.04902	6.7	0.59	33	2	35	3	148	157	22
A-57	1708	718	4	0.47	0	0.00562	4.6	0.03272	9.8	0.04223	8.7	0.47	36	2	33	3	-214	219	-17
A-58	3327	425	7	0.57	0	0.01693	4.8	0.1126	7.6	0.04823	5.9	0.63	108	5	108	8	110	139	**98**

spot number	$^{207}Pb^a$ (cps)	U^b (ppm)	Pb^b (ppm)	$\underline{Th^b}$ U	$\underline{^{206}Pb^c}$ ^{204}Pb	$\underline{^{206}Pb^c}$ ^{238}U	2 s %	$\underline{^{207}Pb^c}$ ^{235}U	2 s %	$\underline{^{207}Pb^c}$ ^{206}Pb	2 s %	rho^d	$\underline{^{206}Pb}$ ^{238}U	2 s (Ma)	$\underline{^{207}Pb}$ ^{235}U	2 s (Ma)	$\underline{^{207}Pb}$ ^{206}Pb	2 s (Ma)	conc %^{c}
A-60	1466	549	3	0.38	0	0.00535	4.8	0.03536	9.2	0.04792	7.8	0.53	34	2	35	3	95	185	36
A-61	1523	614	3	0.31	0	0.00513	5.1	0.02995	8.1	0.04235	6.3	0.62	33	2	30	2	-206	159	-16
A-62	1698	1281	6	0.56	0	0.0051	5	0.03287	10.9	0.04674	9.7	0.46	33	2	33	4	35	232	**93**
A-63	2326	752	4	0.41	0	0.00546	5	0.04151	7.2	0.05516	5.1	0.7	35	2	41	3	418	115	8
A-64	2569	1066	5	0.51	0	0.00526	4.7	0.03287	8.4	0.04532	6.9	0.56	34	2	33	3	-39	168	-86
A-65	842	368	2	0.8	1	0.00478	5.2	0.02641	12.6	0.0401	11	0.42	31	2	26	3	-345	294	-9
A-66	2044	857	4	0.87	0	0.00498	5.9	0.03203	11	0.04668	9.2	0.54	32	2	32	3	32	221	**100**
A-67	962	376	2	0.54	0	0.00565	4.8	0.03774	12.1	0.04848	11	0.4	36	2	38	4	122	261	30
A-68	983	294	2	0.54	1	0.00579	4.9	0.0372	13.7	0.04666	13	0.36	37	2	37	5	31	306	120
A-69	988	352	2	0.32	1	0.00618	5	0.03921	14.2	0.04606	13	0.35	40	2	39	5	0	320	12989
A-70	92901	1661	47	0.56	16	0.02837	5.7	0.2582	16.9	0.06601	16	0.34	180	10	233	35	806	333	22
A-71	969	328	2	0.4	0	0.00629	4.8	0.04482	11.3	0.05169	10	0.42	40	2	45	5	271	235	15
A-72	1764	849	5	0.81	1	0.00607	4.7	0.04186	9.8	0.05006	8.6	0.47	39	2	42	4	197	200	20
A-77	1351	359	2	1.17	1	0.00526	4.9	0.04403	12.7	0.0607	12	0.39	34	2	44	5	628	251	5
A-78	19093	68	19	0.5	0	0.2754	4.6	3.639	5.7	0.09585	3.4	0.81	1568	64	1558	46	1544	64	**102**
A-79	696	273	1	0.5	1	0.00532	5.3	0.03053	15.5	0.04161	15	0.35	34	2	31	5	-251	367	-14
A-80	2984	872	5	1.6	1	0.00578	4.5	0.0469	7.5	0.05885	6	0.6	37	2	47	3	561	131	7
A-81	1831	403	3	0.92	2	0.0065	5.2	0.05785	13.5	0.06453	12	0.39	42	2	57	8	759	263	6
A-82	855	341	2	0.46	1	0.00559	4.9	0.03818	14.6	0.04954	14	0.33	36	2	38	5	173	322	21
A-83	1778	590	4	0.48	0	0.00613	4.7	0.04258	8.5	0.0504	7.1	0.55	39	2	42	4	213	164	18
A-84	1750	739	4	0.5	0	0.00509	4.5	0.0305	7.8	0.0435	6.4	0.57	33	1	31	2	-140	159	-23
A-85	1133	359	2	1.16	2	0.00509	5.2	0.03548	17.2	0.05054	16	0.3	33	2	35	6	219	380	15
A-86	1250	575	3	0.34	1	0.00536	5.4	0.02944	13.3	0.03987	12	0.41	34	2	29	4	-360	313	-10
A-87	1552	626	3	1.49	1	0.00527	4.7	0.03228	12.6	0.04449	12	0.37	34	2	32	4	-84	285	-40
A-88	610	302	1	1.13	5	0.0048	5.6	0.02554	32.6	0.03861	32	0.17	31	2	26	8	-444	846	-7

CM2 (1st & 2nd mount)

spot number	$^{207}Pb^a$ (cps)	U^b (ppm)	Pb^b (ppm)	$\frac{Th^b}{U}$	$\frac{^{206}Pb^c}{^{204}Pb}$	$\frac{^{206}Pb^c}{^{238}U}$	2 s %	$\frac{^{207}Pb^c}{^{235}U}$	2 s %	$\frac{^{207}Pb^c}{^{206}Pb}$	2 s %	rho^d	$\frac{^{206}Pb}{^{238}U}$ (Ma)	2 s (Ma)	$\frac{^{207}Pb}{^{235}U}$ (Ma)	2 s (Ma)	$\frac{^{207}Pb}{^{206}Pb}$ (Ma)	2 s (Ma)	conc %^e
a57	720	619	3	0.83	1239	0.00393	2.3	0.03153	8.9	0.05812	8.6	0.25	25	1	32	3	534	189	5
a58	3639	2405	14	0.1	1550	0.0047	2.1	0.03527	4.5	0.05447	4	0.46	30	1	35	2	390	91	8
a59	1891	200	7	0.95	3071	0.03137	3.3	0.26822	9.1	0.06202	8.4	0.36	199	6	241	20	675	181	30
a60	14000	160	21	0.3	13791	0.11899	4.4	1.67666	6.1	0.1022	4.3	0.72	725	30	1000	40	1664	79	44
a61	5045	66	13	0.56	5782	0.18851	2.3	1.96638	3.9	0.07566	3.2	0.59	1113	24	1104	27	1086	64	103
a62	17073	373	41	0.37	19843	0.10765	2.7	1.16049	3.5	0.07819	2.3	0.77	659	17	782	19	1152	45	57
a63	3470	775	15	0.72	6457	0.0167	3	0.11762	5.6	0.05108	4.8	0.54	107	3	113	6	244	109	44
a64	980	741	2	0.54	701	0.00242	3.2	0.03433	8	0.10273	7.3	0.4	16	1	34	3	1674	135	1
a65	524	223	1	0.72	108	0.00171	6.2	0.08494	23	0.36012	22.2	0.27	11	1	83	18	3750	337	0
a66	13941	109	25	0.7	13666	0.20318	2.6	2.87724	3.1	0.10271	1.8	0.83	1192	28	1376	24	1674	33	71
a67	477	158	1	1.6	400	0.00501	3.2	0.08244	17.1	0.11931	16.8	0.19	32	1	80	13	1946	300	2
a68	2262	126	6	0.93	621	0.04193	2.4	0.47995	9.9	0.08302	9.6	0.24	265	6	398	33	1270	188	21
a69	3318	346	16	2.13	6386	0.03423	1.8	0.24676	3.5	0.05229	3.1	0.51	217	4	224	7	298	70	73
a70	908	104	1	0.5	34	0.00259	7.4	0.20116	10.5	0.56237	7.5	0.71	17	1	186	18	4414	109	0
a71	45028	310	116	2.11	15366	0.26876	1.8	3.68131	2.1	0.09934	1	0.87	1534	25	1567	17	1612	19	95
a72	23033	235	66	0.99	9985	0.24702	2.3	3.06694	3	0.09005	1.9	0.76	1423	29	1424	23	1427	37	100
a73	529	397	2	0.91	750	0.00324	11.1	0.0245	23.4	0.05481	20.6	0.47	21	2	25	6	405	462	5
a74	1939	281	8	0.98	1454	0.02634	2.3	0.17995	5.7	0.04954	5.2	0.41	168	4	168	9	174	122	97
a75	1125	71	1	1.15	101	0.00551	10.2	0.24732	18.2	0.32584	15.1	0.56	35	4	224	37	3598	232	1
a76	4675	240	19	1.47	3596	0.06444	2	0.51322	3.4	0.05776	2.8	0.59	403	8	421	12	521	60	77
a77	894	196	4	1	637	0.01566	2.5	0.12344	12.3	0.05715	12	0.2	100	2	118	14	497	265	20
a78	2526	582	8	0.46	1029	0.0124	2.1	0.09696	5.4	0.05671	5	0.39	79	2	94	5	480	110	17
a79	5377	167	21	1.42	8890	0.09991	2.1	0.84267	3.8	0.06117	3.2	0.56	614	12	621	18	645	68	95
a80	3048	123	10	1.29	1042	0.05784	2.7	0.58743	10.9	0.07366	10.6	0.24	362	9	469	42	1032	215	35
a81	700	160	1	0.76	135	0.00147	5.1	0.10517	8.3	0.51833	6.5	0.61	9	0	102	8	4294	96	0

spot number	$^{207}Pb^a$ (cps)	U^b (ppm)	Pb^b (ppm)	$\frac{Th^b}{U}$	$\frac{^{206}Pb^c}{^{204}Pb}$	$\frac{^{206}Pb^c}{^{238}U}$	2 s %	$\frac{^{207}Pb^c}{^{235}U}$	2 s %	$\frac{^{207}Pb^c}{^{206}Pb}$	2 s %	rho^d	$\frac{^{206}Pb}{^{238}U}$	2 s (Ma)	$\frac{^{207}Pb}{^{235}U}$	2 s (Ma)	$\frac{^{207}Pb}{^{206}Pb}$	2 s (Ma)	conc %^{e}
a83	1056	104	1	0.55	54	0.00266	10.4	0.18948	29	0.51684	27.1	0.36	17	2	176	48	4290	399	0
a84	21545	686	78	1.17	9360	0.0976	1.9	0.83894	3	0.06234	2.4	0.62	600	11	619	14	686	51	88
a85	993	60	2	1.19	831	0.02298	7.8	0.34019	13.7	0.10738	11.3	0.57	146	11	297	36	1755	206	8
a86	4144	58	12	1.06	5474	0.17976	2	1.8903	3.5	0.07627	2.8	0.58	1066	20	1078	23	1102	57	97
a87	2589	193	9	0.71	844	0.04207	2.6	0.3781	9.8	0.06518	9.4	0.27	266	7	326	28	780	198	34
a88	26972	246	60	0.43	16159	0.23513	2	3.07429	2.6	0.09483	1.8	0.74	1361	24	1426	20	1525	33	89
a89	4081	511	10	0.99	327	0.01482	2.3	0.20285	3.9	0.09927	3.1	0.58	95	2	188	7	1610	58	6
a90	27192	254	62	0.46	31713	0.23534	2.4	2.80144	3	0.08633	1.9	0.78	1362	29	1356	23	1346	37	101
a91	3573	829	7	0.52	131	0.00532	2.5	0.12854	7.1	0.1754	6.6	0.35	34	1	123	8	2610	111	1
a92	2084	487	7	0.66	1246	0.01304	2.1	0.10411	5.4	0.0579	5	0.38	84	2	101	5	526	111	16
a93	4125	738	12	0.84	519	0.01337	1.6	0.13638	8.3	0.07396	8.1	0.2	86	1	130	10	1040	164	8
a94	361	180	3	1.26	1477	0.016	2.2	0.0543	11.3	0.02462	11.1	0.2	102	2	54	6			
a95	1615	74	6	0.73	1012	0.07618	2.2	0.58685	4.8	0.05587	4.2	0.45	473	10	469	18	447	94	106
a96	639	244	0	0.48	508	0.00097	19	0.0169	30.1	0.12646	23.3	0.63	6	1	17	5	2049	412	0
a97	242	190	0	0.71	349	0.00153	14.4	-0.01111	482.8	-0.05277	482.6	0.03	10	1	-11	-54			
a98	2653	282	11	0.37	5483	0.03754	2.6	0.25211	5.9	0.04871	5.3	0.43	238	6	228	12	134	126	177
a99	3857	173	12	0.24	2273	0.07332	1.9	0.61145	3.6	0.06048	3.1	0.54	456	9	484	14	621	66	73
a100	10725	493	32	0.08	19238	0.06904	1.9	0.53469	2.8	0.05617	2	0.7	430	8	435	10	459	44	94
A-144	43621	183	79	0.34	b.d.	0.4091	4.7	7.76	5.1	0.1376	2	0.92	2211	87	2204	46	2197	36	**101**
A-149	16030	179	40	0.16	0	0.2209	5.2	2.538	6	0.08333	3.1	0.86	1287	61	1283	44	1276	60	**101**
A-150	1479	378	5	0.21	0	0.01485	5.1	0.09811	10.4	0.04794	9.1	0.49	95	5	95	9	96	215	**99**
A-151	2276	298	10	0.39	0	0.03386	5.2	0.2353	7.8	0.05041	5.8	0.67	215	11	215	15	213	134	**101**
A-152	1032	127	4	0.31	0	0.03466	5.2	0.2425	9.6	0.05076	8.1	0.54	220	11	220	19	229	188	**96**
A-153	8958	5556	30	0.15	1	0.00559	5.4	0.03602	6.7	0.04678	4	0.8	36	2	36	2	37	96	**96**
A-154	933	254	4	0.48	0	0.01733	5.4	0.1155	13.8	0.04834	13	0.39	111	6	111	14	115	298	**96**
A-155	38526	375	79	0.22	0	0.2079	5.8	2.463	6.5	0.08595	2.8	0.9	1218	64	1261	47	1337	54	**91**
A-156	1359	338	4	0.25	0	0.012	6.2	0.08099	11.6	0.04896	9.7	0.54	77	5	79	9	145	228	53

spot number	$^{207}Pb^{a}$ (cps)	U^{b} (ppm)	Pb^{b} (ppm)	$\underline{Th^{b}}$ U	$\underline{^{206}Pb^{c}}$ ^{204}Pb	$\underline{^{206}Pb^{c}}$ ^{238}U	2 s %	$\underline{^{207}Pb^{c}}$ ^{235}U	2 s %	$\underline{^{207}Pb^{c}}$ ^{206}Pb	2 s %	rho^{d}	$\underline{^{206}Pb}$ ^{238}U	2 s (Ma)	$\underline{^{207}Pb}$ ^{235}U	2 s (Ma)	$\underline{^{207}Pb}$ ^{206}Pb	2 s (Ma)	conc %^{e}
A-157	22359	933	57	0.1	1	0.06132	5.5	0.496	21.8	0.05867	21	0.25	384	21	409	73	554	460	69
A-158	747	204	3	0.21	0	0.0173	5.4	0.1149	10.3	0.04821	8.7	0.53	111	6	110	11	109	206	102
A-159	139	360	1	0.68	b.d.	0.00165	47.2	0.1399	48.1	0.6153	9.4	0.98	11	5	133	60	4544	136	0
A-160	8771	2274	34	0.22	0	0.01511	5.1	0.1281	11	0.06152	9.8	0.46	97	5	122	13	657	210	15
A-161	325	413	2	0.53	0	0.00462	5.5	0.02967	19.3	0.04661	19	0.28	30	2	30	6	29	445	104
A-162	3454	94	18	0.64	b.d.	0.1882	6.2	3.175	8.6	0.1224	6	0.72	1112	63	1451	66	1990	106	56
A-163	1440	361	6	0.35	0	0.01634	5.3	0.1101	8.1	0.0489	6.1	0.66	104	5	106	8	142	143	73
A-164	784	218	4	0.16	0	0.01762	5.3	0.1175	10.7	0.04839	9.3	0.49	113	6	113	11	118	219	96
A-165	1617	531	7	0.32	0	0.0139	4.9	0.09227	8.6	0.04815	7.1	0.57	89	4	90	7	106	168	84
A-166	15862	797	60	0.22	0	0.07639	4.8	0.597	6.3	0.0567	4	0.77	475	22	475	24	479	89	99
A-167	6897	1656	29	0.24	0	0.01766	5.2	0.1177	6.1	0.04836	3.2	0.85	113	6	113	7	116	76	97
A-168	5545	1492	18	0.4	1	0.01199	4.8	0.09576	10.1	0.05796	8.9	0.48	77	4	93	9	528	194	15
A-173	896	735	4	0.29	0	0.00512	5	0.03533	10.2	0.05006	8.9	0.49	33	2	35	4	197	206	17
A-174	19914	203	41	0.18	2	0.201	5.2	2.177	7	0.07856	4.6	0.76	1181	57	1174	48	1160	90	102
A-175	5697	168	9	0.23	0	0.05251	6.8	0.4753	7.8	0.06566	3.7	0.88	330	22	395	25	795	77	41
A-176	273	346	2	0.3	b.d.	0.00486	22.7	0.1506	41.4	0.2248	35	0.55	31	7	142	55	3015	556	1
A-177	16508	155	39	0.29	b.d.	0.248	5.1	3.134	8.1	0.09166	6.3	0.64	1428	66	1441	62	1460	119	98
A-178	2552	616	10	0.44	0	0.01641	5.1	0.1089	7.7	0.04817	5.8	0.66	105	5	105	8	107	136	98
A-179	1039	995	5	0.41	1	0.0048	6.3	0.03083	7.1	0.04663	3.3	0.88	31	2	31	2	30	80	104
A-180	1348	333	6	0.31	0	0.01824	5.1	0.1216	11.7	0.04839	10	0.44	117	6	117	13	118	247	99
A-181	7113	1656	29	0.31	0	0.01792	5.5	0.1198	6.9	0.04851	4.1	0.8	114	6	115	8	123	97	93
A-182	512	137	2	0.36	0	0.01615	5.5	0.107	12.4	0.04804	11	0.44	103	6	103	12	101	263	103
A-183	193	140	2	0.42	b.d.	0.01105	14.6	0.3585	50.7	0.2353	49	0.29	71	10	311	136	3088	774	2
A-184	10502	115	26	0.35	b.d.	0.2261	5.2	2.623	14.5	0.08418	14	0.36	1314	62	1307	106	1296	263	101
A-185	1488	1655	7	0.33	0	0.00428	5.3	0.02749	10.9	0.0466	9.5	0.49	28	1	28	3	28	228	99
A-186	705	499	2	0.37	0	0.00479	5.3	0.03964	10.1	0.06003	8.6	0.53	31	2	39	4	604	185	5

spot number	$^{207}Pb^a$ (cps)	U^b (ppm)	Pb^b (ppm)	Th^b/U	$^{206}Pb^c$/^{204}Pb	$^{206}Pb^c$/^{238}U	2 s %	$^{207}Pb^c$/^{235}U	2 s %	$^{207}Pb^c$/^{206}Pb	2 s %	rho^d	^{206}Pb/^{238}U (Ma)	2 s (Ma)	^{207}Pb/^{235}U (Ma)	2 s (Ma)	^{207}Pb/^{206}Pb (Ma)	2 s (Ma)	conc %[e]
A-188	1908	568	8	0.25	0	0.01528	5	0.101	7.6	0.04797	5.7	0.66	98	5	98	7	97	134	**101**
A-189	257	175	1	0.3	0	0.00526	6.3	0.03388	17	0.04677	16	0.37	34	2	34	6	37	379	**92**
A-190	24147	125	47	0.41	b.d.	0.3575	4.8	6.203	5.2	0.1259	1.9	0.93	1970	82	2005	45	2041	34	**97**
A-191	1270	372	6	0.31	0	0.01556	5.4	0.1032	10	0.04812	8.4	0.54	100	5	100	10	105	199	**95**
A-192	8340	1216	35	0.65	b.d.	0.02941	5.2	0.2039	24.1	0.0503	24	0.22	187	10	188	41	208	545	90
A-197	9199	130	29	0.43	0	0.2222	5	2.495	12.5	0.08146	11	0.4	1294	59	1271	90	1232	224	**105**
A-198	7455	1949	39	0.58	0	0.02049	4.9	0.1487	6.9	0.05266	4.8	0.71	131	6	141	9	313	109	42
A-199	1801	493	8	0.24	0	0.01562	5.3	0.1036	10.9	0.04813	9.6	0.49	100	5	100	10	105	226	**95**
A-200	7249	1167	16	0.63	2	0.01368	5.2	0.1505	8.9	0.07981	7.3	0.58	88	4	142	12	1192	144	7
A-201	352	395	1	0.47	0	0.00377	5.8	0.02592	14.8	0.04984	14	0.39	24	1	26	4	187	316	13
A-202	357	405	2	0.36	0	0.00392	6.1	0.02553	20.5	0.04721	20	0.29	25	2	26	5	59	468	43
A-203	286	275	1	0.27	0	0.00435	6.2	0.02802	20	0.0467	19	0.31	28	2	28	6	33	455	85
A-204	114	308	0	0.35	0	0.00063	12.8	0.00457	41.2	0.05245	39	0.31	4	1	5	2	304	893	1
A-205	45390	97	63	0.55	0	0.5796	4.9	16.1	5.5	0.2015	2.5	0.89	2947	116	2883	53	2838	41	**104**
A-206	7695	499	29	0.27	0	0.05909	5	0.4404	26.2	0.05407	26	0.19	370	18	371	81	373	580	**99**
A-207	889	856	3	0.19	0	0.00385	6	0.02883	15.2	0.05439	14	0.4	25	1	29	4	387	314	6
A-208	2633	663	10	0.27	0	0.01582	5.3	0.1102	9.9	0.05053	8.4	0.53	101	5	106	10	219	195	46
A-209	2479	762	11	0.37	0	0.01426	4.8	0.09411	7.7	0.04788	6	0.63	91	4	91	7	93	142	**98**
A-210	1580	408	6	0.24	0	0.01612	5	0.1093	8.7	0.0492	7.1	0.58	103	5	105	9	157	166	66
A-211	1200	312	5	0.39	0	0.0163	5.1	0.1079	8.2	0.04802	6.4	0.63	104	5	104	8	100	150	**105**
A-212	66519	456	139	0.21	0	0.2956	5.5	4.583	6.4	0.1125	3.2	0.86	1669	81	1746	53	1839	58	**91**
A-213	2793	191	10	0.4	0	0.0544	5.1	0.4413	7.9	0.05885	6	0.64	341	17	371	25	561	132	61
A-214	5699	1510	24	0.66	1	0.01604	5	0.1059	17	0.04792	16	0.3	103	5	102	17	94	386	**109**
A-215	845	235	4	0.29	0	0.01605	5.3	0.1064	11.2	0.04812	9.9	0.47	103	5	103	11	104	233	**98**
A-216	1519	454	6	0.38	0	0.01335	5.1	0.09212	9.4	0.05005	7.8	0.55	86	4	89	8	196	182	44
A-221	2560	761	10	0.14	0	0.01305	4.8	0.09848	6.8	0.05476	4.8	0.7	84	4	95	6	402	108	21
A-222	54279	349	111	0.32	0	0.3099	4.9	4.604	5.5	0.1078	2.5	0.89	1740	74	1750	46	1761	46	**99**

spot number	$^{207}Pb^a$ (cps)	U^b (ppm)	Pb^b (ppm)	$\underline{Th^b}$ U	$\underline{^{206}Pb^c}$ ^{204}Pb	$\underline{^{206}Pb^c}$ ^{238}U	2 s %	$\underline{^{207}Pb^c}$ ^{235}U	2 s %	$\underline{^{207}Pb^c}$ ^{206}Pb	2 s %	rho^d	$\underline{^{206}Pb}$ ^{238}U	2 s (Ma)	$\underline{^{207}Pb}$ ^{235}U	2 s (Ma)	$\underline{^{207}Pb}$ ^{206}Pb	2 s (Ma)	conc %^{e}
A-223	281	796	1	0.34	0	0.00075	10.3	0.01075	30.5	0.1046	29	0.34	5	0	11	3	1707	528	0
A-224	2138	567	7	0.53	0	0.01331	5	0.09728	8.6	0.05303	6.9	0.59	85	4	94	8	329	157	26
A-225	3225	780	13	0.22	0	0.01682	5.3	0.1288	6.4	0.05555	3.6	0.82	108	6	123	7	434	81	25
A-226	15392	962	70	0.08	0	0.07352	5.1	0.5714	9	0.05638	7.5	0.56	457	22	459	33	467	165	**98**

CM3 (1st & 2nd mount)

spot number	$^{207}Pb^a$ (cps)	U^b (ppm)	Pb^b (ppm)	$\underline{Th^b}$ U	$\underline{^{206}Pb^c}$ ^{204}Pb	$\underline{^{206}Pb^c}$ ^{238}U	2 s %	$\underline{^{207}Pb^c}$ ^{235}U	2 s %	$\underline{^{207}Pb^c}$ ^{206}Pb	2 s %	rho^d	$\underline{^{206}Pb}$ ^{238}U	2 s (Ma)	$\underline{^{207}Pb}$ ^{235}U	2 s (Ma)	$\underline{^{207}Pb}$ ^{206}Pb	2 s (Ma)	conc %^{e}
a47	1130	1133	5	0.49	2462	0.00409	2.4	0.02624	6.2	0.04654	5.7	0.39	26	1	26	2	26	138	102
a48	27233	410	72	1.01	24049	0.14821	1.9	1.50021	2.4	0.07341	1.6	0.77	891	16	930	15	1025	32	87
a49	11932	519	38	0.75	2266	0.06681	1.9	0.59553	2.6	0.06464	1.8	0.74	417	8	474	10	763	37	55
A-5	4675	924	17	0.18	0	0.01923	5.8	0.1335	7.1	0.05035	4	0.82	123	7	127	8	210	94	58
A-6	4360	232	16	0.3	0	0.07208	5	0.5508	7	0.05543	4.9	0.72	449	22	445	25	429	108	**105**
A-7	3203	769	14	0.37	0	0.01837	4.9	0.1228	6.3	0.0485	4.1	0.76	117	6	118	7	123	96	**95**
A-8	5156	1258	21	0.4	0	0.01693	5.2	0.1173	6.1	0.05029	3.3	0.85	108	6	113	7	208	75	52
A-9	2070	67	10	0.49	0	0.1515	7.4	2.223	35.5	0.1064	35	0.21	910	63	1188	249	1738	637	52
A-10	31181	205	57	0.27	0	0.2687	4.7	3.809	5.8	0.1028	3.3	0.82	1534	65	1595	47	1675	62	**92**
A-11	4750	241	17	0.28	0	0.0724	4.9	0.5682	6.9	0.05693	4.8	0.72	451	21	457	25	488	105	**92**
A-12	1255	154	1	0.34	0	0.00308	8.4	0.3188	22	0.7502	20	0.38	20	2	281	54	4830	291	0
A-13	7129	342	28	0.25	0	0.08238	5	0.6525	6.4	0.05747	4	0.78	510	24	510	26	509	88	**100**
A-14	31694	178	58	0.35	1	0.3157	5.1	5.013	7.5	0.1152	5.6	0.67	1769	79	1822	64	1883	100	**94**
A-15	471	7	2	0.5	0	0.2206	7.3	2.461	21.4	0.08092	20	0.34	1285	86	1261	154	1219	395	**105**
A-16	6417	368	25	0.08	0	0.06938	5	0.5339	5.9	0.05583	3.1	0.85	432	21	434	21	445	69	**97**
A-17	46109	1025	147	0.26	0	0.1443	5.4	1.372	6.4	0.06898	3.4	0.85	869	44	877	38	898	70	**97**
A-18	12164	144	32	0.13	0	0.2179	5.3	2.449	6.2	0.08154	3.3	0.85	1271	61	1257	45	1234	65	**103**
A-19	30583	403	86	0.39	b.d.	0.2105	4.8	2.538	5.8	0.08745	3.3	0.83	1232	54	1283	43	1370	63	90
A-20	1901	93	7	0.46	0	0.07493	5.3	0.6055	8.6	0.05863	6.8	0.61	466	24	481	33	553	149	84
A-21	20489	316	56	0.07	0	0.1769	4.9	1.864	5.1	0.07646	1.6	0.95	1050	47	1068	34	1107	32	**95**

spot number	$^{207}Pb^a$ (cps)	U^b (ppm)	Pb^b (ppm)	$\frac{Th^b}{U}$	$\frac{^{206}Pb^c}{^{204}Pb}$	$\frac{^{206}Pb^c}{^{238}U}$	2 s %	$\frac{^{207}Pb^c}{^{235}U}$	2 s %	$\frac{^{207}Pb^c}{^{206}Pb}$	2 s %	rho^d	$\frac{^{206}Pb}{^{238}U}$ (Ma)	2 s (Ma)	$\frac{^{207}Pb}{^{235}U}$ (Ma)	2 s (Ma)	$\frac{^{207}Pb}{^{206}Pb}$ (Ma)	2 s (Ma)	conc %^{e}
A-22	75599	150	91	0.24	b.d.	0.5467	4.7	14.48	5.2	0.1921	2.1	0.92	2811	108	2782	49	2760	34	102
A-23	287	354	-1	0.97	b.d.	-0.00185	44.8	-0.1781	47.3	0.6993	15	0.95	-12	-5	-199	-104	4729	221	0
A-24	18355	268	47	0.26	b.d.	0.172	5.1	2.077	5.5	0.08762	2	0.93	1023	49	1141	38	1374	38	74
A-29	42580	273	86	0.24	b.d.	0.3048	4.8	4.647	5.5	0.1106	2.6	0.88	1715	72	1758	46	1809	47	95
A-30	71	125	0	0.26	0	0.00237	8.5	0.0169	43.7	0.0518	43	0.19	15	1	17	7	276	981	6
A-31	54463	566	125	0.28	0	0.2185	5.4	2.599	6.2	0.08628	3.1	0.87	1274	63	1300	46	1344	60	95
A-32	17341	286	18	0.1	0	0.06261	7.7	0.9021	8.2	0.1045	2.9	0.94	391	29	653	39	1706	53	23
A-33	32	160	0	0.36	0	0.00069	13.7	0.00471	149.4	0.04955	150	0.09	4	1	5	7	173	3471	3
A-34	2725	271	9	1.84	0	0.03491	5.4	0.2438	10.8	0.05067	9.3	0.5	221	12	222	21	225	215	98
A-35	18599	245	48	0.4	0	0.1971	5.4	2.133	9.7	0.07852	8.1	0.56	1160	58	1160	67	1159	160	100
A-36	70427	911	190	0.06	b.d.	0.2066	5	2.379	5.4	0.08355	2.1	0.92	1211	55	1236	39	1281	41	94
A-37	6210	268	22	0.15	0	0.08463	5.5	0.6713	24.8	0.05755	24	0.22	524	27	522	101	512	531	102
A-38	4279	172	15	0.28	b.d.	0.08535	5.3	0.7	32.7	0.0595	32	0.16	528	27	539	137	585	700	90
A-39	73625	194	109	0.37	0	0.5163	4.9	12.83	5.7	0.1803	2.9	0.86	2683	108	2667	53	2655	47	101
A-40	44895	304	97	0.41	0	0.3097	4.9	4.453	5.7	0.1043	2.9	0.86	1739	74	1722	47	1702	54	102
A-41	9938	133	23	0.65	0	0.1755	5.2	1.81	18.2	0.07481	17	0.28	1042	50	1049	119	1063	351	98
A-42	4644	210	13	0.33	0	0.06297	5.6	0.5099	7.5	0.05874	5	0.74	394	21	418	26	557	109	71
A-43	784	220	3	0.2	0	0.01521	5.2	0.1006	9.7	0.04798	8.2	0.54	97	5	97	9	98	193	100
A-44	9830	530	36	0.16	1	0.06916	5	0.5347	12.7	0.05609	12	0.39	431	21	435	45	455	260	95
A-45	4648	266	17	0.46	b.d.	0.06358	5	0.6721	21.2	0.07668	21	0.24	397	19	522	86	1112	411	36
A-46	6013	193	16	0.21	b.d.	0.08334	5.5	0.703	27.8	0.0612	27	0.2	516	27	541	116	646	585	80
A-47	7039	1819	32	0.13	b.d.	0.01769	5.4	0.1571	17.2	0.06446	16	0.31	113	6	148	24	756	345	15
A-48	159	589	0	0.24	7	0.00063	8.7	0.0045	90.6	0.05147	90	0.1	4	0	5	4	261	2071	2
A-53	9319	446	31	0.48	1	0.06978	6	0.5369	18.4	0.05582	17	0.32	435	25	436	65	445	387	98
A-54	19436	933	61	0.57	0	0.06588	5.7	0.5102	9.4	0.05618	7.5	0.61	411	23	419	32	459	165	90
A-55	43805	934	128	0.08	0	0.1374	6.7	1.318	10.1	0.06962	7.6	0.66	830	52	854	58	916	156	91
A-56	2287	678	11	0.53	b.d.	0.0163	5.5	0.1098	49.4	0.04889	49	0.11	104	6	106	50	142	1153	74

spot number	$^{207}Pb^a$ (cps)	U^b (ppm)	Pb^b (ppm)	Th^b/U	$^{206}Pb^c/^{204}Pb$	$^{206}Pb^c/^{238}U$	2 s %	$^{207}Pb^c/^{235}U$	2 s %	$^{207}Pb^c/^{206}Pb$	2 s %	rho^d	$^{206}Pb/^{238}U$ (Ma)	2 s (Ma)	$^{207}Pb/^{235}U$ (Ma)	2 s (Ma)	$^{207}Pb/^{206}Pb$ (Ma)	2 s (Ma)	conc %[e]
A-57	845	175	4	0.25	b.d.	0.01975	7.7	0.584	9.8	0.2145	6.1	0.79	126	10	467	37	2939	98	4
A-58	14716	103	16	0.26	2	0.1569	7	1.584	10.5	0.07322	7.8	0.67	940	62	964	65	1019	157	**92**
A-59	20389	1674	77	0.23	0	0.04707	5.1	0.3399	8.8	0.05239	7.2	0.58	297	15	297	23	302	164	**98**
A-60	38388	267	80	0.37	0	0.2943	4.9	4.147	6	0.1022	3.5	0.82	1663	73	1664	49	1665	64	**100**
A-61	45995	716	100	0.13	0	0.1384	5.2	1.619	6.8	0.08483	4.3	0.77	836	41	977	43	1311	84	64
A-62	25234	352	68	0.2	b.d.	0.1928	5	2.041	7.2	0.07678	5.2	0.69	1137	52	1129	49	1115	104	**102**
A-63	919	90	3	0.38	4	0.03464	5.8	0.1491	25.4	0.03122	25	0.23	220	13	141	33	-1039	739	-21
A-64	9202	357	21	0.78	b.d.	0.05859	7.4	0.5502	14.5	0.06812	12	0.51	367	26	445	52	872	258	42
A-65	4264	271	24	0.36	b.d.	0.08903	5.7	0.6656	50.2	0.05424	50	0.11	550	30	518	204	380	1121	145
A-66	9515	99	22	0.36	b.d.	0.2152	7.4	3.88	13.7	0.1308	12	0.54	1256	84	1610	111	2108	203	60
A-67	88761	859	211	0.58	0	0.2432	5	2.85	5.4	0.08501	2.1	0.92	1403	63	1369	41	1315	40	**107**
A-68	7383	140	22	0.35	b.d.	0.156	5.1	1.729	10.6	0.08041	9.3	0.48	934	44	1019	68	1207	184	77
A-69	25794	221	62	0.35	0	0.2719	5.2	3.748	9	0.1	7.3	0.58	1550	72	1582	72	1624	136	**95**
A-70	204	270	0	0.46	21	0.00066	17.2	0.1376	200.9	1.503	200	0.09	4	1	131	247	5803	2754	0
A-71	1520	977	5	0.32	4	0.0049	5.3	0.03151	6.5	0.04668	3.6	0.83	31	2	31	2	32	87	**99**
A-72	21358	267	51	0.25	b.d.	0.1906	5	2.093	7.8	0.07969	6	0.64	1124	51	1147	53	1189	118	**95**
A-77	25434	286	57	0.11	4	0.1978	5.1	2.201	6.8	0.08074	4.5	0.75	1163	55	1181	48	1215	89	**96**
A-78	9475	537	34	0.03	0	0.06346	5	0.4782	6.3	0.05468	3.8	0.79	397	19	397	21	398	86	**100**
A-79	4103	1081	18	0.4	b.d.	0.01703	5.1	0.1206	32	0.05138	32	0.16	109	6	116	35	257	726	42
A-80	5597	86	16	0.21	0	0.1816	5.1	1.867	6.6	0.0746	4.1	0.78	1076	51	1070	44	1057	83	**102**
A-81	2222	149	8	0.63	0	0.05375	5.1	0.4164	7.6	0.0562	5.6	0.68	337	17	353	23	460	124	73
A-82	10378	139	28	0.53	0	0.197	5	2.174	6.5	0.08004	4.2	0.77	1159	53	1173	45	1198	83	**97**
A-83	1667	482	9	0.24	0	0.01852	5.4	0.1236	9.4	0.04843	7.7	0.57	118	6	118	10	119	181	**99**
A-84	6330	1569	23	0.59	1	0.0151	5.5	0.1001	26.7	0.04808	26	0.2	97	5	97	25	103	618	**94**
A-85	4390	234	15	0.17	0	0.06547	5.3	0.4991	6.6	0.05531	3.9	0.8	409	21	411	22	424	88	**96**
A-86	34285	530	93	0.03	0	0.1758	5.1	1.809	6.2	0.07467	3.6	0.82	1044	49	1049	41	1059	72	**99**
A-87	2144	485	7	0.14	b.d.	0.01385	6.8	0.2013	17.5	0.1055	16	0.39	89	6	186	30	1722	296	5

spot number	$^{207}Pb^a$ (cps)	U^b (ppm)	Pb^b (ppm)	Th^b/U	$^{206}Pb^c$/^{204}Pb	$^{206}Pb^c$/^{238}U	2 s %	$^{207}Pb^c$/^{235}U	2 s %	$^{207}Pb^c$/^{206}Pb	2 s %	rho^d	^{206}Pb/^{238}U	2 s (Ma)	^{207}Pb/^{235}U	2 s (Ma)	^{207}Pb/^{206}Pb	2 s (Ma)	conc %[e]
A-88	10560	411	35	0.76	0	0.08674	5	0.706	6.1	0.05904	3.5	0.82	536	26	542	26	568	76	**94**
A-89	3543	899	12	0.35	0	0.01359	5.4	0.09754	8	0.05206	5.9	0.67	87	5	94	7	287	135	30
A-90	3195	246	12	0.59	0	0.04819	4.8	0.352	7.3	0.05299	5.4	0.67	303	14	306	19	328	123	**93**
A-91	5011	181	16	0.54	b.d.	0.08543	5.9	0.9245	16.1	0.07851	15	0.37	528	30	665	78	1159	297	46
A-92	11297	165	33	0.29	1	0.1992	5.7	2.174	19.9	0.07917	19	0.28	1171	61	1173	138	1176	377	**100**
A-93	6168	212	19	0.31	b.d.	0.0905	5.3	0.8011	19.3	0.06422	19	0.27	558	28	597	87	748	392	75
A-94	7148	805	28	0.43	1	0.03513	5.3	0.2459	21.4	0.05078	21	0.25	223	12	223	43	230	480	**97**
A-95	2423	781	11	0.21	0	0.01382	5.9	0.09289	9.6	0.04876	7.6	0.62	88	5	90	8	136	178	65
A-96	6063	416	21	0.25	b.d.	0.05082	6.3	0.5785	18.3	0.08258	17	0.34	320	20	464	68	1259	335	25
A-101	3551	891	14	0.33	0	0.01653	5.2	0.1102	6.6	0.04837	4	0.8	106	5	106	7	117	93	**90**
A-102	1357	80	5	0.36	0	0.06583	5.2	0.5073	10	0.05591	8.5	0.52	411	21	417	34	448	189	**92**
A-103	22905	164	13	0.18	4	0.07696	14.2	0.8793	15.5	0.08288	6.1	0.92	478	65	641	73	1266	120	38
A-104	21357	67	31	0.4	0	0.4339	5.5	8.288	6.6	0.1386	3.6	0.83	2323	107	2263	60	2209	63	**105**
A-105	33135	294	80	0.26	b.d.	0.2662	5	3.756	5.5	0.1024	2.2	0.92	1522	68	1583	44	1667	41	**91**
A-106	21068	113	43	0.21	b.d.	0.3617	5.2	6.076	7.4	0.1219	5.3	0.7	1990	89	1987	65	1983	94	**100**
A-107	3507	365	12	0.42	0	0.03315	5.4	0.2451	6.7	0.05365	4	0.8	210	11	223	13	356	90	59
A-108	254	439	0	0.43	0	0.00068	10.4	0.01599	46.5	0.1717	45	0.22	4	0	16	7	2574	757	0
A-109	48	336	0	0.42	0	0.00063	14	0.00555	111.9	0.06376	110	0.13	4	1	6	6	733	2352	1
A-110	2088	66	6	0.39	0	0.09171	5	0.7664	6.8	0.06063	4.6	0.73	566	27	578	30	625	100	**90**
A-111	2679	83	10	0.25	b.d.	0.118	6	1.688	16.6	0.1038	15	0.36	719	41	1004	106	1692	286	42
A-112	3235	148	6	0.4	0	0.04309	9.4	0.4078	12.4	0.06865	8.1	0.76	272	25	347	36	888	168	31
A-113	10778	195	34	0.31	2	0.1726	5	1.823	12.5	0.07663	11	0.4	1026	48	1054	82	1111	229	**92**
A-114	280	240	1	0.47	0	0.00249	6.4	0.03699	27.6	0.108	27	0.23	16	1	37	10	1765	491	1
A-115	993	634	3	0.31	0	0.00391	5.3	0.04719	9.2	0.08748	7.5	0.58	25	1	47	4	1371	145	2
A-116	683	593	2	0.23	0	0.00357	5.7	0.02846	14.8	0.05787	14	0.39	23	1	28	4	524	299	4
A-117	41963	531	109	0.08	0	0.204	4.9	2.257	5.7	0.08025	2.9	0.86	1197	54	1199	40	1202	58	**100**
A-118	38104	1228	132	0.08	0	0.1083	4.9	0.9124	6.2	0.0611	3.8	0.79	663	31	658	30	642	83	**103**

spot number	^{207}Pb[a] (cps)	U[b] (ppm)	Pb[b] (ppm)	Th[b]/U	^{206}Pb[c]/^{204}Pb	^{206}Pb[c]/^{238}U	2 s %	^{207}Pb[c]/^{235}U	2 s %	^{207}Pb[c]/^{206}Pb	2 s %	rho[d]	^{206}Pb/^{238}U (Ma)	2 s (Ma)	^{207}Pb/^{235}U (Ma)	2 s (Ma)	^{207}Pb/^{206}Pb (Ma)	2 s (Ma)	conc %[e]
A-119	106108	672	211	0.24	0	0.3057	5	4.476	5.1	0.1062	1.1	0.97	1720	75	1727	42	1735	21	99
A-120	2769	802	12	0.21	0	0.01565	5.2	0.1035	7.7	0.048	5.6	0.68	100	5	100	7	98	133	102
A-125	6357	186	18	0.14	b.d.	0.09823	5.5	0.8822	25.2	0.06515	25	0.22	604	32	642	120	779	518	78
A-126	42232	379	94	0.3	0	0.2424	4.9	3.147	5.5	0.09421	2.7	0.87	1399	61	1444	43	1512	51	93
A-127	524	128	4	0.32	b.d.	0.02126	34.7	2.172	40	0.7413	20	0.87	136	47	1172	278	4813	286	3
A-128	6905	215	23	0.35	1	0.1064	5.1	0.908	18.4	0.06193	18	0.28	652	32	656	89	671	377	97
A-129	7506	201	21	0.07	0	0.1023	6.6	1.247	13.4	0.08843	12	0.49	628	39	822	75	1391	224	45
A-130	17846	1657	50	0.25	3	0.03063	5.1	0.2113	29.4	0.05004	29	0.17	195	10	195	52	196	672	99
A-131	1721	153	6	0.47	0	0.03685	5.2	0.3053	10.6	0.06009	9.3	0.49	233	12	270	25	606	201	38
A-132	5239	51	13	0.38	0	0.2477	5.6	3.106	28.5	0.09096	28	0.2	1427	72	1434	219	1445	532	99
A-133	332	608	0	0.37	7	0.00062	8.4	0.01171	88	0.1379	88	0.1	4	0	12	10	2200	1521	0
A-134	123379	1129	308	0.23	1	0.2682	5	3.469	5.9	0.09383	3.1	0.85	1532	68	1520	46	1504	59	102
A-135	8452	293	25	0.53	0	0.08738	5	0.7876	6	0.06538	3.4	0.83	540	26	590	27	786	71	69
A-136	1450	13	2	0.54	b.d.	0.1856	8.4	1.931	118.8	0.07548	120	0.07	1098	85	1092	795	1081	2377	102
A-137	85025	344	140	0.4	b.d.	0.3853	4.8	7.333	5	0.1381	1.4	0.96	2101	86	2153	45	2203	25	95
A-138	1414	567	4	0.94	13	0.00664	12.3	0.1933	23.2	0.2114	20	0.53	43	5	179	38	2916	319	1
A-139	720	38	3	1.12	0	0.07007	5.8	0.5372	11.7	0.05562	10	0.49	437	24	437	42	437	228	100
A-140	1452	117	4	1.14	0	0.03731	5.7	0.4029	12	0.07834	11	0.47	236	13	344	35	1155	211	20
A-141	11674	591	43	0.17	0	0.07407	5.1	0.597	6	0.05847	3.1	0.86	461	23	475	23	547	67	84
A-142	848	887	3	0.25	0	0.00349	5.2	0.02768	12.5	0.05758	11	0.41	22	1	28	3	513	251	4
A-143	9622	193	41	0.26	0	0.2097	5.3	2.286	20.6	0.07911	20	0.26	1227	59	1208	146	1174	394	104

CM5

spot number	^{207}Pb[a] (cps)	U[b] (ppm)	Pb[b] (ppm)	Th[b]/U	^{206}Pb[c]/^{204}Pb	^{206}Pb[c]/^{238}U	2 s %	^{207}Pb[c]/^{235}U	2 s %	^{207}Pb[c]/^{206}Pb	2 s %	rho[d]	^{206}Pb/^{238}U (Ma)	2 s (Ma)	^{207}Pb/^{235}U (Ma)	2 s (Ma)	^{207}Pb/^{206}Pb (Ma)	2 s (Ma)	conc %[e]
A-4	1087	396	6	0.42	0	0.01636	4.6	0.1226	10.2	0.05435	9.1	0.45	105	5	117	11	385	204	27
A-5	759	1065	5	0.44	0	0.0052	4.3	0.03351	10.7	0.04677	9.8	0.4	33	1	33	4	37	234	91
A-6	8457	205	36	0.58	0	0.1766	4.9	1.791	7	0.07359	5	0.7	1048	47	1042	45	1030	101	102

spot number	$^{207}Pb^a$ (cps)	U^b (ppm)	Pb^b (ppm)	$\underline{Th^b}$ U	$\frac{^{206}Pb^c}{^{204}Pb}$	$\frac{^{206}Pb^c}{^{238}U}$	2 s %	$\frac{^{207}Pb^c}{^{235}U}$	2 s %	$\frac{^{207}Pb^c}{^{206}Pb}$	2 s %	rho^d	$\frac{^{206}Pb}{^{238}U}$ (Ma)	2 s (Ma)	$\frac{^{207}Pb}{^{235}U}$ (Ma)	2 s (Ma)	$\frac{^{207}Pb}{^{206}Pb}$ (Ma)	2 s (Ma)	conc %[e]
A-8	10537	588	57	0.58	0	0.09778	5	0.8276	6.2	0.06141	3.7	0.8	601	28	612	28	653	79	92
A-9	10701	572	55	0.24	0	0.09801	5.4	0.8078	16.7	0.05979	16	0.32	603	31	601	76	596	343	101
A-10	9894	616	38	0.72	b.d.	0.06199	11.2	0.6717	15.2	0.07862	10	0.73	388	42	522	62	1162	204	33
A-11	21691	1208	115	0.02	0	0.096	5.5	0.7773	6.6	0.05874	3.7	0.83	591	31	584	29	557	81	106
A-13	526	845	3	0.34	0	0.0037	4.7	0.02368	11.4	0.0465	10	0.41	24	1	24	3	23	251	105
A-14	74717	1282	295	0.09	b.d.	0.2274	4	2.733	5	0.08719	3.1	0.79	1321	48	1337	37	1364	59	97
A-15	571	893	4	0.21	0	0.00466	7.3	0.03051	20.6	0.04755	19	0.35	30	2	31	6	76	457	39
A-16	12143	566	25	0.16	2	0.04525	11.3	0.3245	12.8	0.05203	6	0.88	285	31	285	32	286	136	100
A-17	22395	82	43	1.29	0	0.487	4.2	11.23	6.9	0.1673	5.5	0.6	2558	88	2542	64	2530	92	101
A-18	6776	1133	44	0.07	0	0.03951	5.3	0.3415	10.8	0.06271	9.4	0.49	250	13	298	28	698	199	36
A-19	11456	201	44	0.29	b.d.	0.2151	3.9	2.535	7.4	0.0855	6.2	0.54	1256	45	1282	54	1326	120	95
A-20	503	183	3	0.38	0	0.01766	4.2	0.1174	13	0.04823	12	0.32	113	5	113	14	110	290	103
A-21	4217	4	2	0.39	74	0.2711	30.8	29.78	85.3	0.7968	80	0.36	1547	423	3480	838	4916	1134	31
A-22	39328	410	120	0.61	0	0.2855	4.4	4.011	7.4	0.1019	5.9	0.6	1619	63	1636	60	1659	110	98
A-23	3639	103	17	0.65	0	0.1689	4.6	1.708	7.2	0.07335	5.5	0.64	1006	43	1011	46	1023	112	98
A-27	42082	509	140	0.04	0	0.2651	4	4.206	7.4	0.1151	6.2	0.54	1516	54	1675	60	1881	111	81
A-28	505	201	3	0.43	0	0.01452	4.8	0.09588	12	0.0479	11	0.4	93	4	93	11	94	262	99
A-29	4658	212	27	0.29	0	0.13	5.2	1.153	28.3	0.06435	28	0.18	788	39	779	154	753	587	105
A-30	15422	663	78	0.16	2	0.1179	4.5	1.046	8.9	0.06434	7.7	0.5	719	30	727	46	752	163	96
A-31	1483	90	9	0.29	b.d.	0.09058	5.4	1.467	20.4	0.1175	20	0.26	559	29	917	123	1918	352	29
A-32	115	294	0	0.35	0	0.00048	15.6	0.01036	48.8	0.1567	46	0.32	3	0	10	5	2420	786	0
A-33	6239	154	24	0.25	0	0.1579	4.2	1.667	6.1	0.07663	4.5	0.68	945	37	996	39	1111	90	85
A-34	20782	413	72	0.31	b.d.	0.1742	6.5	1.802	12.8	0.07508	11	0.51	1035	62	1046	84	1070	222	97
A-35	261	473	2	0.39	0	0.0038	4.7	0.02436	15.3	0.04652	15	0.3	24	1	24	4	24	351	101
A-36	31662	627	116	0.2	0	0.1846	3.9	2.02	6.5	0.0794	5.1	0.61	1092	39	1122	44	1182	101	92
A-37	2620	409	17	0.36	b.d.	0.04086	4.1	0.5423	5.8	0.09629	4.1	0.71	258	10	440	21	1553	77	17
A-38	4651	161	23	0.25	0	0.143	3.8	1.387	6	0.07038	4.6	0.63	862	31	884	35	939	94	92

175

spot number	$^{207}Pb^a$ (cps)	U^b (ppm)	Pb^b (ppm)	$\frac{Th^b}{U}$	$\frac{^{206}Pb^c}{^{204}Pb}$	$\frac{^{206}Pb^c}{^{238}U}$	2 s %	$\frac{^{207}Pb^c}{^{235}U}$	2 s %	$\frac{^{207}Pb^c}{^{206}Pb}$	2 s %	rho^d	$\frac{^{206}Pb}{^{238}U}$	2 s (Ma)	$\frac{^{207}Pb}{^{235}U}$	2 s (Ma)	$\frac{^{207}Pb}{^{206}Pb}$	2 s (Ma)	conc %^{e}
A-39	937	1528	1	0.64	0	0.00067	5.8	0.02063	16.1	0.2248	15	0.36	4	0	21	3	3015	241	0
A-40	565	705	1	0.61	0	0.00158	10.1	0.03314	28.7	0.152	27	0.35	10	1	33	9	2368	457	0
A-41	1216	60	6	0.99	0	0.1039	3.9	0.8817	8.6	0.06156	7.6	0.46	637	24	642	41	658	163	97
A-42	2038	27	7	0.29	2	0.2443	4.1	2.838	9.1	0.08426	8.1	0.46	1409	52	1366	68	1298	157	109
A-43	2432	67	11	0.95	0	0.1605	3.8	1.578	7.2	0.07132	6	0.54	960	34	962	44	966	123	99
A-44	14021	138	44	0.37	0	0.3074	4.5	4.474	5.9	0.1056	3.8	0.77	1728	69	1726	49	1724	69	100
A-45	2275	869	15	0.26	0	0.01731	4.4	0.1299	6.8	0.05447	5.1	0.66	111	5	124	8	390	115	28
A-46	6996	180	29	0.16	0	0.1625	6.1	1.642	21.7	0.07334	21	0.28	970	55	987	137	1023	422	95
A-50	2473	235	16	0.54	0	0.06954	4.2	0.5432	6	0.05667	4.3	0.7	433	18	441	22	478	96	91
A-51	6107	620	34	0.36	b.d.	0.05497	6.7	0.4026	30.7	0.05313	30	0.22	345	22	344	89	334	679	103
A-52	1355	226	7	0.72	0	0.03305	4.8	0.2331	9.2	0.05116	7.9	0.52	210	10	213	18	248	181	85
A-53	16854	92	39	0.54	b.d.	0.3937	5	7.914	10.8	0.1458	9.6	0.46	2140	90	2221	97	2297	165	93
A-54	22594	302	68	0.25	0	0.2212	7.4	2.671	11.7	0.08759	9.1	0.63	1288	87	1320	86	1373	174	94
A-55	441	74	2	1.2	0	0.03323	5	0.2296	17	0.05013	16	0.29	211	10	210	32	200	376	105
A-56	52206	421	131	0.17	0	0.2945	3.8	5.447	4.7	0.1342	2.8	0.8	1664	55	1892	40	2153	49	77
A-57	2317	174	10	0.12	1	0.05684	5	0.4251	12	0.05426	11	0.42	356	17	360	36	381	246	93
A-58	1636	255	9	0.58	0	0.0341	4.8	0.237	10.2	0.05043	9	0.47	216	10	216	20	214	209	101
A-59	3363	251	16	0.17	2	0.06477	5	0.4629	10.8	0.05185	9.6	0.46	405	19	386	35	278	220	145
A-60	724	115	4	0.68	0	0.0363	4.7	0.255	10.6	0.05096	9.5	0.44	230	11	231	22	238	219	96
A-61	8998	231	38	0.27	0	0.1658	5.3	1.721	8.2	0.07532	6.3	0.65	989	48	1017	53	1076	126	92
A-62	34796	346	108	0.17	0	0.3019	4	4.45	5.9	0.1069	4.3	0.67	1701	59	1722	49	1747	80	97
A-63	2177	414	13	0.42	0	0.03076	4.1	0.2158	8.8	0.0509	7.8	0.47	195	8	198	16	235	179	83
A-64	1100	461	7	0.17	0	0.01533	4.8	0.1013	7.8	0.04796	6.1	0.62	98	5	98	7	97	144	101
A-65	48	895	1	0.69	0	0.00077	7.5	0.00499	42.5	0.04701	42	0.18	5	0	5	2	49	998	10
A-66	3535	273	21	0.2	0	0.07935	4	0.616	6.3	0.05632	4.9	0.64	492	19	487	24	464	108	106
A-67	4256	1359	26	0.3	0	0.0198	4.7	0.1331	6.1	0.04874	3.9	0.77	126	6	127	7	135	91	94

spot number	$^{207}Pb^a$ (cps)	U^b (ppm)	Pb^b (ppm)	$\underline{Th^b}$ U	$\underline{^{206}Pb^c}$ ^{204}Pb	$\underline{^{206}Pb^c}$ ^{238}U	2 s %	$\underline{^{207}Pb^c}$ ^{235}U	2 s %	$\underline{^{207}Pb^c}$ ^{206}Pb	2 s %	rho^d	$\underline{^{206}Pb}$ ^{238}U	2 s (Ma)	$\underline{^{207}Pb}$ ^{235}U	2 s (Ma)	$\underline{^{207}Pb}$ ^{206}Pb	2 s (Ma)	conc $\%^e$
A-69	847	109	5	0.95	0	0.04756	4.3	0.3428	8.1	0.0523	6.8	0.54	300	13	299	21	298	155	**101**
A-73	9653	1033	58	0.24	0	0.05689	5	0.4204	14	0.0536	13	0.36	357	18	356	42	354	294	**101**
A-74	3241	238	15	0.42	0	0.06498	10.2	0.5017	14.1	0.05601	9.8	0.72	406	40	413	48	452	217	90
A-75	3811	310	22	0.44	0	0.07227	4.6	0.5645	8	0.05667	6.6	0.57	450	20	454	29	478	146	**94**
A-76	3160	83	13	1.16	0	0.161	4	1.659	6.9	0.07477	5.7	0.57	962	35	993	44	1062	114	**91**
A-77	6909	138	22	0.16	2	0.1563	4.6	1.56	9.2	0.07239	8	0.5	936	40	954	57	996	162	**94**
A-78	961	8655	6	1.04	0	0.00076	4.8	0.0049	10.5	0.04682	9.4	0.46	5	0	5	1	39	224	12
A-79	8107	170	30	0.42	0	0.1758	4.9	1.88	6.9	0.07758	4.8	0.72	1044	48	1074	45	1136	95	**92**
A-80	792	1078	5	0.45	0	0.00434	4.6	0.02789	10.7	0.04659	9.7	0.42	28	1	28	3	27	233	**102**
A-81	5018	728	25	0.54	0	0.03412	5	0.29	9.9	0.06166	8.5	0.5	216	11	259	23	662	183	33
A-82	48898	449	135	0.2	1	0.2932	4	4.17	5.7	0.1032	4.1	0.69	1657	58	1668	47	1682	76	**99**
A-83	5687	430	33	0.26	b.d.	0.07538	4	0.8374	10.2	0.08059	9.4	0.39	469	18	618	47	1211	185	39
A-84	50094	508	123	0.15	0	0.2323	5	3.776	5.5	0.1179	2.2	0.91	1347	61	1588	44	1924	40	70
A-85	1425	651	9	0.21	0	0.01433	6	0.1035	15	0.05238	14	0.4	92	5	100	14	301	313	30
A-86	39014	900	153	0.15	2	0.1692	4.2	1.784	11.7	0.07647	11	0.36	1008	39	1040	76	1107	219	**91**
A-87	1252	584	7	0.23	0	0.01222	4.1	0.08347	8.9	0.04957	7.9	0.46	78	3	81	7	174	183	45
A-88	248	563	1	0.57	0	0.00208	6.5	0.02199	24.4	0.07665	24	0.27	13	1	22	5	1112	469	1
A-89	6207	414	31	0.53	0	0.07499	5.4	0.5832	7.6	0.05642	5.3	0.72	466	24	467	28	468	117	**100**
A-90	12231	362	58	0.24	1	0.1595	5.3	1.604	18.2	0.07294	17	0.29	954	47	972	114	1012	353	**94**
A-91	9654	122	36	0.24	b.d.	0.289	4.7	4.044	9.7	0.1015	8.5	0.48	1637	68	1643	79	1651	157	**99**
A-92	38167	466	127	0.48	b.d.	0.2649	4.6	3.804	8	0.1042	6.5	0.58	1515	62	1594	64	1699	121	89
A-96	1511	695	9	0.33	0	0.01333	5.6	0.08781	9.4	0.0478	7.5	0.6	85	5	85	8	89	178	**96**
A-97	15383	380	64	0.12	b.d.	0.1676	3.8	1.744	7.9	0.07548	7	0.48	999	35	1025	51	1081	140	**92**
A-98	8887	409	39	0.15	0	0.09614	4.8	0.9069	6.8	0.06844	4.8	0.71	592	27	655	33	881	98	67
A-99	389	639	2	0.3	0	0.0035	4.8	0.02243	15.3	0.04647	14	0.32	23	1	23	3	21	347	**105**
A-100	1551	665	9	0.25	0	0.01372	4.9	0.09553	10.6	0.05053	9.4	0.46	88	4	93	9	219	217	40
A-101	138	203	0	0.53	0	0.00214	8.9	0.03367	32	0.1142	31	0.28	14	1	34	11	1866	554	1

spot number	$^{207}Pb^a$ (cps)	U^b (ppm)	Pb^b (ppm)	$\frac{Th^b}{U}$	$\frac{^{206}Pb^c}{^{204}Pb}$	$\frac{^{206}Pb^c}{^{238}U}$	2 s %	$\frac{^{207}Pb^c}{^{235}U}$	2 s %	$\frac{^{207}Pb^c}{^{206}Pb}$	2 s %	rho^d	$\frac{^{206}Pb}{^{238}U}$	2 s (Ma)	$\frac{^{207}Pb}{^{235}U}$	2 s (Ma)	$\frac{^{207}Pb}{^{206}Pb}$	2 s (Ma)	conc %[e]
A-102	5393	258	29	0.61	0	0.1117	3.9	0.9584	5.6	0.06225	4	0.69	683	25	682	28	682	86	100
A-103	1575	121	8	0.32	0	0.06635	4.7	0.539	11	0.05894	10	0.42	414	19	438	39	564	217	73
A-104	11542	283	52	0.45	b.d.	0.1835	4.2	1.983	10.1	0.07839	9.2	0.42	1086	42	1110	68	1156	182	94
A-105	65283	831	241	0.01	0	0.2837	4.8	3.842	5.2	0.09825	2.1	0.91	1610	68	1602	42	1591	40	101
A-106	8349	167	37	0.17	0	0.2194	4.3	2.51	14.3	0.08301	14	0.3	1279	50	1275	104	1269	266	101
A-107	11526	269	57	0.31	b.d.	0.2124	3.8	2.384	8.8	0.08142	7.9	0.43	1241	43	1238	63	1231	155	101
A-108	24982	1307	106	0.2	0	0.08128	4.5	0.7401	8.3	0.06606	7	0.54	504	22	562	36	808	146	62
A-109	3439	1393	22	0.57	0	0.01585	4.5	0.111	6.6	0.05079	4.9	0.68	101	4	107	7	230	113	44
A-110	64	1054	1	0.32	0	0.0005	8.5	0.00325	36.6	0.04734	36	0.23	3	0	3	1	66	848	5
A-111	14618	410	68	0.57	0	0.1648	4.5	1.721	5.5	0.07573	3.2	0.82	984	41	1016	36	1087	64	90
A-112	2583	133	13	0.07	0	0.1011	4.8	0.8538	9.9	0.06124	8.7	0.48	621	28	627	46	647	186	96
A-113	4833	112	20	1.57	0	0.1795	4.4	1.875	6.9	0.07577	5.3	0.64	1064	43	1072	46	1088	106	98
A-114	2911	132	14	0.45	0	0.1043	5.4	0.8937	6.5	0.06218	3.6	0.83	639	33	648	31	680	77	94
A-115	275	404	1	0.39	0	0.00376	6	0.02429	28	0.04686	27	0.22	24	1	24	7	41	655	59
A-119	263	434	0	0.37	0	0.00103	9.5	0.02907	24	0.2042	22	0.39	7	1	29	7	2860	359	0
A-120	2823	469	14	0.32	0	0.03122	4.6	0.2367	6.7	0.05499	4.9	0.68	198	9	216	13	411	109	48
A-121	14113	289	54	0.21	0	0.1853	4.4	1.904	5.7	0.07453	3.6	0.77	1096	44	1082	38	1055	72	104
A-122	1088	1580	6	0.25	0	0.0041	4.6	0.02627	10.3	0.04652	9.3	0.44	26	1	26	3	24	222	109
A-123	2402	1155	15	0.32	0	0.0136	3.9	0.08974	10.6	0.04787	9.9	0.36	87	3	87	9	92	235	94
A-124	4374	317	10	1.09	0	0.03193	12.6	0.2508	14.5	0.05698	7.1	0.87	203	25	227	29	490	158	41
A-125	12366	205	42	0.28	0	0.2052	4.7	2.401	9.7	0.0849	8.5	0.49	1203	52	1243	70	1313	165	92
A-126	35917	1304	188	0.2	0	0.145	4.9	1.384	7.6	0.06924	5.8	0.64	873	40	882	45	905	120	96
A-127	873	344	5	0.23	0	0.01608	6.8	0.1067	12.6	0.04814	11	0.54	103	7	103	12	105	250	98
A-129	2720	1126	19	0.17	0	0.01736	4.7	0.1153	7.7	0.04819	6.2	0.6	111	5	111	8	108	146	103
A-130	9527	119	35	0.28	b.d.	0.2872	4	3.923	14	0.09911	13	0.28	1627	57	1618	113	1607	250	101
A-131	15937	354	62	0.28	0	0.1754	4.9	1.887	11.3	0.07803	10	0.43	1042	47	1076	75	1147	203	91
A-132	398	504	3	0.22	0	0.00386	7.2	0.02695	22.7	0.05063	22	0.32	25	2	27	6	223	497	11

spot number	$^{207}Pb^a$ (cps)	U^b (ppm)	Pb^b (ppm)	$\underline{Th^b}$ U	$\underline{^{206}Pb^c}$ ^{204}Pb	$\underline{^{206}Pb^c}$ ^{238}U	2 s %	$\underline{^{207}Pb^c}$ ^{235}U	2 s %	$\underline{^{207}Pb^c}$ ^{206}Pb	2 s %	rho^d	$\underline{^{206}Pb}$ ^{238}U	2 s (Ma)	$\underline{^{207}Pb}$ ^{235}U	2 s (Ma)	$\underline{^{207}Pb}$ ^{206}Pb	2 s (Ma)	conc $\%^e$
A-133	1065	488	4	0.24	0	0.00912	5.8	0.06078	17.2	0.04833	16	0.34	59	3	60	10	115	382	51
A-134	2063	853	15	0.26	0	0.01742	4	0.1157	9.1	0.04816	8.2	0.44	111	4	111	10	107	194	104
A-135	622	1051	4	0.61	0	0.00428	5	0.02749	19.5	0.04662	19	0.26	28	1	28	5	29	451	95
A-136	165	532	0	0.36	0	0.00063	11.2	0.01189	32.2	0.1375	30	0.35	4	0	12	4	2195	524	0
A-137	22329	547	96	0.28	0	0.1743	4.3	1.859	9.5	0.07741	8.5	0.45	1036	41	1067	63	1131	169	92
A-138	262	1747	2	1.03	0	0.00089	5	0.00835	20.2	0.06834	20	0.25	6	0	8	2	878	405	1
A-142	1289	103	7	0.42	0	0.07096	4.7	0.5535	9.8	0.05659	8.5	0.48	442	20	447	35	475	189	93
A-143	38	292	0	0.35	0	0.0008	13.1	0.00652	79.3	0.0588	78	0.17	5	1	7	5	559	1706	1
A-144	5	283	0	0.33	9	0.00061	15.7	-0.00337	489.7	-0.04022	490	0.03	4	1	-3	-17	0	0	
A-145	12415	354	57	0.14	1	0.1605	4	1.544	5.3	0.0698	3.5	0.75	960	35	948	33	922	71	104
A-146	5593	106	22	0.35	0	0.2048	4.3	2.326	5.6	0.08239	3.5	0.77	1201	47	1220	40	1254	69	96
A-147	1575	641	6	0.24	0	0.01027	6.3	0.07452	8.3	0.05265	5.3	0.76	66	4	73	6	313	122	21
A-148	61746	1726	196	0.12	2	0.1147	4.7	0.9745	6.2	0.06165	4	0.76	700	31	691	31	661	87	106
A-149	8482	101	31	0.91	0	0.2956	3.9	4.048	5.9	0.09933	4.5	0.66	1670	57	1644	48	1611	83	104
A-150	2162	196	11	0.34	0	0.0593	4.1	0.4796	8	0.05868	6.9	0.52	371	15	398	26	554	150	67
A-151	3397	285	18	0.29	b.d.	0.06201	4.5	0.8054	6	0.09422	4	0.74	388	17	600	27	1512	75	26
A-152	2858	56	12	0.3	0	0.2063	4.2	2.339	7.3	0.08224	6	0.57	1209	46	1224	52	1251	117	97
A-153	1019	420	7	0.23	0	0.01608	5.3	0.1069	10.1	0.04824	8.6	0.53	103	5	103	10	110	202	93

PL2

spot number	$^{207}Pb^a$ (cps)	U^b (ppm)	Pb^b (ppm)	$\underline{Th^b}$ U	$\underline{^{206}Pb^c}$ ^{204}Pb	$\underline{^{206}Pb^c}$ ^{238}U	2 s %	$\underline{^{207}Pb^c}$ ^{235}U	2 s %	$\underline{^{207}Pb^c}$ ^{206}Pb	2 s %	rho^d	$\underline{^{206}Pb}$ ^{238}U	2 s (Ma)	$\underline{^{207}Pb}$ ^{235}U	2 s (Ma)	$\underline{^{207}Pb}$ ^{206}Pb	2 s (Ma)	conc $\%^e$
A-96	698	243	1	0.63	16	0.00323	6.9	0.02987	32.7	0.06713	32	0.21	21	1	30	10	841	665	2
A-101	3349	2932	13	0.6	1	0.00442	3.1	0.0284	6.7	0.04658	5.9	0.46	28	1	28	2	27	142	106
A-102	883	284	5	0.38	1	0.01719	3.4	0.09772	11.5	0.04125	11	0.3	110	4	95	10	-273	280	-40
A-103	882	470	2	0.65	6	0.00436	5.2	0.02803	23.8	0.04662	23	0.22	28	1	28	7	29	558	97
A-104	702	329	1	0.54	10	0.00412	4.8	0.02899	31.7	0.05107	31	0.15	26	1	29	9	243	721	11
A-105	260	246	1	0.66	2	0.00482	4.6	0.02821	29.8	0.04248	29	0.15	31	1	28	8	-199	739	-16
A-106	51	66	0	0.6	18	0.00187	16.8	0.01471	200.8	0.05719	200	0.08	12	2	15	30	498	4408	2

spot number	$^{207}Pb^a$ (cps)	U^b (ppm)	Pb^b (ppm)	$\underline{Th^b}$ U	$\underline{^{206}Pb^c}$ ^{204}Pb	$\underline{^{206}Pb^c}$ ^{238}U	2 s %	$\underline{^{207}Pb^c}$ ^{235}U	2 s %	$\underline{^{207}Pb^c}$ ^{206}Pb	2 s %	rho^d	$\underline{^{206}Pb}$ ^{238}U	2 s (Ma)	$\underline{^{207}Pb}$ ^{235}U	2 s (Ma)	$\underline{^{207}Pb}$ ^{206}Pb	2 s (Ma)	conc %^{e}
A-107	1089	431	2	0.86	9	0.00443	4.2	0.03818	21	0.0626	21	0.2	28	1	38	8	694	438	4
A-108	314	212	1	0.49	3	0.0046	6.2	0.03168	21.6	0.04994	21	0.29	30	2	32	7	191	480	15
A-109	412	527	2	0.91	0	0.00455	3.8	0.0252	14.3	0.04021	14	0.26	29	1	25	4	-338	356	-9
A-110	523	572	3	0.67	1	0.00497	3.7	0.03019	13.9	0.04406	13	0.27	32	1	30	4	-108	329	-30
A-111	153	205	1	0.4	3	0.00549	4.2	0.01739	52.9	0.02299	53	0.08	35	1	18	9	-2068	2010	-2
A-112	213	246	1	0.57	2	0.00526	4.1	0.02639	46.4	0.03638	46	0.09	34	1	26	12	-603	1257	-6
A-113	516	553	3	0.4	1	0.00499	3.1	0.02794	16	0.04065	16	0.19	32	1	28	4	-310	403	-10
A-114	325	334	1	0.65	1	0.00442	5.4	0.02549	25.1	0.04182	25	0.21	28	2	26	6	-238	620	-12
A-115	178	219	1	0.65	3	0.00499	5.2	0.01864	40.5	0.02712	40	0.13	32	2	19	8	-1482	1328	-2
A-116	155	181	1	0.77	3	0.00466	5.8	0.01896	52	0.02951	52	0.11	30	2	19	10	-1211	1608	-2
A-117	354	372	2	0.39	1	0.00503	3.2	0.02821	20.4	0.04073	20	0.16	32	1	28	6	-305	516	-11
A-118	260	366	2	0.39	2	0.00445	4.5	0.01944	36	0.03167	36	0.12	29	1	20	7	-996	1058	-3
A-119	115	164	1	0.31	2	0.00366	7.3	0.01321	70.9	0.02619	71	0.1	24	2	13	9	-1599	2400	-1
A-120	325	435	2	0.86	1	0.00419	4.4	0.02688	16.6	0.04659	16	0.26	27	1	27	4	28	384	**98**
A-125	994	512	3	0.5	8	0.00499	4.5	0.03786	22.4	0.05509	22	0.2	32	1	38	8	415	489	8
A-126	845	1038	4	0.74	1	0.00425	3.8	0.02937	14.1	0.05019	14	0.27	27	1	29	4	203	316	13
A-127	747	816	3	1.06	1	0.00425	3.1	0.02493	12.6	0.04259	12	0.25	27	1	25	3	-192	305	-14
A-128	728	872	4	1.22	1	0.00469	3.1	0.02763	14	0.04277	14	0.22	30	1	28	4	-182	341	-17
A-129	426	524	2	0.8	1	0.0048	3.3	0.02412	21	0.03644	21	0.16	31	1	24	5	-599	563	-5
A-130	2348	861	12	0.65	1	0.01398	2.5	0.08365	6.9	0.04339	6.4	0.37	90	2	82	5	-146	158	-61
A-131	874	833	3	0.15	3	0.00396	3.1	0.02758	16.8	0.05049	16	0.19	25	1	28	5	217	382	12
A-132	615	680	3	1.24	1	0.0044	3.6	0.02302	15.7	0.03793	15	0.23	28	1	23	4	-491	408	-6
A-133	596	854	3	0.59	1	0.00376	3	0.02097	12.3	0.04044	12	0.24	24	1	21	3	-323	306	-7
A-134	317	453	2	0.62	1	0.00366	4.8	0.02075	22.3	0.04119	22	0.21	24	1	21	5	-277	554	-9
A-135	250508	609	1	0.67	87	0.00219	208	0.02489	260	0.08238	160	0.8	14	29	25	64	1254	3050	1
A-136	301	381	2	0.64	2	0.00442	3.9	0.01897	26.1	0.03113	26	0.15	28	1	19	5	-1047	772	-3

PL3

spot number	$^{207}Pb^a$ (cps)	U^b (ppm)	Pb^b (ppm)	$\underline{Th^b}$ U	$\underline{^{206}Pb^c}$ ^{204}Pb	$\underline{^{206}Pb^c}$ ^{238}U	2 s %	$\underline{^{207}Pb^c}$ ^{235}U	2 s %	$\underline{^{207}Pb^c}$ ^{206}Pb	2 s %	rho^d	$\underline{^{206}Pb}$ ^{238}U	2 s (Ma)	$\underline{^{207}Pb}$ ^{235}U	2 s (Ma)	$\underline{^{207}Pb}$ ^{206}Pb	2 s (Ma)	conc %^{e}
A-5	168	150	0.73	0	1.09528	0.0	4.09631	0.0	21.71018	0.0	21.32	0	32	1					
A-6	583	567	2.58	0	0.63066	0.0	2.98801	0.0	9.63033	0.0	9.16	0	30	1					
A-7	248	131	0.22	0	14.88067	0.0	9.10807	0.0	62.80236	0.1	62.14	0	11	1	14	9	693	1	2
A-8	254	249	0.95	1	2.08812	0.0	3.38029	0.0	24.61334	0.0	24.38	0	25	1					
A-9	190	243	0.81	0	0.82776	0.0	2.90564	0.0	19.75623	0.0	19.54	0	22	1					
A-10	5483	690	23.19	0	0.90506	0.0	2.19903	0.2	3.65119	0.0	2.91	1	218	5					
A-11	225	276	1.07	0	0.48717	0.0	3.46974	0.0	11.38228	0.0	10.84	0	26	1					
A-12	346	374	1.44	0	1.56734	0.0	3.08128	0.0	13.19154	0.0	12.83	0	26	1					
A-13	670	811	2.98	0	1.07859	0.0	2.73097	0.0	8.16469	0.0	7.69	0	24	1					
A-14	250	279	1.06	0	0.49757	0.0	2.47132	0.0	18.17629	0.0	18.01	0	25	1					
A-15	2507	857	10.34	1	1.19880	0.0	2.13487	0.1	10.87963	0.0	10.67	0	80	2					
A-16	463	519	1.93	1	1.39610	0.0	2.69527	0.0	21.07274	0.0	20.90	0	25	1					
A-17	122	243	0.39	1	2.67955	0.0	5.51859	0.0	23.72270	0.0	23.07	0	11	1					
A-18	376	474	1.82	0	1.14576	0.0	2.72523	0.0	11.46430	0.0	11.14	0	25	1					
A-19	379	56	0.14	0	49.25673	0.0	13.74502	0.0	74.13372	0.1	72.85	0	16	2	36	26	1692	2	1
A-20	138	159	0.62	0	0.73531	0.0	3.42823	0.0	26.12899	0.0	25.90	0	26	1					
A-21	3004	556	13.00	1	0.94693	0.0	2.31660	0.1	9.70215	0.0	9.42	0	153	4					
A-22	26	41	0.10	1	6.23780	0.0	6.43790	0.0	460.25469	0.0	460.21	0	16	1					
A-23	594	189	0.33	1	27.88026	0.0	4.70199	0.0	85.81769	0.0	85.69	0	12	1					
A-24	392	334	1.52	1	1.83655	0.0	3.34792	0.0	15.94266	0.0	15.59	0	30	1					
A-29	308	283	0.99	1	1.48535	0.0	3.00023	0.0	12.42226	0.1	12.05	0	23	1	25	3	245	1	9
A-30	289	374	1.22	0	0.78102	0.0	3.34438	0.0	13.56273	0.0	13.14	0	22	1					
A-31	260	606	1.14	1	1.49690	0.0	3.91542	0.0	23.34516	0.0	23.01	0	13	0					
A-32	176	185	0.73	1	0.98387	0.0	4.05526	0.0	15.39638	0.0	14.85	0	26	1					
A-33	665	441	1.38	1	7.67288	0.0	3.14009	0.0	88.46755	0.0	88.41	0	21	1					
A-34	190	185	0.70	1	1.60540	0.0	3.01045	0.0	21.57396	0.0	21.36	0	25	1					

spot number	$^{207}Pb^a$ (cps)	U^b (ppm)	Pb^b (ppm)	$\underline{Th^b}$ U	$\underline{^{206}Pb^c}$ ^{204}Pb	$\underline{^{206}Pb^c}$ ^{238}U	2 s %	$\underline{^{207}Pb^c}$ ^{235}U	2 s %	$\underline{^{207}Pb^c}$ ^{206}Pb	2 s %	rho^d	$\underline{^{206}Pb}$ ^{238}U	2 s (Ma)	$\underline{^{207}Pb}$ ^{235}U	2 s (Ma)	$\underline{^{207}Pb}$ ^{206}Pb	2 s (Ma)	conc %^{e}
A-35	329	383	1.34	1	1.27255	0.0	2.72268	0.0	16.38634	0.0	16.16	0	23	1					
A-36	297	400	1.46	1	0.83062	0.0	2.92812	0.0	11.71086	0.0	11.34	0	24	1					
A-37	97	115	0.47	0	-0.17663	0.0	3.65544	0.0	24.84593	0.0	24.58	0	27	1	27	7	69	1	39
A-38	748	695	3.13	1	0.83764	0.0	2.25490	0.0	7.01294	0.0	6.64	0	30	1	30	2	29	1	101
A-39	339	387	1.40	0	0.76989	0.0	2.75919	0.0	11.52564	0.0	11.19	0	24	1					
A-40	291	252	0.96	1	3.79533	0.0	3.50304	0.0	32.27582	0.0	32.09	0	26	1					
A-41	3523	62	0.68	0	89.48728														
A-42	39	46	0.09	0	4.05504	0.0	6.15874	0.0	53.24185	0.0	52.88	0	13	1	13	7	12	1	113
A-43	163	148	0.71	0	0.79633	0.0	3.33130	0.0	15.18661	0.0	14.82	0	32	1					
A-44	228	276	0.98	0	1.16850	0.0	2.78722	0.0	21.13471	0.0	20.95	0	24	1					
A-45	597	641	2.12	1	2.79775	0.0	2.73710	0.0	20.55307	0.0	20.37	0	22	1					
A-46	344	339	0.94	0	0.98900	0.0	5.29986	0.0	20.82500	0.0	20.14	0	18	1					
A-47	381	457	1.64	1	0.84707	0.0	2.75347	0.0	15.74561	0.0	15.50	0	24	1					
A-48	980	278	4.62	0	0.73806	0.0	2.23193	0.1	8.82257	0.0	8.54	0	109	2					
A-53	825	853	3.65	1	1.51978	0.0	2.56118	0.0	14.24798	0.0	14.02	0	28	1					
A-54	1608	1811	5.76	1	2.54081	0.0	2.34281	0.0	18.88907	0.0	18.74	0	21	0					
A-55	975	855	3.67	0	2.73794	0.0	2.50722	0.0	15.02843	0.0	14.82	0	28	1					
A-56	714	1002	3.38	1	0.77195	0.0	2.39708	0.0	11.22497	0.0	10.97	0	22	1					
A-57	434	540	2.09	0	1.52859	0.0	2.94300	0.0	16.13908	0.0	15.87	0	26	1					
A-58	252	334	1.05	0	1.71456	0.0	3.99553	0.0	25.03369	0.0	24.71	0	21	1					
A-59	226	233	1.06	1	1.04900	0.0	3.26669	0.0	20.94982	0.0	20.69	0	30	1					
A-60	41	89	0.20	1	0.81266	0.0	4.44139	0.0	27.98469	0.0	27.63	0	15	1	15	4	16	1	91
A-61	765	799	3.46	0	0.63514	0.0	2.37830	0.0	9.26370	0.0	8.95	0	29	1					
A-62	736	670	3.05	0	0.81656	0.0	2.78368	0.0	7.22254	0.0	6.66	0	30	1					
A-63	636	918	3.00	0	1.11334	0.0	2.77659	0.0	12.75064	0.0	12.44	0	22	1					

spot number	$^{207}Pb^a$ (cps)	U^b (ppm)	Pb^b (ppm)	Th^b / U	$\frac{^{206}Pb^c}{^{204}Pb}$	$\frac{^{206}Pb^c}{^{238}U}$	2 s %	$\frac{^{207}Pb^c}{^{235}U}$	2 s %	$\frac{^{207}Pb^c}{^{206}Pb}$	2 s %	rho^d	$\frac{^{206}Pb}{^{238}U}$	2 s (Ma)	$\frac{^{207}Pb}{^{235}U}$	2 s (Ma)	$\frac{^{207}Pb}{^{206}Pb}$	2 s (Ma)	conc %^{e}
A-65	85	186	0.45	1	1.45644	0.0	3.61854	0.0	32.02249	0.0	31.82	0	16	1					
A-66	523	619	2.31	0	1.11700	0.0	2.83584	0.0	9.81348	0.0	9.39	0	25	1					
A-67	415	504	1.91	0	0.63916	0.0	2.58611	0.0	13.86805	0.0	13.62	0	25	1					
A-68	788	611	2.74	1	3.92008	0.0	2.78065	0.0	37.68475	0.0	37.58	0	30	1					
A-69	926	1149	4.32	1	1.30035	0.0	2.84791	0.0	14.63825	0.0	14.36	0	25	1					
A-70	292	361	1.38	0	1.05935	0.0	2.87857	0.0	13.56378	0.0	13.25	0	25	1					
A-71	174	69	0.12	1	16.75754	0.0	6.79522	0.0	71.76814	0.1	71.45	0	11	1	15	11	734	1	2
A-72	324	274	0.94	0	2.75585	0.0	2.72671	0.0	14.89913	0.0	14.65	0	23	1	23	3	24	1	93
A-77	55	136	0.31	1	1.00405	0.0	4.85657	0.0	37.95640	0.0	37.64	0	15	1					
A-78	1577	2066	7.32	0	0.29951	0.0	2.24006	0.0	4.75105	0.0	4.19	0	23	1					
A-79	176	194	0.69	0	1.13184	0.0	3.14761	0.0	12.34894	0.0	11.94	0	23	1	23	3	22	1	106
A-80	234	146	0.78	1	3.45131	0.0	3.72240	0.0	22.92089	0.0	22.62	0	35	1					
A-81	1382	101	0.29	1	64.70951	0.0	18.72780	0.0	156.82721	0.1	155.70	0	18	3	30	47	1129	3	2
A-82	305	72	0.18	1	28.93518	0.0	6.98364	0.0	450.19248	0.0	450.14	0	17	1					
A-83	902	1851	4.25	1	1.20953	0.0	2.44914	0.0	10.90906	0.0	10.63	0	15	0					
A-84	1001	199	3.97	0	0.75644	0.0	2.54760	0.1	9.23405	0.0	8.88	0	130	3	131	11	138	3	94
A-85	180	126	0.28	0	9.71199	0.0	6.67653	0.0	53.85922	0.0	53.44	0	15	1					
A-86	566	138	1.76	0	2.92367	0.0	3.42374	0.1	20.18628	0.0	19.89	0	85	3					
A-87	1245	320	1.33	0	17.37980	0.0	3.36177	0.0	39.15971	0.0	39.02	0	28	1					
A-88	835	248	0.88	0	16.54048	0.0	4.14696	0.0	33.38123	0.1	33.12	0	23	1	27	9	406	1	6
A-89	4448	511	1.95	0	32.90849	0.0	4.24637	0.0	40.80459	0.0	40.58	0	25	1					
A-90	337	398	1.50	0	1.00615	0.0	3.13426	0.0	13.85587	0.0	13.50	0	25	1					
A-91	970	1345	4.35	1	2.09087	0.0	2.28695	0.0	17.00723	0.0	16.85	0	22	0					
A-92	181	171	0.82	1	0.79186	0.0	3.12162	0.0	13.38086	0.0	13.01	0	32	1					
A-93	425	372	1.59	0	2.82915	0.0	2.66435	0.0	14.45811	0.0	14.21	0	29	1					
A-94	370	372	1.79	0	0.68168	0.0	3.05917	0.0	8.69110	0.0	8.13	0	32	1					
A-95	46	42	0.17	0	3.27789	0.0	6.54412	0.0	77.63480	0.0	77.36	0	27	2					

spot number	$^{207}Pb^a$ (cps)	U^b (ppm)	Pb^b (ppm)	$\underline{Th^b}$ U	$\underline{^{206}Pb^c}$ ^{204}Pb	$\underline{^{206}Pb^c}$ ^{238}U	2 s %	$\underline{^{207}Pb^c}$ ^{235}U	2 s %	$\underline{^{207}Pb^c}$ ^{206}Pb	2 s %	rho^d	$\underline{^{206}Pb}$ ^{238}U	2 s (Ma)	$\underline{^{207}Pb}$ ^{235}U	2 s (Ma)	$\underline{^{207}Pb}$ ^{206}Pb	2 s (Ma)	conc %^{e}
A-96	42	104	0.21	0	-0.34613	0.0	4.57316	0.0	30.18399	0.0	29.84	0	13	1	13	4	14	1	**99**
A-101	347	306	1.60	1	1.34026	0.0	3.32728	0.0	16.45395	0.0	16.11	0	35	1					
A-102	411	360	1.44	1	2.75137	0.0	3.70564	0.0	29.62245	0.0	29.39	0	26	1					
A-103	397	522	1.89	1	0.50619	0.0	2.70410	0.0	12.62243	0.0	12.33	0	24	1					
A-104	872	52	0.27	0	67.59308	0.0	10.73431	0.1	70.23505	0.1	69.41	0	32	3	94	63	2270	3	1
A-105	1040	1168	4.38	0	1.30911	0.0	2.57426	0.0	11.49074	0.0	11.20	0	25	1					
A-106	415	465	1.59	0	1.23859	0.0	2.62664	0.0	10.81706	0.0	10.49	0	23	1	23	2	22	1	**104**
A-107	685	679	2.97	1	1.44310	0.0	2.38837	0.0	11.51801	0.0	11.27	0	29	1					
A-108																			
A-109	256	230	1.11	0	1.38578	0.0	2.95967	0.0	11.00982	0.0	10.60	0	32	1	32	3	33	1	**96**
A-110	235	251	1.15	1	0.97009	0.0	2.69847	0.0	19.80543	0.0	19.62	0	30	1					
A-111	183	98	0.18	0	17.23284	0.0	6.91301	0.0	120.83426	0.0	120.64	0	12	1					
A-112	67	70	0.16	1	3.92533	0.0	6.12293	0.0	45.91976	0.1	45.51	0	15	1	17	8	305	1	5
A-113	2396	2800	10.96	1	1.22368	0.0	2.26525	0.0	8.04984	0.0	7.72	0	26	1					
A-114	251	188	0.92	1	0.79894	0.0	3.53083	0.0	13.48163	0.1	13.01	0	32	1	36	5	301	1	11
A-115	327	258	0.93	1	3.43976	0.0	3.00988	0.0	31.67385	0.0	31.53	0	24	1					
A-116	253	303	1.15	1	1.40935	0.0	2.72048	0.0	36.09132	0.0	35.99	0	25	1	25	9	26	1	**95**
A-117	552	294	1.36	1	3.66578	0.0	3.55012	0.0	9.93198	0.1	9.28	0	30	1	36	3	416	1	7
A-118	339	144	0.52	0	10.84566	0.0	3.32440	0.0	66.31986	0.0	66.24	0	24	1					
A-119	282	274	1.34	1	1.42395	0.0	3.32862	0.0	20.37507	0.0	20.10	0	32	1					
A-120	56	110	0.26	1	1.92860	0.0	5.69425	0.0	21.52240	0.0	20.76	0	16	1	16	3	16	1	**97**
A-125	430	483	1.78	0	0.76866	0.0	2.65936	0.0	10.28396	0.0	9.93	0	24	1					
A-126	765	1010	3.19	1	1.06130	0.0	2.76344	0.0	10.30309	0.0	9.93	0	21	1					
A-127	155	185	0.71	0	1.73236	0.0	3.23038	0.0	19.70344	0.0	19.44	0	25	1					
A-128	4257	35	9.00	2	2.72072	0.3	2.30494	3.1	5.81249	0.1	5.34	0	1470	30	1442	45	1401	30	**105**
A-129	257	248	1.23	0	1.97339	0.0	3.46443	0.0	26.43748	0.0	26.21	0	33	1					

spot number	$^{207}Pb^{a}$ (cps)	U^{b} (ppm)	Pb^{b} (ppm)	$\underline{Th^{b}}$ U	$\underline{^{206}Pb^{c}}$ ^{204}Pb	$\underline{^{206}Pb^{c}}$ ^{238}U	2 s %	$\underline{^{207}Pb^{c}}$ ^{235}U	2 s %	$\underline{^{207}Pb^{c}}$ ^{206}Pb	2 s %	rho^{d}	$\underline{^{206}Pb}$ ^{238}U	2 s (Ma)	$\underline{^{207}Pb}$ ^{235}U	2 s (Ma)	$\underline{^{207}Pb}$ ^{206}Pb	2 s (Ma)	conc %^{e}
A-131	303	347	1.39	1	0.82082	0.0	2.94000	0.0	14.92783	0.0	14.64	0	26	1					
A-132	198	245	0.90	0	0.15270	0.0	3.76563	0.0	15.86340	0.0	15.41	0	24	1	24	4	34	1	71
A-133	334	254	0.88	0	5.52373	0.0	3.27256	0.0	43.41811	0.0	43.29	0	23	1					
A-134	651	769	2.81	0	1.04474	0.0	2.39348	0.0	11.59253	0.0	11.34	0	24	1					
A-135	524	705	2.70	0	0.62357	0.0	2.43225	0.0	11.51783	0.0	11.26	0	25	1					
A-136	271	345	1.35	1	1.01056	0.0	3.34050	0.0	12.29809	0.0	11.84	0	26	1					
A-137	447	216	0.85	0	7.30104	0.0	3.01267	0.0	27.24903	0.0	27.08	0	26	1					
A-138	300	342	1.37	0	0.66112	0.0	2.56186	0.0	9.64574	0.0	9.30	0	26	1	26	3	26	1	101
A-139	116	133	0.50	0	0.96332	0.0	5.68599	0.0	25.77801	0.0	25.14	0	25	1					
A-140	211	289	1.01	0	1.34799	0.0	4.13372	0.0	20.71727	0.0	20.30	0	23	1					
A-141	78	72	0.29	0	-0.26718	0.0	3.89129	0.0	26.07550	0.1	25.78	0	27	1	35	9	681	1	4
A-142	57	133	0.28	1	1.72915	0.0	4.68639	0.0	44.64239	0.0	44.40	0	14	1					
A-143	86	154	0.32	1	3.78827	0.0	4.51461	0.0	49.13421	0.0	48.93	0	14	1					
A-144	24	72	0.16	1	1.25009	0.0	5.99882	0.0	77.43033	0.0	77.20	0	15	1					
A-149	371	486	1.67	1	1.32431	0.0	2.60032	0.0	17.40218	0.0	17.21	0	23	1					
A-150	467	138	0.63	0	14.03211	0.0	4.79494	0.0	82.75225	0.0	82.61	0	31	1					
A-151	342	384	1.34	1	2.05952	0.0	3.37818	0.0	22.95211	0.0	22.70	0	23	1					
A-152	272	277	1.08	0	1.11153	0.0	3.21598	0.0	13.18410	0.0	12.79	0	26	1					
A-153	1609	2297	7.37	1	1.46833	0.0	3.74074	0.0	19.76007	0.0	19.40	0	21	1					
A-154	131	122	0.58	0	0.73968	0.0	3.51432	0.0	23.78063	0.0	23.52	0	31	1					
A-155	595	728	2.33	1	4.01869	0.0	3.45860	0.0	48.46147	0.0	48.34	0	22	1					
A-156	107	199	0.67	1	5.16876	0.0	4.92445	0.0	77.56317	0.0	77.41	0	22	1					
A-157	336	438	1.68	0	0.72227	0.0	2.28986	0.0	19.84871	0.0	19.72	0	25	1					
A-158	1501	996	4.11	0	3.10928	0.0	2.13072	0.0	10.20699	0.1	9.98	0	27	1	33	3	511	1	5
A-159	549	582	2.40	0	0.62805	0.0	3.08558	0.0	12.21181	0.0	11.82	0	27	1					
A-160	217	291	1.09	0	0.45810	0.0	3.37868	0.0	10.65061	0.0	10.10	0	25	1					
A-161	91	111	0.40	0	1.33004	0.0	3.78618	0.0	29.16793	0.0	28.92	0	24	1					

spot number	^{207}Pb[a] (cps)	U[b] (ppm)	Pb[b] (ppm)	Th[b] / U	^{206}Pb[c] / ^{204}Pb	^{206}Pb[c] / ^{238}U	2 s %	^{207}Pb[c] / ^{235}U	2 s %	^{207}Pb[c] / ^{206}Pb	2 s %	rho[d]	^{206}Pb / ^{238}U (Ma)	2 s (Ma)	^{207}Pb / ^{235}U (Ma)	2 s (Ma)	^{207}Pb / ^{206}Pb (Ma)	2 s (Ma)	conc %[e]
A-162	503	627	2.41	0	1.25661	0.0	2.36095	0.0	10.80532	0.0	10.54	0	26	1					
A-163	1109	1149	5.29	1	0.48898	0.0	2.54804	0.0	6.19414	0.0	5.65	0	30	1					
A-164	87	123	0.26	1	4.47531	0.0	12.25874	0.0	41.82691	0.1	39.99	0	14	2	17	7	461	2	3
A-165	847	241	3.68	0	0.65655	0.0	2.33624	0.1	5.28341	0.0	4.74	0	100	2	100	5	100	2	**100**
A-166	57	155	0.31	1	2.05273	0.0	5.12825	0.0	68.23141	0.0	68.04	0	13	1					
A-167	552	650	2.51	0	1.10290	0.0	2.61565	0.0	14.20109	0.0	13.96	0	26	1					
A-168	2043	2553	9.73	0	0.64469	0.0	2.10651	0.0	7.22071	0.0	6.91	0	25	1					
A-173	1651	281	7.60	0	0.72929	0.0	2.11033	0.2	7.81498	0.0	7.52	0	177	4					
A-174	15055	1846	60.06	0	0.51486	0.0	2.12588	0.2	4.12398	0.1	3.53	1	210	4	211	8	217	4	**97**
A-175	1009	927	4.09	0	1.94452	0.0	2.35249	0.0	12.05302	0.0	11.82	0	29	1					
A-176	487	526	2.00	0	1.60310	0.0	2.49600	0.0	10.62032	0.0	10.32	0	25	1					
A-177	443	508	2.39	1	0.62485	0.0	2.66638	0.0	13.04785	0.0	12.77	0	31	1					
A-178	554	640	2.69	1	0.97406	0.0	2.45703	0.0	13.28785	0.0	13.06	0	28	1					
A-179	666	742	2.98	1	1.16598	0.0	2.62238	0.0	15.29317	0.0	15.07	0	27	1					
A-180	978	1450	4.97	1	1.08254	0.0	2.52404	0.0	10.61885	0.0	10.31	0	23	1					
A-181	318	311	1.40	1	1.07314	0.0	2.88194	0.0	15.68528	0.0	15.42	0	30	1					
A-182	1020	1228	4.47	0	1.45312	0.0	2.47848	0.0	9.76920	0.0	9.45	0	24	1					
A-183	245	252	1.14	1	1.16392	0.0	3.30197	0.0	16.42119	0.0	16.09	0	30	1					
A-184	1491	1887	6.91	0	0.77511	0.0	2.24846	0.0	9.47060	0.0	9.20	0	24	1					
A-185	324	307	1.39	0	1.65840	0.0	3.32455	0.0	21.43482	0.0	21.18	0	30	1					
A-186	552	595	2.48	1	2.52355	0.0	3.69147	0.0	21.23265	0.0	20.91	0	27	1					
A-187	197	196	0.89	0	0.73129	0.0	3.31491	0.0	11.67981	0.0	11.20	0	30	1	30	3	28	1	**105**
A-188	1721	1720	7.98	0	1.03317	0.0	2.29514	0.0	4.91435	0.0	4.35	0	31	1					
A-189	188	191	0.81	0	1.06896	0.0	3.06314	0.0	12.76929	0.0	12.40	0	28	1					
A-190	396	717	1.40	1	2.58960	0.0	3.01630	0.0	17.31470	0.0	17.05	0	13	0					
A-191	842	990	4.04	1	1.31007	0.0	2.73525	0.0	13.62149	0.0	13.34	0	27	1					

spot number	$^{207}Pb^a$ (cps)	U^b (ppm)	Pb^b (ppm)	Th^b/U	$^{206}Pb^c$/^{204}Pb	$^{206}Pb^c$/^{238}U	2 s %	$^{207}Pb^c$/^{235}U	2 s %	$^{207}Pb^c$/^{206}Pb	2 s %	rho^d	^{206}Pb/^{238}U (Ma)	2 s (Ma)	^{207}Pb/^{235}U (Ma)	2 s (Ma)	^{207}Pb/^{206}Pb (Ma)	2 s (Ma)	conc %^{e}
A-197	268	253	1.15	1	2.28456	0.0	3.24264	0.0	22.30514	0.0	22.07	0	30	1					
A-198	321	190	0.89	0	5.83222	0.0	3.43540	0.0	26.78934	0.0	26.57	0	31	1					
A-199	199	208	0.97	1	0.91358	0.0	3.01764	0.0	13.14792	0.0	12.80	0	31	1					
A-200	350	431	1.45	1	1.74269	0.0	2.58722	0.0	14.66600	0.0	14.44	0	22	1					
A-201	259	250	0.87	1	1.73170	0.0	3.72106	0.0	21.40802	0.0	21.08	0	23	1					
A-202	91	96	0.33	0	2.77104	0.0	3.96950	0.0	29.42045	0.0	29.15	0	23	1					
A-203	544	340	2.01	1	4.33654	0.0	2.76191	0.0	23.42621	0.0	23.26	0	39	1					
A-204	1037	1419	5.30	0	0.66999	0.0	2.21569	0.0	6.28878	0.0	5.89	0	25	1					
A-205	394	539	1.84	1	1.27883	0.0	3.09505	0.0	11.06138	0.0	10.62	0	23	1					
A-206	285	213	1.01	0	4.88521	0.0	3.63491	0.0	35.58029	0.0	35.39	0	32	1					
A-207	382	400	1.56	0	2.20253	0.0	2.61210	0.0	15.56544	0.0	15.34	0	26	1					
A-208	69	67	0.17	1	4.70393	0.0	9.75128	0.0	99.50428	0.0	99.03	0	17	2					
A-209	39	101	0.22	0	0.94956	0.0	4.89908	0.0	37.26763	0.0	36.94	0	14	1					
A-210	353	430	1.45	1	1.29235	0.0	2.72625	0.0	10.92032	0.0	10.57	0	22	1	22	2	21	1	104
A-211	169	221	0.92	0	1.19657	0.0	3.79689	0.0	24.35332	0.0	24.06	0	28	1					
A-212	41	68	0.07	0	0.77683	0.0	28.93779	0.0	57.17984	0.1	49.32	1	7	2	8	5	450	2	2
A-213	349	415	1.71	0	0.54654	0.0	3.46327	0.0	14.74553	0.0	14.33	0	27	1					
A-214	601	879	3.10	1	0.36621	0.0	2.86629	0.0	7.53830	0.0	6.97	0	23	1					
A-215	236	333	1.08	1	1.84013	0.0	3.51762	0.0	25.56865	0.0	25.33	0	22	1					
A-216	246	295	1.11	0	1.01513	0.0	3.64223	0.0	21.29904	0.0	20.99	0	25	1					
A-221	177	396	0.78	1	2.05964	0.0	3.61546	0.0	20.49999	0.0	20.18	0	13	0					
A-222	405	577	1.84	1	0.54388	0.0	2.42846	0.0	12.34562	0.0	12.10	0	21	1					
A-223	4221	41	0.31	1	83.25722														
A-224	177	230	0.79	0	1.45168	0.0	4.13511	0.0	16.25140	0.0	15.72	0	23	1					
A-225	224	302	1.08	0	0.76658	0.0	3.31722	0.0	14.42882	0.0	14.04	0	24	1					
A-226	357	448	1.71	0	1.29752	0.0	2.76838	0.0	12.92772	0.0	12.63	0	25	1					
A-227	427	562	2.14	0	0.74120	0.0	2.74339	0.0	8.89426	0.0	8.46	0	25	1					

spot number	$^{207}Pb^a$ (cps)	U^b (ppm)	Pb^b (ppm)	$\frac{Th^b}{U}$	$\frac{^{206}Pb^c}{^{204}Pb}$	$\frac{^{206}Pb^c}{^{238}U}$	2 s %	$\frac{^{207}Pb^c}{^{235}U}$	2 s %	$\frac{^{207}Pb^c}{^{206}Pb}$	2 s %	rho^d	$\frac{^{206}Pb}{^{238}U}$	2 s (Ma)	$\frac{^{207}Pb}{^{235}U}$	2 s (Ma)	$\frac{^{207}Pb}{^{206}Pb}$	2 s (Ma)	conc %[e]
A-228	119	104	0.47	1	1.65102	0.0	4.09780	0.0	20.62159	0.0	20.21	0	30	1					
A-229	898	932	4.54	0	0.59392	0.0	2.28809	0.0	7.21393	0.0	6.84	0	32	1					
A-230	314	185	0.65	0	6.80635	0.0	5.87891	0.0	31.67398	0.1	31.12	0	23	1	25	8	252	1.3533	9
A-231	557	14	0.05	1	42.17497	0.0	102.27836	0.1	149.59549	0.2	109.17	1	20	21	100	143	3106	20.787	1
A-232	431	518	1.73	0	1.42998	0.0	2.75833	0.0	16.34638	0.0	16.11	0	22	1					
A-233	722	673	2.55	0	1.56377	0.0	2.94430	0.0	8.79352	0.1	8.29	0	25	1	28	2	346	0.7288	7
A-234	2731	800	11.70	1	1.32866	0.0	2.20554	0.1	10.67552	0.0	10.45	0	97	2					
A-235	1491	1907	7.55	0	0.40639	0.0	2.15115	0.0	5.77111	0.0	5.36	0	26	1					
A-236	308	312	1.36	0	2.46369	0.0	2.99916	0.0	27.47283	0.0	27.31	0	29	1					
A-237	71	85	0.18	0	4.40330	0.0	10.62049	0.0	56.69838	0.0	55.69	0	14	2					
A-238	214	247	0.85	1	1.66065	0.0	3.38704	0.0	11.41936	0.0	10.91	0	23	1					
A-239	403	387	1.66	1	1.74769	0.0	3.59763	0.0	10.71548	0.0	10.09	0	28	1					
A-240	1347	439	5.96	0	0.57117	0.0	2.22146	0.1	6.24445	0.0	5.84	0	89	2					
A-245	20963	177	1.26	1	82.43951														
A-246	347	439	1.61	0	1.30012	0.0	2.79734	0.0	10.37598	0.0	9.99	0	24	1					
A-247	229	266	0.92	1	2.00462	0.0	3.04162	0.0	27.29766	0.0	27.13	0	23	1					
A-248	27	65	0.14	1	1.97152	0.0	6.68195	0.0	29.21842	0.0	28.44	0	14	1					
A-249	326	428	1.46	0	1.23685	0.0	2.88851	0.0	12.52240	0.0	12.18	0	23	1					
A-250	207	311	1.10	1	1.08745	0.0	2.93075	0.0	26.05272	0.0	25.89	0	24	1					
A-251	592	186	2.90	0	0.75827	0.0	2.80976	0.1	10.81978	0.0	10.45	0	103	3					
A-252	46	91	0.20	0	2.03780	0.0	5.86590	0.0	34.41951	0.0	33.92	0	14	1					
A-253	33	91	0.18	0	1.89766	0.0	6.44557	0.0	71.08972	0.0	70.80	0	13	1					
A-254	2669	3759	14.10	0	0.34327	0.0	2.14885	0.0	5.09848	0.0	4.62	0	25	1					
A-255	38	68	0.18	1	2.19809	0.0	6.61493	0.0	30.58827	0.1	29.86	0	17	1	23	7	651	1.137	3
A-256	519	536	2.54	0	0.86033	0.0	2.72918	0.0	12.59939	0.0	12.30	0	31	1					
A-257	107	132	0.59	1	0.80069	0.0	3.50288	0.0	29.35575	0.0	29.15	0	30	1					

RM3 (1st & 2nd mount)

| spot | $^{207}Pb^a$ | U^b | Pb^b | Th^b | $^{206}Pb^c$ | $^{206}Pb^c$ | 2 s | $^{207}Pb^c$ | 2 s | $^{207}Pb^c$ | 2 s | rho^d | ^{206}Pb | 2 s | ^{207}Pb | 2 s | ^{207}Pb | 2 s | |
number	(cps)	(ppm)	(ppm)	U	^{204}Pb	^{238}U	%	^{235}U	%	^{206}Pb	%		^{238}U	(Ma)	^{235}U	(Ma)	^{206}Pb	(Ma)	conc %^{e}
a41	2072	521	7	0.1	4056	0.01361	2.8	0.09683	5.5	0.05158	4.7	0.51	87	2	94	5	267	108	33
a42	3096	2061	11	0.12	3628	0.00594	1.8	0.03829	3.7	0.04673	3.3	0.48	38	1	38	1	35	78	**108**
a43	3660	395	14	0.04	6208	0.03673	2.1	0.30004	5	0.05924	4.5	0.42	233	5	266	12	576	98	40
a44	1053	720	6	0.27	2269	0.00554	2.8	0.03574	7.7	0.0468	7.1	0.37	36	1	36	3	39	170	**91**
a45	92	122	0	0.86	167	0.00246	3	-0.02442	278.1	-0.07194	278.1	0.01	16	0	-25	-68			
a46	2334	30	5	0.39	2063	0.16857	3	1.7234	6.6	0.07415	5.9	0.45	1004	28	1017	43	1046	118	**96**
A-5	1616	523	9.39	0.001	0.21293	0.018	5.12401	0.123	8.66374	0.049	7	0.59	117	6	118	10	131	164	90
A-6	1245	672	7.07	0.015	1.13682	0.011	5.94834	0.071	9.09091	0.048	6.9	0.65	69	4	70	6	99	163	70
A-7	586	1037	4.31	0.251	0	0.004	4.95191	0.027	12.6077	0.047	12	0.39	27	1	27	3	29	278	**95**
A-8	5140	353	29.79	0.303	0.08394	0.085	4.74938	0.69	13.1965	0.059	12	0.36	528	24	533	55	554	269	**95**
A-9	51348	1396	257.88	0.061	0.64295	0.185	4.80162	1.858	5.40864	0.073	2.5	0.89	1092	48	1066	36	1014	50	**108**
A-10	3143	816	25.53	0.047	1.42738	0.032	4.6034	0.222	8.80916	0.051	7.5	0.52	202	9	204	16	222	174	**91**
A-11	3716	1030	23.07	0.159	1.53598	0.023	4.99662	0.146	8.37713	0.046	6.7	0.6	146	7	138	11	8	162	1892
A-12	6412	143	26.67	0.27	2.68174	0.186	5.66147	1.893	10.1814	0.074	8.5	0.56	1101	57	1079	68	1034	171	**106**
A-15	14858	740	30.87	0.033	2.36726	0.04	6.27761	0.702	6.98279	0.128	3.1	0.9	251	15	540	29	2071	54	12
A-16	24930	280	48.6	0.177	4.07177	0.165	5.06648	3.003	6.05744	0.132	3.3	0.84	983	46	1408	46	2128	58	46
A-17	580	149	0.75	0.626	19.4871	0.005	11.2718	0.034	76.5745	0.049	76	0.15	33	4	34	26	134	1780	25
A-18	4079	62	13.6	0.184	1.58723	0.216	5.98349	2.447	8.56418	0.082	6.1	0.7	1262	69	1257	62	1247	120	**101**
A-21	460	806	0.02	0.551	62.4081	0	188.601	0.001	202.899	0.292	75	0.93	0	0	1	2	3425	1163	0
A-22	4575	2140	28.47	0.002	0.03477	0.014	8.29517	0.09	11.7296	0.048	8.3	0.71	87	7	87	10	97	196	90
A-23	1333	803	10.21	0.024	0.48974	0.013	5.20947	0.085	13.0919	0.048	12	0.4	83	4	83	10	80	285	**105**
A-24	5493	147	18.61	0.528	0	0.126	5.43723	1.364	6.46939	0.079	3.5	0.84	763	39	874	38	1166	69	65
A-29	2356	1375	14.27	0.048	0	0.011	5.75735	0.07	7.98747	0.048	5.5	0.72	68	4	69	5	96	131	71
A-30	8135	218	27.71	0.059	0.27759	0.124	4.39605	1.737	5.05253	0.102	2.5	0.87	752	31	1022	33	1656	46	45

189

spot number	$^{207}Pb^{[a]}$ (cps)	$U^{[b]}$ (ppm)	$Pb^{[b]}$ (ppm)	$Th^{[b]}/U$	$^{206}Pb^{[c]}/^{204}Pb$	$^{206}Pb^{[c]}/^{238}U$	2 s %	$^{207}Pb^{[c]}/^{235}U$	2 s %	$^{207}Pb^{[c]}/^{206}Pb$	2 s %	$rho^{[d]}$	$^{206}Pb/^{238}U$	2 s (Ma)	$^{207}Pb/^{235}U$	2 s (Ma)	$^{207}Pb/^{206}Pb$	2 s (Ma)	conc %[e]
A-31	14997	1304	57.95	0.071	4.00603	0.044	4.70412	0.457	7.67189	0.075	6.1	0.61	280	13	382	24	1060	122	26
A-32	14044	382	67.75	0.129	0	0.177	4.61255	1.822	5.52211	0.075	3	0.84	1050	45	1053	36	1060	61	99
A-33	2099	859	8.72	0.005	0.49085	0.01	5.36801	0.068	9.1787	0.047	7.4	0.58	67	4	67	6	69	177	96
A-34	1787	2025	10.74	0.31	0	0.005	4.89954	0.036	8.5222	0.048	7	0.57	35	2	36	3	82	165	42
A-35	3634	4062	23.1	0.021	0.26416	0.006	4.73284	0.038	6.98888	0.047	5.1	0.68	37	2	37	3	42	123	90
A-36	1403	888	11.71	0.008	0	0.013	6.01461	0.089	10.142	0.048	8.2	0.59	86	5	87	8	94	193	92
A-37	25046	416	28.06	0.128	5.91469	0.065	8.99554	0.991	10.3721	0.11	5.2	0.87	408	36	699	52	1798	94	23
A-38	2709	1304	18.36	0.05	0.60149	0.014	4.93289	0.095	7.24039	0.048	5.3	0.68	92	5	92	6	102	125	90
A-39	2685	1503	18.81	0.083	0	0.013	5.00073	0.084	6.99056	0.048	4.9	0.72	82	4	82	6	81	116	101
A-40	4660	82	14.14	0.194	1.3434	0.173	5.69941	1.718	8.39144	0.072	6.2	0.68	1026	54	1015	54	992	125	103
A-41	605	528	3.09	0.427	0	0.006	5.66186	0.042	10.6061	0.051	9	0.53	38	2	41	4	221	207	17
A-42	8130	216	26.6	0.108	2.3883	0.124	5.10902	1.119	6.78638	0.066	4.5	0.75	753	36	762	36	791	94	95
A-43	14908	310	20.65	0.149	1.80239	0.067	13.4399	0.599	15.5953	0.065	7.9	0.86	419	55	477	59	765	167	55
A-45	895	440	4.51	0.031	0.71569	0.01	4.89762	0.068	12.2143	0.047	11	0.4	67	3	67	8	69	266	97
A-46	463	198	2.51	0.009	0.33797	0.013	6.73029	0.106	18.2581	0.06	17	0.37	82	5	102	18	599	367	14
A-47	29573	323	92.63	0.098	0.85903	0.28	4.56614	3.942	5.34829	0.102	2.8	0.85	1589	64	1622	43	1666	52	95
A-48	13288	137	44.23	0.415	b.d.	0.313	4.9945	4.552	8.94582	0.105	7.4	0.56	1757	77	1741	74	1721	136	102
A-53	3501	4185	23.82	0.261	0	0.006	4.5035	0.038	6.28935	0.047	4.4	0.72	37	2	37	2	35	105	106
A-54	848	424	5	0.102	0	0.012	5.27417	0.079	10.1384	0.047	8.7	0.52	77	4	77	8	72	206	107
A-55	1800	47	6.16	0.386	0	0.13	5.54914	1.568	7.57601	0.087	5.2	0.73	789	41	958	47	1368	99	58
A-56	975	513	6.11	0.042	0.83999	0.012	6.0541	0.081	19.5338	0.048	19	0.31	78	5	79	15	100	439	78
A-57	7219	115	23.39	0.76	1.08563	0.203	5.68336	2.247	20.2482	0.08	19	0.28	1190	62	1196	142	1207	383	99
A-58	7998	456	26.63	0.102	2.9451	0.06	5.29234	0.385	8.86551	0.047	7.1	0.6	374	19	331	25	37	170	1011
A-59	2031	42	7.25	0.31	0	0.174	7.17467	1.838	9.28337	0.077	5.9	0.77	1033	68	1059	61	1113	118	93
A-60	5031	104	20.6	0.197	0.66941	0.198	7.05854	2.129	41.9233	0.078	41	0.17	1164	75	1158	290	1147	821	101
A-61	3209	79	12.57	0.26	2.78159	0.161	4.87523	1.296	7.53225	0.058	5.7	0.65	964	44	844	43	541	126	178
A-62	1590	1492	11.56	0.527	b.d.	0.007	5.18358	0.12	12.3905	0.116	11	0.42	48	2	115	13	1899	202	3

spot number	$^{207}Pb^a$ (cps)	U^b (ppm)	Pb^b (ppm)	$\frac{Th^b}{U}$	$\frac{^{206}Pb^c}{^{204}Pb}$	$\frac{^{206}Pb^c}{^{238}U}$	2 s %	$\frac{^{207}Pb^c}{^{235}U}$	2 s %	$\frac{^{207}Pb^c}{^{206}Pb}$	2 s %	rho^d	$\frac{^{206}Pb}{^{238}U}$	2 s (Ma)	$\frac{^{207}Pb}{^{235}U}$	2 s (Ma)	$\frac{^{207}Pb}{^{206}Pb}$	2 s (Ma)	conc %[e]
A-63	12763	460	36.5	0.034	0.01598	0.078	5.81379	1.001	6.64419	0.093	3.2	0.88	484	27	705	34	1492	61	32
A-67	8165	129	9.56	0.38	0.31152	0.072	10.2943	1.08	18.5428	0.109	15	0.56	448	45	744	98	1780	281	25
A-68	1508	575	7.41	0.384	b.d.	0.013	7.74772	0.122	87.7714	0.068	87	0.09	83	6	117	97	882	1808	9
A-69	2821	2659	15.58	0.031	0.37517	0.006	6.14537	0.039	9.81422	0.047	7.7	0.63	39	2	38	4	36	183	**108**
A-70	4274	1138	25.04	0.019	0.85857	0.022	4.87717	0.151	8.34919	0.049	6.8	0.58	143	7	143	11	140	159	**102**
A-71	1409	783	7.35	0.054	0.81269	0.01	5.52876	0.062	10.6478	0.047	9.1	0.52	62	3	62	6	58	217	**106**
A-72	4241	1230	30.71	0.02	b.d.	0.025	6.11501	0.271	16.2909	0.079	15	0.38	158	10	244	35	1178	299	13
A-77	5797	766	32.62	0.081	0	0.043	5.09668	0.388	6.82998	0.066	4.5	0.75	270	13	333	19	799	95	34
A-78	7584	175	32.39	0.263	b.d.	0.184	4.95797	2.009	10.492	0.079	9.2	0.47	1086	50	1118	71	1182	183	**92**
A-79	763	453	4.61	0.081	1.40601	0.01	6.04131	0.068	17.3643	0.047	16	0.35	67	4	67	11	67	388	**100**
A-80	40357	416	97.42	0.121	3.88081	0.22	4.97482	4.396	6.79542	0.145	4.6	0.73	1282	58	1712	56	2286	80	56
A-81	19332	473	52.13	0.233	2.13765	0.111	6.86381	0.936	9.07953	0.061	5.9	0.76	681	44	671	45	637	128	**107**
A-82	30377	341	83.51	0.181	1.73592	0.24	5.0455	3.109	5.76568	0.094	2.8	0.88	1388	63	1435	44	1505	53	**92**
A-83	1839	2001	11.84	0.349	0	0.006	4.87787	0.039	7.23863	0.047	5.3	0.67	39	2	39	3	37	128	**106**
A-84	149	207	0.56	0.458	0	0.003	9.28951	0.035	28.3401	0.096	27	0.33	17	2	35	10	1550	503	1
A-85	1011	1286	8.09	0.055	0.63278	0.006	4.88421	0.042	8.86144	0.047	7.4	0.55	41	2	41	4	41	177	**102**
A-86	404	670	5.03	0.38	b.d.	0.007	14.4741	0.203	38.1594	0.221	35	0.38	43	6	187	65	2988	568	1
A-88	1021	1190	6.42	0.17	0	0.006	5.25867	0.036	10.5245	0.047	9.1	0.5	35	2	35	4	35	218	**101**
A-89	1517	798	9.63	0.047	0.69836	0.012	5.03569	0.081	8.5139	0.048	6.9	0.59	79	4	79	6	76	163	**104**
A-90	2799	43	5.16	0.503	6.15447	0.121	7.2443	0.808	21.2851	0.048	20	0.34	737	50	601	97	116	472	636
A-91	1651	1864	10.03	0.048	0.90771	0.006	5.22957	0.035	7.37011	0.047	5.2	0.71	35	2	35	3	33	124	**107**
A-92	383	141	2.04	0.622	0	0.015	6.3805	0.098	13.6944	0.048	12	0.47	95	6	95	12	89	287	**106**
A-93	5268	78	16.1	0.245	2.61531	0.208	5.476	1.616	18.5174	0.056	18	0.3	1220	61	977	116	464	392	263
A-94	691	677	4.11	0.319	0	0.006	5.72813	0.04	10.2912	0.047	8.5	0.56	40	2	40	4	43	204	**92**
A-95	1917	249	1.18	1.023	0	0.004	12.0025	0.195	17.6937	0.378	13	0.68	24	3	181	29	3821	197	1
A-96	585	151	3.32	0.315	0	0.022	5.97738	0.152	15.4042	0.049	14	0.39	143	8	143	21	153	333	**93**
A-101	2016	2351	13.68	0.062	0.63761	0.006	4.52591	0.038	7.77638	0.047	6.3	0.58	38	2	38	3	35	151	**109**

spot number	$^{207}Pb^a$ (cps)	U^b (ppm)	Pb^b (ppm)	Th^b/U	$^{206}Pb^c$/^{204}Pb	$^{206}Pb^c$/^{238}U	2 s %	$^{207}Pb^c$/^{235}U	2 s %	$^{207}Pb^c$/^{206}Pb	2 s %	rho^d	^{206}Pb/^{238}U (Ma)	2 s (Ma)	^{207}Pb/^{235}U (Ma)	2 s (Ma)	^{207}Pb/^{206}Pb (Ma)	2 s (Ma)	conc %^{e}
A-102	805	472	3.04	0.021	b.d.	0.006	16.7368	0.089	44.1107	0.103	41	0.38	40	7	87	37	1672	754	2
A-103	14267	337	59.89	0.203	b.d.	0.177	5.07977	1.852	15.5808	0.076	15	0.33	1051	49	1064	103	1091	295	96
A-104	2279	1036	19.75	0.015	0.24585	0.019	4.44362	0.13	6.92447	0.049	5.3	0.64	124	5	124	8	129	125	96
A-105	1658	822	8.74	0.061	0.71453	0.011	5.82982	0.071	8.78078	0.048	6.6	0.66	70	4	70	6	77	156	90
A-107	6494	1270	36.87	0.041	2.1277	0.03	5.03573	0.204	8.85268	0.05	7.3	0.57	188	9	188	15	188	169	100
A-108	12509	1217	46.05	0.045	2.80886	0.038	5.18037	0.322	8.32447	0.061	6.5	0.62	242	12	283	21	643	140	38
A-109	1334	624	7.47	0.007	0.31593	0.012	5.60377	0.08	9.44962	0.048	7.6	0.59	78	4	78	7	74	181	106
A-111	2804	1799	18.68	0.314	0	0.011	5.59349	0.069	9.10705	0.047	7.2	0.61	68	4	68	6	72	171	94
A-112	4537	64	16.38	0.311	0	0.25	5.49675	3.184	8.76654	0.093	6.8	0.63	1437	71	1453	68	1478	130	97
A-113	7102	454	16.6	0.05	2.24835	0.037	8.10578	0.294	10.599	0.058	6.8	0.76	234	19	261	24	512	150	46
A-114	2140	1069	12.81	0.372	0	0.012	4.94956	0.081	7.41511	0.048	5.5	0.67	78	4	79	6	84	131	93
A-115	8892	1984	57.02	0.283	b.d.	0.029	4.99283	0.238	8.77515	0.059	7.2	0.57	185	9	217	17	582	157	32
A-116	2180	2288	14.1	0.063	b.d.	0.006	5.43741	0.081	21.5617	0.097	21	0.25	39	2	79	16	1574	391	2
A-117	2774	707	4.33	0.125	b.d.	0.006	15.2353	0.083	17.6798	0.101	9	0.86	38	6	81	14	1632	167	2
A-118	163	238	1.41	0.282	b.d.	0.005	46.0591	0.251	73.4568	0.395	57	0.63	30	14	227	150	3888	862	1
A-119	483	906	2.3	1.054	0	0.003	7.82298	0.021	20.795	0.058	19	0.38	17	1	21	4	547	421	3
A-120	1756	987	11.89	0.054	0.43698	0.012	4.93707	0.081	8.1577	0.048	6.5	0.61	79	4	79	6	82	154	96
A-125	13416	539	37.88	0.074	0.73809	0.069	7.62121	0.872	8.69236	0.092	4.2	0.88	431	32	637	41	1457	79	30
A-126	22631	349	83.07	0.019	0.41428	0.232	4.49397	3.351	6.7298	0.105	5	0.67	1342	54	1493	53	1714	92	78
A-127	529	805	2.81	0.801	0	0.004	5.94653	0.029	13.9012	0.06	13	0.43	23	1	29	4	589	273	4
A-128	209	241	1.41	0.357	0	0.006	7.40445	0.039	30.361	0.047	29	0.24	39	3	39	12	47	703	82
A-129	1260	826	7.26	0.007	0.31023	0.009	5.10122	0.059	7.82469	0.047	5.9	0.65	58	3	58	4	63	141	92
A-130	13518	269	31.25	0.154	1.6453	0.117	11.7498	1.018	14.8047	0.063	9	0.79	713	79	713	76	714	191	100
A-131	1868	33	2.03	0.348	6.13815	0.062	24.0992	0.475	44.4753	0.056	37	0.54	386	90	394	145	445	831	87
A-132	74629	888	227.1	0.141	1.42499	0.251	4.67751	3.301	5.06109	0.095	1.9	0.92	1442	60	1481	39	1537	36	94

RM4 (1st mount)

spot number	$^{207}Pb^a$ (cps)	U^b (ppm)	Pb^b (ppm)	Th^b/U	$\frac{^{206}Pb^c}{^{204}Pb}$	$\frac{^{206}Pb^c}{^{238}U}$	2 s %	$\frac{^{207}Pb^c}{^{235}U}$	2 s %	$\frac{^{207}Pb^c}{^{206}Pb}$	2 s %	rho^d	$\frac{^{206}Pb}{^{238}U}$ (Ma)	2 s (Ma)	$\frac{^{207}Pb}{^{235}U}$ (Ma)	2 s (Ma)	$\frac{^{207}Pb}{^{206}Pb}$ (Ma)	2 s (Ma)	conc %^{e}
a50	596	85	2	0.64	1164	0.02403	2.9	0.17166	9	0.0518	8.5	0.32	153	4	161	13	277	195	55
a51	1375	313	6	0.73	2344	0.01639	1.7	0.10887	5.3	0.04818	5	0.33	105	2	105	5	108	118	97

Sample set "Colorado Plateau"

CV2 (1st mount)

spot number	$^{207}Pb^a$ (cps)	U^b (ppm)	Pb^b (ppm)	Th^b/U	$\frac{^{206}Pb^c}{^{204}Pb}$	$\frac{^{206}Pb^c}{^{238}U}$	2 s %	$\frac{^{207}Pb^c}{^{235}U}$	2 s %	$\frac{^{207}Pb^c}{^{206}Pb}$	2 s %	rho^d	$\frac{^{206}Pb}{^{238}U}$ (Ma)	2 s (Ma)	$\frac{^{207}Pb}{^{235}U}$ (Ma)	2 s (Ma)	$\frac{^{207}Pb}{^{206}Pb}$ (Ma)	2 s (Ma)	conc %^{e}
a1	3096	393	13	1.41	2836	0.02648	2	0.1956	4.4	0.05358	3.9	0.45	168	3	181	7	354	89	48
a2	2055	23	-5	0.87	31	-0.19016	103.2	-16.7828	105	0.64008	19.5	0.98	-1360	-1399			4602	281	-30
a3	2175	233	8	0.99	3112	0.03108	2	0.22231	5.1	0.05188	4.7	0.39	197	4	204	9	280	107	70
a4	4947	72	13	0.68	6771	0.16656	2.4	1.69793	3.7	0.07394	2.8	0.66	993	22	1008	24	1040	56	96
a5	6491	256	18	0.28	2475	0.06851	2	0.54562	3.9	0.05776	3.4	0.5	427	8	442	14	521	74	82
a6	11612	122	27	0.56	14431	0.20536	1.8	2.29485	2.6	0.08105	1.8	0.69	1204	20	1211	18	1223	36	98
a7	22550	254	39	0.58	823	0.1395	3.7	2.02431	4.2	0.10525	1.9	0.89	842	29	1124	29	1719	34	49
a8	6420	96	7	0.06	3418	0.07575	8	0.89049	8.7	0.08526	3.5	0.92	471	36	647	43	1321	67	36
a9	77317	322	104	0.29	66160	0.31318	1.8	5.06424	2.2	0.11728	1.2	0.83	1756	28	1830	18	1915	22	92
a10	2183	27	6	0.83	2892	0.18546	2.1	1.94542	4.1	0.07608	3.5	0.52	1097	21	1097	28	1097	70	100
a11	5768	79	16	1.15	7706	0.16672	1.7	1.73286	2.4	0.07538	1.6	0.74	994	16	1021	15	1079	32	92
a12	3662	239	9	1.01	1422	0.03338	4.6	0.28389	7.9	0.06169	6.4	0.58	212	10	254	18	663	138	32
a13	10846	94	22	0.53	3176	0.22185	2	2.58165	3.3	0.0844	2.7	0.59	1292	23	1295	25	1302	52	99
a14	12647	30	12	1.88	34	0.16129	6.2	10.65971	8.9	0.47933	6.4	0.69	964	55	2494	86	4179	95	23
a15	4392	115	15	1.25	7243	0.10775	1.8	0.90944	3.6	0.06122	3.1	0.51	660	12	657	18	647	67	102
a16	2293	53	7	1.1	1405	0.10383	2.4	1.017	5.7	0.07104	5.1	0.42	637	14	712	29	959	105	66
a17	4013	55	12	1.42	3578	0.17418	2.3	1.84555	3.6	0.07685	2.8	0.64	1035	22	1062	24	1117	55	93
a18	9645	536	26	0.01	13295	0.05309	2	0.40006	3	0.05465	2.3	0.67	333	7	342	9	398	51	84
a19	955	96	3	1.02	1174	0.03142	3.3	0.23894	10.2	0.05515	9.7	0.32	199	6	218	20	418	216	48

spot number	$^{207}Pb^a$ (cps)	U^b (ppm)	Pb^b (ppm)	Th^b/U	$^{206}Pb^c$/^{204}Pb	$^{206}Pb^c$/^{238}U	2 s %	$^{207}Pb^c$/^{235}U	2 s %	$^{207}Pb^c$/^{206}Pb	2 s %	rho^d	^{206}Pb/^{238}U	2 s (Ma)	^{207}Pb/^{235}U	2 s (Ma)	^{207}Pb/^{206}Pb	2 s (Ma)	conc $\%^e$
a20	29114	2753	605	0.07	29917	0.23142	4	2.68592	4.1	0.08417	0.7	0.99	1342	49	1325	31	1297	14	103
a21	69043	392	122	0.32	40494	0.30448	2	4.28664	2.3	0.10211	1.2	0.86	1713	29	1691	19	1663	22	103
a22	11142	92	19	0.53	2955	0.19616	2.4	2.78093	3.5	0.10282	2.4	0.71	1155	26	1350	26	1676	45	69
a23	3560	141	11	0.59	6056	0.07166	2.1	0.58223	4.4	0.05893	3.8	0.49	446	9	466	16	565	83	79
a24	24645	89	41	0.98	1879	0.39952	2.2	7.44282	4.2	0.13511	3.6	0.51	2167	40	2166	38	2165	63	100
a25	3630	198	15	1.61	6765	0.06007	1.8	0.44851	4	0.05415	3.5	0.45	376	7	376	13	377	80	100
a26	7561	104	18	0.39	5594	0.1725	1.9	1.74253	2.8	0.07326	2.1	0.67	1026	18	1024	18	1021	43	100
a27	3060	101	11	1	2621	0.09276	2.1	0.76147	3.6	0.05954	2.9	0.59	572	12	575	16	587	63	97
a28	11684	337	42	1.35	1588	0.10282	2	0.93283	3.2	0.0658	2.5	0.63	631	12	669	16	800	52	79
a29	30400	160	49	0.97	12156	0.25509	2.3	4.10152	2.5	0.11661	1	0.91	1465	30	1655	20	1905	18	77
a30	25792	163	52	1.09	23965	0.26449	2.3	3.70254	2.8	0.10153	1.6	0.82	1513	31	1572	23	1652	30	92
a31	26060	158	47	1.02	14755	0.24522	2.3	3.40789	3.1	0.10079	2.1	0.75	1414	29	1506	24	1639	38	86
a32	24985	919	300	0.33	117	0.22284	2.6	6.38641	3.5	0.20786	2.3	0.75	1297	31	2030	31	2889	37	45
a33	30865	168	51	0.21	28778	0.30075	2	4.48104	2.3	0.10806	1.1	0.88	1695	30	1727	19	1767	20	96
a34	6430	90	15	0.16	8966	0.1687	2.2	1.67993	4.2	0.07222	3.6	0.53	1005	21	1001	27	992	72	101
a35	1378	108	5	0.55	1363	0.04176	2.3	0.29613	4.4	0.05143	3.7	0.52	264	6	263	10	260	86	101
a36	20112	111	35	0.49	7012	0.29323	2.1	4.25226	2.9	0.10517	1.9	0.73	1658	31	1684	24	1717	36	97
a37	714	9	2	1.22	835	0.18279	2.7	2.19943	8.3	0.08727	7.8	0.33	1082	27	1181	60	1366	151	79
a38	16881	465	31	0.51	433	0.05816	2.7	0.7095	5.7	0.08847	5	0.48	364	10	544	24	1393	95	26
a39	55170	863	77	0.64	212	0.06281	2.4	1.57297	2.7	0.18164	1.3	0.87	393	9	960	17	2668	22	15
a40	2725	118	9	0.61	4940	0.07061	1.9	0.54028	4.3	0.0555	3.9	0.44	440	8	439	16	432	87	102

CV3

spot number	$^{207}Pb^a$ (cps)	U^b (ppm)	Pb^b (ppm)	Th^b/U	$^{206}Pb^c$/^{204}Pb	$^{206}Pb^c$/^{238}U	2 s %	$^{207}Pb^c$/^{235}U	2 s %	$^{207}Pb^c$/^{206}Pb	2 s %	rho^d	^{206}Pb/^{238}U	2 s (Ma)	^{207}Pb/^{235}U	2 s (Ma)	^{207}Pb/^{206}Pb	2 s (Ma)	conc $\%^e$
a52	38877	370	50	1.14	443	0.11432	2.8	2.31771	4	0.14703	2.9	0.7	698	18	1218	29	2312	49	30
a53	1092	274	4	0.49	2331	0.01557	1.9	0.10083	6.4	0.04698	6.1	0.3	100	2	98	6	48	145	206
a54	32262	531	59	1.33	150	0.08261	2.1	1.78277	4	0.15652	3.4	0.52	512	10	1039	26	2418	58	21

spot number	$^{207}Pb^a$ (cps)	U^b (ppm)	Pb^b (ppm)	$\frac{Th^b}{U}$	$\frac{^{206}Pb^c}{^{204}Pb}$	$\frac{^{206}Pb^c}{^{238}U}$	2 s %	$\frac{^{207}Pb^c}{^{235}U}$	2 s %	$\frac{^{207}Pb^c}{^{206}Pb}$	2 s %	rho^d	$\frac{^{206}Pb}{^{238}U}$ (Ma)	2 s (Ma)	$\frac{^{207}Pb}{^{235}U}$ (Ma)	2 s (Ma)	$\frac{^{207}Pb}{^{206}Pb}$ (Ma)	2 s (Ma)	conc %^{e}
a55	5752	300	21	1.38	3954	0.05897	2	0.49598	3.5	0.061	3	0.55	369	7	409	12	639	64	58
a56	5899	72	11	1.02	1141	0.12393	3.8	1.39765	6.8	0.0818	5.7	0.55	753	27	888	41	1241	111	61

CV5

spot number	$^{207}Pb^a$ (cps)	U^b (ppm)	Pb^b (ppm)	$\frac{Th^b}{U}$	$\frac{^{206}Pb^c}{^{204}Pb}$	$\frac{^{206}Pb^c}{^{238}U}$	2 s %	$\frac{^{207}Pb^c}{^{235}U}$	2 s %	$\frac{^{207}Pb^c}{^{206}Pb}$	2 s %	rho^d	$\frac{^{206}Pb}{^{238}U}$ (Ma)	2 s (Ma)	$\frac{^{207}Pb}{^{235}U}$ (Ma)	2 s (Ma)	$\frac{^{207}Pb}{^{206}Pb}$ (Ma)	2 s (Ma)	conc %^{e}
A-133	127383	4028	260	0.81	0	0.05757	5.9	1.67500	6.9	0.21110	3.5	0.86	361	21	999	44	2914	56	12
A-134	23604	468	75	0.44	3	0.16120	5.9	1.62300	20.7	0.07305	20.0	0.28	963	53	979	130	1015	402	95
A-135	773	74	5	0.39	0	0.06910	4.9	0.53660	10.7	0.05633	9.5	0.46	431	20	436	38	465	210	93
A-136	26194	389	102	0.30	b.d.	0.25820	4.8	3.32300	6.6	0.09335	4.5	0.73	1481	64	1486	52	1494	86	99
A-137	14641	389	62	0.11	b.d.	0.15830	4.7	1.60100	15.5	0.07337	15.0	0.30	947	41	971	97	1024	299	93
A-139	35146	674	115	0.27	1	0.16780	4.5	2.08900	7.4	0.09030	5.9	0.61	1000	42	1145	51	1431	113	70
A-140	1119	523	7	0.30	0	0.01458	5.0	0.09605	8.7	0.04778	7.1	0.57	93	5	93	8	88	169	106
A-141	10003	215	50	0.26	0	0.22840	4.7	2.84400	11.3	0.09031	10.0	0.41	1326	56	1367	85	1432	196	93
A-142	1747	427	11	0.64	0	0.02731	4.8	0.19960	7.0	0.05301	5.1	0.68	174	8	185	12	329	116	53
A-143	247	490	2	0.31	0	0.00439	5.7	0.02822	17.8	0.04668	17.0	0.32	28	2	28	5	32	403	88
A-144	153289	2891	457	0.34	5	0.15680	5.0	1.75600	7.1	0.08125	5.0	0.71	939	44	1029	46	1227	98	77
A-150	855	1330	6	0.26	0	0.00424	5.6	0.02719	11.9	0.04655	10.0	0.47	27	2	27	3	25	251	107
A-151	93642	966	297	0.27	0	0.29880	4.8	4.41900	5.6	0.10730	2.9	0.86	1685	71	1716	47	1754	53	96
A-152	29660	591	126	0.39	b.d.	0.21190	4.7	2.44000	6.5	0.08356	4.4	0.73	1239	53	1255	47	1282	87	97
A-153	2736	150	16	0.52	0	0.10710	5.4	0.89860	9.7	0.06084	8.0	0.56	656	34	651	46	633	172	104
A-154	4853	384	29	0.46	b.d.	0.07644	5.0	0.80210	17.3	0.07612	17.0	0.29	475	23	598	78	1098	332	43
A-155	38451	421	114	0.32	1	0.26160	6.1	4.21900	16.8	0.11700	16.0	0.36	1498	81	1678	138	1910	282	78
A-156	5744	116	24	0.27	0	0.20580	5.2	2.36500	6.7	0.08336	4.2	0.78	1206	58	1232	48	1277	83	94
A-157	70299	680	256	0.19	b.d.	0.36350	4.9	5.74400	5.5	0.11460	2.4	0.90	1999	84	1938	47	1874	43	107
A-158	75235	2623	72	0.08	24	0.02799	5.8	0.21990	15.2	0.05700	14.0	0.38	178	10	202	28	491	310	36
A-159	15259	178	45	0.22	2	0.24970	5.1	3.04800	7.5	0.08854	5.6	0.67	1437	65	1420	58	1394	107	103
A-160	51699	169	109	0.29	b.d.	0.57730	4.9	16.37000	7.0	0.20570	5.0	0.70	2938	115	2899	67	2871	81	102

spot number	$^{207}Pb^a$ (cps)	U^b (ppm)	Pb^b (ppm)	$\frac{Th^b}{U}$	$\frac{^{206}Pb^c}{^{204}Pb}$	$\frac{^{206}Pb^c}{^{238}U}$	2 s %	$\frac{^{207}Pb^c}{^{235}U}$	2 s %	$\frac{^{207}Pb^c}{^{206}Pb}$	2 s %	rho^d	$\frac{^{206}Pb}{^{238}U}$	2 s (Ma)	$\frac{^{207}Pb}{^{235}U}$	2 s (Ma)	$\frac{^{207}Pb}{^{206}Pb}$	2 s (Ma)	conc %[e]
A-161	90580	1285	322	0.07	1	0.24610	4.6	3.22800	5.1	0.09515	2.2	0.90	1418	58	1464	39	1530	42	**93**
A-162	21270	343	77	0.29	b.d.	0.22240	5.3	2.63600	9.0	0.08598	7.3	0.59	1294	62	1311	66	1337	141	**97**
A-163	14196	719	48	0.96	5	0.06775	5.9	0.61010	27.6	0.06533	27.0	0.21	423	24	484	106	784	567	54
A-164	3254	246	18	0.09	1	0.07381	4.7	0.55640	8.1	0.05468	6.6	0.58	459	21	449	29	399	147	115
A-165	27648	228	81	0.28	1	0.34430	4.9	5.27700	8.5	0.11120	6.9	0.58	1908	81	1865	72	1818	125	**105**
A-166	10329	948	60	0.14	b.d.	0.06425	5.1	0.49130	17.4	0.05548	17.0	0.29	401	20	406	58	431	370	**93**
A-167	6286	653	38	0.75	b.d.	0.05889	4.8	0.53430	13.2	0.06583	12.0	0.36	369	17	435	47	800	258	46
A-168	22977	299	84	0.29	0	0.27710	5.2	3.62600	10.5	0.09493	9.1	0.49	1577	73	1555	83	1526	172	**103**
A-173	1728	759	12	0.28	0	0.01656	4.6	0.11530	10.2	0.05052	9.1	0.45	106	5	111	11	218	210	49
A-174	39913	1136	132	0.09	5	0.11600	4.6	1.12400	5.7	0.07029	3.5	0.79	708	31	765	31	936	72	76
A-175	5070	59	17	0.48	b.d.	0.28670	7.0	4.15700	37.1	0.10520	36.0	0.19	1625	100	1666	303	1718	669	**95**
A-176	21191	261	59	0.24	3	0.22440	6.1	2.53500	6.9	0.08195	3.2	0.89	1305	72	1282	50	1244	63	**105**
A-177	1676	53	4	0.41	b.d.	0.07065	13.1	1.27200	35.0	0.13060	32.0	0.37	440	56	833	199	2105	570	21
A-178	2413	593	14	0.38	b.d.	0.02294	7.4	0.39090	9.4	0.12360	5.8	0.79	146	11	335	27	2009	103	7
A-179	3424	285	20	0.51	3	0.07258	6.4	0.55080	45.1	0.05505	45.0	0.14	452	28	446	163	414	999	**109**
A-180	7519	162	32	0.19	b.d.	0.19330	5.3	2.18300	18.4	0.08193	18.0	0.29	1139	55	1176	128	1243	345	**92**
A-182	1081	71	7	0.60	0	0.09332	5.1	0.74860	11.4	0.05820	10.0	0.45	575	28	567	50	537	223	**107**
A-183	181	220	1	0.26	0	0.00555	6.8	0.03576	28.7	0.04673	28.0	0.24	36	2	36	10	35	669	**102**
A-185	52419	528	140	0.17	2	0.25510	4.7	4.09300	5.6	0.11640	3.1	0.84	1465	62	1653	46	1901	56	77
A-186	6159	151	30	0.15	0	0.19530	5.0	2.06900	6.4	0.07688	4.1	0.77	1150	52	1139	44	1118	81	**103**
A-187	22438	208	75	0.22	b.d.	0.34880	4.8	5.64800	6.5	0.11750	4.4	0.74	1929	80	1924	56	1917	79	**101**
A-189	4512	394	26	0.18	b.d.	0.06691	5.6	0.63020	27.8	0.06833	27.0	0.20	418	23	496	109	878	564	48
A-190	4245	345	19	0.68	b.d.	0.05422	6.1	0.59800	16.9	0.08001	16.0	0.36	340	20	476	64	1197	312	28
A-191	2948	242	18	0.45	b.d.	0.07256	5.1	0.82050	23.4	0.08203	23.0	0.22	452	22	608	107	1246	446	36
A-192	12022	170	46	0.21	b.d.	0.26350	4.9	3.45000	12.3	0.09497	11.0	0.40	1508	66	1516	97	1527	214	**99**

spot number	$^{207}Pb^a$ (cps)	U^b (ppm)	Pb^b (ppm)	$\frac{Th^b}{U}$	$\frac{^{206}Pb^c}{^{204}Pb}$	$\frac{^{206}Pb^c}{^{238}U}$	2 s %	$\frac{^{207}Pb^c}{^{235}U}$	2 s %	$\frac{^{207}Pb^c}{^{206}Pb}$	2 s %	rho^d	$\frac{^{206}Pb}{^{238}U}$	2 s (Ma)	$\frac{^{207}Pb}{^{235}U}$	2 s (Ma)	$\frac{^{207}Pb}{^{206}Pb}$	2 s (Ma)	conc %^{e}
A-198	3566	95	18	0.27	b.d.	0.18690	5.7	1.94700	41.9	0.07556	42.0	0.14	1105	58	1097	281	1083	833	**102**
A-199	1297	234	9	0.67	0	0.03864	4.8	0.28640	8.5	0.05377	7.0	0.56	244	11	256	19	361	158	68
A-200	19070	197	59	0.20	1	0.29310	4.7	4.15200	9.4	0.10280	8.2	0.50	1657	69	1665	77	1674	151	**99**
A-201	5178	118	21	0.24	3	0.18030	5.0	1.43100	11.2	0.05759	10.0	0.45	1068	50	902	67	513	219	208
A-202	3282	43	10	0.41	b.d.	0.23070	7.7	2.95400	51.1	0.09289	51.0	0.15	1338	94	1396	388	1485	958	**90**
A-203	9121	120	33	0.43	1	0.27010	4.9	3.35700	19.2	0.09017	19.0	0.26	1541	67	1494	150	1429	355	**108**
A-204	6047	85	21	0.30	b.d.	0.23960	5.7	3.15900	20.0	0.09567	19.0	0.29	1385	71	1447	154	1541	360	90
A-205	5444	304	31	0.32	0	0.10230	4.8	0.85780	6.3	0.06083	4.2	0.75	628	28	629	30	632	90	**99**
A-206	16061	356	67	0.33	b.d.	0.18640	4.7	2.04100	9.6	0.07946	8.3	0.49	1102	47	1129	65	1183	165	**93**
A-207	33161	1459	136	0.21	2	0.09367	6.6	0.91150	7.9	0.07059	4.3	0.84	577	36	658	38	945	88	61
A-208	6303	365	30	1.04	b.d.	0.08100	4.9	0.83500	15.8	0.07479	15.0	0.31	502	24	616	73	1062	303	47

LSS2

spot number	$^{207}Pb^a$ (cps)	U^b (ppm)	Pb^b (ppm)	$\frac{Th^b}{U}$	$\frac{^{206}Pb^c}{^{204}Pb}$	$\frac{^{206}Pb^c}{^{238}U}$	2 s %	$\frac{^{207}Pb^c}{^{235}U}$	2 s %	$\frac{^{207}Pb^c}{^{206}Pb}$	2 s %	rho^d	$\frac{^{206}Pb}{^{238}U}$	2 s (Ma)	$\frac{^{207}Pb}{^{235}U}$	2 s (Ma)	$\frac{^{207}Pb}{^{206}Pb}$	2 s (Ma)	conc %^{e}
a1	15010	604	36	0.11	27395	0.06358	1.7	0.48566	2.6	0.05540	2.0	0.65	397	7	402	9	428	45	93
a2	4223	141	11	0.90	2284	0.07206	2.5	0.55711	4.3	0.05607	3.5	0.58	449	11	450	16	455	78	99
a3	130464	547	189	0.64	39946	0.31981	2.0	4.63902	2.4	0.10520	1.3	0.83	1789	31	1756	20	1718	24	104
a4	1268	102	4	0.83	2536	0.03286	2.2	0.22902	5.1	0.05054	4.6	0.43	208	5	209	10	220	106	95
a5	16247	543	23	0.91	181	0.03281	2.4	0.22864	8.9	0.05054	8.6	0.27	208	5	209	17	220	199	95
a6	10771	335	23	0.11	3391	0.07325	1.8	0.60595	2.8	0.05999	2.1	0.66	456	8	481	11	603	45	76
a7	76063	360	57	0.71	1807	0.12161	3.6	1.66128	4.2	0.09908	2.0	0.88	740	25	994	27	1607	37	46
a8	29960	200	47	0.34	25690	0.22829	2.3	2.71226	2.7	0.08617	1.4	0.85	1326	27	1332	20	1342	27	99
a9	5215	52	10	0.61	7014	0.18118	2.1	1.87300	3.5	0.07498	2.8	0.60	1073	21	1072	23	1068	56	101
a10	41861	151	54	0.48	7994	0.33433	1.8	5.19099	2.2	0.11261	1.2	0.82	1859	29	1851	19	1842	23	101
a11	7629	58	13	0.54	9477	0.21401	1.9	2.40271	3.3	0.08143	2.7	0.58	1250	22	1243	24	1232	53	101
a12	8881	185	20	0.65	14681	0.10000	2.0	0.84430	3.2	0.06123	2.5	0.62	614	12	622	15	647	54	95
a13	54968	249	68	0.37	11539	0.26152	2.1	3.61782	2.4	0.10033	1.2	0.87	1498	28	1553	19	1630	22	92

spot number	$^{207}Pb^a$ (cps)	U^b (ppm)	Pb^b (ppm)	$\frac{Th^b}{U}$	$\frac{^{206}Pb^c}{^{204}Pb}$	$\frac{^{206}Pb^c}{^{238}U}$	2 s %	$\frac{^{207}Pb^c}{^{235}U}$	2 s %	$\frac{^{207}Pb^c}{^{206}Pb}$	2 s %	rho^d	$\frac{^{206}Pb}{^{238}U}$ (Ma)	2 s (Ma)	$\frac{^{207}Pb}{^{235}U}$ (Ma)	2 s (Ma)	$\frac{^{207}Pb}{^{206}Pb}$ (Ma)	2 s (Ma)	conc %^{e}
a14	54959	156	63	0.44	8540	0.37882	1.8	6.74249	1.9	0.12909	0.7	0.92	2071	32	2078	17	2086	13	99
a15	6846	65	13	0.53	9122	0.18712	2.0	1.96226	3.1	0.07606	2.4	0.64	1106	20	1103	21	1097	48	101
a16	5520	139	15	1.21	9465	0.09089	2.0	0.73833	3.3	0.05891	2.6	0.62	561	11	561	14	564	56	99
a17	6364	259	15	0.07	11750	0.06131	1.9	0.46290	3.5	0.05476	2.9	0.54	384	7	386	11	402	65	95
a18	1581	80	4	1.85	1247	0.04071	2.7	0.38784	15.4	0.06909	15.1	0.18	257	7	333	45	901	312	29
a19	19336	175	36	0.70	25813	0.18503	1.9	1.93302	2.3	0.07577	1.2	0.83	1094	19	1093	15	1089	25	101
a20	20930	564	18	0.35	205	0.02492	4.0	0.16917	13.5	0.04923	12.9	0.30	159	6	159	20	159	302	100
a21	6962	150	21	1.98	4704	0.10149	1.9	0.85237	3.9	0.06091	3.4	0.48	623	11	626	18	636	74	98
a22	50760	272	85	0.94	21871	0.27097	2.0	3.48624	2.3	0.09331	1.0	0.90	1546	28	1524	18	1494	19	103
a23	1568	484	2	0.15	371	0.00464	2.0	0.02982	45.6	0.04665	45.5	0.04	30	1	30	13	31	1091	95
a24	7836	65	14	0.42	2825	0.20737	1.9	2.30969	2.7	0.08078	1.9	0.71	1215	21	1215	19	1216	37	100
a25	31648	269	62	0.86	40319	0.20501	1.6	2.24542	2.1	0.07944	1.4	0.75	1202	17	1195	15	1183	27	102
a26	9438	338	29	1.44	17151	0.06717	1.9	0.51461	2.6	0.05557	1.8	0.73	419	8	422	9	435	39	96
a27	215422	1295	386	0.20	3264	0.29605	1.8	4.19280	2.2	0.10271	1.2	0.84	1672	27	1673	18	1674	22	100
a28	189978	1783	333	0.30	1695	0.18495	3.2	2.46889	3.3	0.09682	0.9	0.96	1094	32	1263	24	1564	16	70
a29	14076	58	22	1.32	13642	0.30970	1.8	4.46594	2.8	0.10459	2.1	0.67	1739	28	1725	23	1707	38	102
a30	15547	82	26	0.95	16717	0.26959	2.0	3.48912	2.7	0.09387	1.8	0.75	1539	28	1525	22	1505	34	102
a31	31341	126	42	0.52	9796	0.31724	1.7	4.73252	2.1	0.10819	1.2	0.80	1776	26	1773	18	1769	23	100
a32	24906	920	46	0.30	367	0.04611	1.9	0.67121	3.3	0.10556	2.6	0.59	291	5	521	13	1724	48	17
a33	8033	80	10	0.33	139	0.09403	3.5	1.80846	18.6	0.13949	18.2	0.19	579	19	1049	129	2221	316	26
a34	4012	154	11	1.11	4061	0.06205	2.0	0.48565	4.0	0.05676	3.4	0.51	388	8	402	13	482	76	80
a35	1993	38	4	0.80	2707	0.09439	3.0	0.96636	6.3	0.07425	5.6	0.47	581	17	687	32	1048	112	55
a36	45157	216	49	0.55	4614	0.20839	4.9	3.66540	5.4	0.12757	2.3	0.90	1220	54	1564	44	2065	41	59
a37	67532	343	84	0.37	9760	0.23641	3.7	3.43677	3.8	0.10543	1.1	0.96	1368	45	1513	30	1722	20	79
a38	45041	192	60	0.33	11940	0.30713	2.2	4.39675	2.7	0.10383	1.6	0.80	1727	33	1712	23	1694	30	102

spot number	$^{207}Pb^a$ (cps)	U^b (ppm)	Pb^b (ppm)	$\frac{Th^b}{U}$	$\frac{^{206}Pb^c}{^{204}Pb}$	$\frac{^{206}Pb^c}{^{238}U}$	2 s %	$\frac{^{207}Pb^c}{^{235}U}$	2 s %	$\frac{^{207}Pb^c}{^{206}Pb}$	2 s %	rho^d	$\frac{^{206}Pb}{^{238}U}$	2 s (Ma)	$\frac{^{207}Pb}{^{235}U}$	2 s (Ma)	$\frac{^{207}Pb}{^{206}Pb}$	2 s (Ma)	conc $\%^e$
a40	5376	176	13	0.58	5748	0.07021	1.8	0.54748	3.5	0.05656	3.0	0.51	437	8	443	13	474	67	92
a41	29139	329	63	0.92	25056	0.16912	1.9	1.68008	2.5	0.07205	1.7	0.73	1007	17	1001	16	987	35	102
a42	53399	183	70	0.54	19549	0.35123	1.7	5.60558	2.0	0.11575	1.0	0.86	1940	29	1917	18	1892	19	103
a43	8859	104	17	0.24	12524	0.16306	1.9	1.60964	2.9	0.07159	2.2	0.64	974	17	974	18	974	46	100
a44	68055	805	48	0.28	255	0.04701	14.2	0.69372	14.5	0.10704	2.9	0.98	296	41	535	62	1750	52	17
a45	16006	603	30	0.34	1214	0.04884	2.1	0.35669	4.3	0.05297	3.7	0.49	307	6	310	11	327	84	94
a46	9310	84	17	0.47	11715	0.19395	1.6	2.13219	3.0	0.07973	2.6	0.54	1143	17	1159	21	1190	50	96
a47	76674	555	139	0.60	5600	0.23580	2.6	2.88785	3.0	0.08882	1.6	0.85	1365	32	1379	23	1400	31	97
a48	6100	613	18	0.64	5833	0.02753	2.1	0.18821	6.1	0.04958	5.7	0.35	175	4	175	10	175	133	100
a49	29372	491	41	0.40	487	0.07333	3.6	1.00935	4.3	0.09983	2.3	0.84	456	16	709	22	1621	43	28
a50	48242	230	74	0.68	48767	0.29522	1.6	4.07566	2.0	0.10013	1.1	0.83	1668	24	1649	16	1626	20	103
a51	2982	100	7	0.42	5341	0.07306	1.7	0.56909	4.0	0.05649	3.6	0.42	455	7	457	15	472	80	96
a52	9176	190	22	0.93	2277	0.10179	2.3	0.97186	2.8	0.06925	1.6	0.83	625	14	689	14	906	32	69
a53	18356	140	30	0.33	22620	0.21425	1.9	2.42510	2.5	0.08209	1.5	0.79	1251	22	1250	18	1248	29	100
a54	21980	96	32	0.68	14429	0.30213	1.9	4.24135	2.7	0.10182	1.8	0.72	1702	29	1682	22	1657	34	103
a55	4414	107	11	0.56	5029	0.09432	2.0	0.78199	3.1	0.06013	2.4	0.63	581	11	587	14	608	53	96
a56	11612	286	27	0.55	18825	0.09201	1.8	0.77099	2.7	0.06077	2.1	0.66	567	10	580	12	631	44	90
a57	1433	52	4	0.61	2434	0.07003	2.5	0.53639	5.8	0.05555	5.3	0.43	436	11	436	21	434	117	100
a58	30948	656	72	0.42	14398	0.10938	1.9	0.92858	2.4	0.06157	1.5	0.78	669	12	667	12	659	32	102
a59	43014	334	59	1.04	1578	0.16196	2.4	1.95510	2.6	0.08755	1.0	0.92	968	21	1100	17	1373	19	70
a60	13013	108	23	0.21	15951	0.21457	1.6	2.44131	2.3	0.08252	1.7	0.69	1253	18	1255	17	1258	32	100
a61	72175	639	115	0.74	792	0.16191	2.5	2.03498	3.2	0.09115	1.9	0.79	967	23	1127	22	1450	37	67
a62	19076	447	42	0.53	1281	0.08814	2.4	0.78183	3.2	0.06433	2.2	0.74	545	12	587	14	753	46	72
a63	178466	299	159	0.55	40611	0.46876	1.9	11.36140	2.0	0.17578	0.7	0.94	2478	39	2553	19	2613	12	95
a64	45312	389	102	0.40	6286	0.26429	2.0	2.86222	2.4	0.07855	1.4	0.83	1512	27	1372	19	1161	27	130
a65	4285	329	14	1.18	8507	0.03660	2.3	0.25675	3.4	0.05088	2.6	0.66	232	5	232	7	235	60	99

spot number	$^{207}Pb^a$ (cps)	U^b (ppm)	Pb^b (ppm)	$\underline{Th^b}$ U	$\underline{^{206}Pb^c}$ ^{204}Pb	$\underline{^{206}Pb^c}$ ^{238}U	2 s %	$\underline{^{207}Pb^c}$ ^{235}U	2 s %	$\underline{^{207}Pb^c}$ ^{206}Pb	2 s %	rho^d	$\underline{^{206}Pb}$ ^{238}U	2 s (Ma)	$\underline{^{207}Pb}$ ^{235}U	2 s (Ma)	$\underline{^{207}Pb}$ ^{206}Pb	2 s (Ma)	conc %^{e}
a66	4920	276	13	0.64	9256	0.04589	1.6	0.34051	3.3	0.05382	2.9	0.48	289	5	298	9	363	66	80
a67	25277	111	39	0.93	2594	0.30404	1.9	4.64757	2.3	0.11086	1.4	0.81	1711	28	1758	19	1814	25	94
a68	55181	248	96	1.80	53345	0.29197	2.7	4.21452	2.9	0.10469	1.0	0.93	1651	40	1677	24	1709	19	97
a69	7238	288	18	0.32	6604	0.06452	1.8	0.50849	3.0	0.05716	2.5	0.58	403	7	417	10	498	54	81
a70	76079	467	121	0.44	9561	0.24983	1.8	3.45358	2.1	0.10026	1.0	0.88	1438	23	1517	16	1629	18	88
a71	68753	296	80	0.22	4134	0.27120	3.7	4.02101	3.8	0.10753	1.1	0.96	1547	51	1638	32	1758	20	88
a72	11166	119	24	0.95	15090	0.17435	1.7	1.79485	2.9	0.07466	2.4	0.57	1036	16	1044	19	1059	49	98
a73	61423	269	101	1.18	59447	0.31333	1.6	4.51465	1.9	0.10450	1.0	0.85	1757	25	1734	16	1706	18	103
a74	3450	609	10	0.72	4587	0.01549	1.9	0.10223	9.5	0.04786	9.3	0.20	99	2	99	9	92	221	108
a75	7919	826	19	0.48	991	0.02141	1.9	0.19880	3.6	0.06733	3.1	0.51	137	3	184	6	848	65	16
a76	69387	333	92	0.36	67400	0.26488	3.0	3.80259	3.3	0.10412	1.5	0.89	1515	40	1593	27	1699	28	89
a77	19676	253	28	0.24	3885	0.10942	4.7	1.27911	6.7	0.08478	4.8	0.70	669	30	836	39	1311	93	51
a78	34711	251	49	0.32	2375	0.18164	3.9	2.57298	4.1	0.10274	1.2	0.95	1076	38	1293	30	1674	22	64
a79	14618	117	25	0.34	18048	0.21152	1.9	2.38439	2.5	0.08176	1.7	0.73	1237	21	1238	18	1240	34	100
a80	16398	344	30	0.54	659	0.07918	1.9	0.89328	3.6	0.08183	3.0	0.54	491	9	648	17	1241	60	40
a81	37614	173	58	0.76	11662	0.29637	2.0	4.22953	2.2	0.10350	1.0	0.90	1673	29	1680	18	1688	18	99
a82	35301	472	64	0.58	3174	0.12492	2.0	1.24302	2.4	0.07217	1.2	0.86	759	14	820	13	991	24	77
a83	46551	339	77	0.97	2072	0.21438	3.0	2.84716	3.3	0.09632	1.2	0.93	1252	35	1368	25	1554	22	81
a84	3237	34	6	0.43	4324	0.18735	1.7	1.94060	4.1	0.07512	3.7	0.42	1107	18	1095	28	1072	74	103
a85	1367	39	6	5.67	1963	0.06734	2.4	0.64295	5.9	0.06925	5.4	0.41	420	10	504	24	906	110	46
a86	120801	1077	185	0.46	4053	0.15866	5.4	2.00219	5.5	0.09152	0.9	0.99	949	48	1116	38	1458	17	65
a87	4716	531	17	1.30	6086	0.02694	1.7	0.18422	3.1	0.04960	2.6	0.55	171	3	172	5	176	60	97
a88	860	160	4	2.99	659	0.01666	2.5	0.11426	7.0	0.04975	6.6	0.36	106	3	110	7	183	153	58
a89	27229	111	38	1.22	3102	0.28503	3.8	4.40936	4.1	0.11220	1.6	0.92	1617	55	1714	35	1835	30	88
a90	17341	90	28	0.79	18088	0.27770	2.3	3.71408	3.0	0.09700	1.9	0.76	1580	32	1574	24	1567	36	101

spot number	$^{207}Pb^a$ (cps)	U^b (ppm)	Pb^b (ppm)	$\underline{Th^b}$ U	$\underline{^{206}Pb^c}$ ^{204}Pb	$\underline{^{206}Pb^c}$ ^{238}U	2 s %	$\underline{^{207}Pb^c}$ ^{235}U	2 s %	$\underline{^{207}Pb^c}$ ^{206}Pb	2 s %	rho^d	$\underline{^{206}Pb}$ ^{238}U	2 s (Ma)	$\underline{^{207}Pb}$ ^{235}U	2 s (Ma)	$\underline{^{207}Pb}$ ^{206}Pb	2 s (Ma)	conc %^{e}
a92	42661	197	65	0.55	9113	0.30704	1.7	4.35442	2.0	0.10286	1.0	0.87	1726	26	1704	16	1676	18	103
a93	49101	269	81	0.57	6622	0.28254	1.7	3.76384	2.0	0.09662	1.0	0.86	1604	25	1585	16	1560	19	103
a94	20425	100	26	0.60	10363	0.23752	2.2	3.47290	2.9	0.10605	1.9	0.76	1374	27	1521	23	1733	34	79
a95	29545	194	59	1.28	6217	0.26016	1.8	3.29887	2.3	0.09196	1.3	0.81	1491	24	1481	18	1467	25	102
a96	87014	69	49	0.71	23	0.20125	7.7	18.98966	7.9	0.68437	1.9	0.97	1182	84	3041	80	4698	27	25
a97	5463	578	21	0.62	8878	0.02760	1.9	0.18972	3.6	0.04986	3.0	0.53	175	3	176	6	188	70	93
a98	31622	164	44	0.67	32553	0.24407	2.5	3.31196	2.8	0.09842	1.3	0.89	1408	31	1484	22	1594	24	88
a99	87637	734	159	0.27	59085	0.21664	2.0	2.42025	2.2	0.08103	0.9	0.91	1264	23	1249	16	1222	18	103
a100	9268	69	16	0.65	10378	0.22366	2.0	2.78442	2.7	0.09029	1.8	0.75	1301	24	1351	21	1432	35	91
a101	10520	333	22	0.42	461	0.06456	2.1	0.71016	6.0	0.07978	5.6	0.35	403	8	545	26	1191	111	34
a102	18290	198	39	0.82	24928	0.17903	1.8	1.83160	2.4	0.07420	1.6	0.74	1062	17	1057	16	1047	33	101
a103	23108	452	58	1.07	2396	0.11000	2.0	1.03054	2.7	0.06795	1.8	0.73	673	13	719	14	867	38	78
a104	61365	630	100	0.13	9616	0.16505	2.3	1.66714	2.6	0.07326	1.3	0.87	985	21	996	17	1021	26	96
a105	34645	325	58	0.20	12177	0.18430	1.7	1.98583	2.1	0.07815	1.1	0.84	1090	17	1111	14	1151	22	95
a106	57462	237	86	1.01	41345	0.30644	2.0	4.64502	2.2	0.10994	1.0	0.90	1723	30	1757	19	1798	17	96
a107	2921	241	7	0.55	425	0.02894	2.6	0.28141	6.4	0.07052	5.8	0.40	184	5	252	14	944	120	19

LSS3

spot number	$^{207}Pb^a$ (cps)	U^b (ppm)	Pb^b (ppm)	$\underline{Th^b}$ U	$\underline{^{206}Pb^c}$ ^{204}Pb	$\underline{^{206}Pb^c}$ ^{238}U	2 s %	$\underline{^{207}Pb^c}$ ^{235}U	2 s %	$\underline{^{207}Pb^c}$ ^{206}Pb	2 s %	rho^d	$\underline{^{206}Pb}$ ^{238}U	2 s (Ma)	$\underline{^{207}Pb}$ ^{235}U	2 s (Ma)	$\underline{^{207}Pb}$ ^{206}Pb	2 s (Ma)	conc %^{e}
a108	4555	150	12	0.79	5032	0.07472	1.8	0.58395	3.1	0.05668	2.6	0.56	465	8	467	12	479	58	97
a109	21427	180	45	1.17	7335	0.20778	1.9	2.32044	2.3	0.08099	1.3	0.83	1217	21	1219	17	1221	26	100
a110	7780	318	20	0.62	5784	0.05935	2.0	0.45031	3.5	0.05503	2.8	0.59	372	7	377	11	413	63	90
a111	4867	119	16	2.21	7922	0.09879	2.1	0.84294	3.6	0.06188	3.0	0.58	607	12	621	17	670	64	91
a112	95928	182	87	0.13	20654	0.46286	1.7	10.99658	2.0	0.17231	1.0	0.85	2452	34	2523	19	2580	18	95
a113	436	348	2	1.14	945	0.00403	2.5	0.02587	9.6	0.04656	9.3	0.26	26	1	26	2	27	222	97
a114	8668	159	21	0.81	13585	0.11932	1.9	1.06269	3.0	0.06459	2.4	0.62	727	13	735	16	761	50	95
a115	786	90	3	1.53	1622	0.02360	2.6	0.15991	8.6	0.04914	8.2	0.30	150	4	151	12	154	193	97

spot number	$^{207}Pb^a$ (cps)	U^b (ppm)	Pb^b (ppm)	$\frac{Th^b}{U}$	$\frac{^{206}Pb^c}{^{204}Pb}$	$\frac{^{206}Pb^c}{^{238}U}$	2 s %	$\frac{^{207}Pb^c}{^{235}U}$	2 s %	$\frac{^{207}Pb^c}{^{206}Pb}$	2 s %	rho^d	$\frac{^{206}Pb}{^{238}U}$	2 s (Ma)	$\frac{^{207}Pb}{^{235}U}$	2 s (Ma)	$\frac{^{207}Pb}{^{206}Pb}$	2 s (Ma)	conc %^{e}
a116	11808	259	36	1.94	3805	0.10088	1.9	0.89622	3.4	0.06443	2.8	0.55	620	11	650	16	756	59	82

LSS4

spot number	$^{207}Pb^a$ (cps)	U^b (ppm)	Pb^b (ppm)	$\frac{Th^b}{U}$	$\frac{^{206}Pb^c}{^{204}Pb}$	$\frac{^{206}Pb^c}{^{238}U}$	2 s %	$\frac{^{207}Pb^c}{^{235}U}$	2 s %	$\frac{^{207}Pb^c}{^{206}Pb}$	2 s %	rho^d	$\frac{^{206}Pb}{^{238}U}$	2 s (Ma)	$\frac{^{207}Pb}{^{235}U}$	2 s (Ma)	$\frac{^{207}Pb}{^{206}Pb}$	2 s (Ma)	conc %^{e}
A-189	10537	392	34	1.00	0	0.08742	4.3	0.74400	5.4	0.06174	3.2	0.81	540	22	565	23	665	68	81
A-190	3291	97	9	0.64	1	0.09839	4.6	0.82440	7.5	0.06079	5.9	0.62	605	27	611	34	631	127	96
A-191	10362	58	13	0.67	0	0.22510	4.5	2.64000	5.2	0.08509	2.7	0.86	1309	53	1312	38	1317	52	99
A-192	28583	374	37	1.00	1	0.09834	4.8	1.00100	5.6	0.07388	3.0	0.85	605	27	705	29	1038	61	58
A-197	3539	154	7	0.30	0	0.04583	4.9	0.33230	8.0	0.05259	6.3	0.61	289	14	291	20	311	145	93
A-198	2312	111	5	0.68	0	0.04169	4.7	0.29650	8.7	0.05160	7.3	0.54	263	12	264	20	267	168	99
A-199	77428	259	86	0.57	0	0.32440	4.7	4.70900	5.2	0.10530	2.3	0.90	1811	74	1769	43	1719	41	105
A-200	114979	388	119	0.94	0	0.29850	4.7	4.35300	4.9	0.10580	1.3	0.96	1684	70	1704	41	1728	24	97
A-201	4851	148	15	1.89	0	0.09998	4.7	0.93710	6.6	0.06800	4.7	0.71	614	28	671	33	868	97	71
A-202	20991	382	35	1.12	0	0.09323	4.6	0.82800	5.9	0.06443	3.6	0.79	575	25	612	27	755	76	76
A-203	47032	161	45	0.40	0	0.27540	4.7	3.94300	5.1	0.10390	2.0	0.92	1568	65	1622	41	1694	37	93
A-204	23333	161	17	1.19	6	0.10260	4.6	1.31700	6.2	0.09312	4.2	0.73	630	27	853	36	1490	80	42
A-205	5045	206	10	2.01	0	0.04907	4.6	0.36590	5.7	0.05409	3.3	0.81	309	14	317	15	374	75	83
A-206	25589	158	44	0.57	0	0.27370	4.5	3.55800	5.2	0.09432	2.7	0.86	1559	62	1540	41	1514	50	103
A-207	12120	283	27	0.43	0	0.09483	4.4	0.81430	5.1	0.06229	2.5	0.87	584	25	605	23	684	52	85
A-208	20130	553	37	0.02	0	0.06711	4.4	0.50760	5.1	0.05488	2.6	0.86	419	18	417	17	407	58	103
A-209	26013	508	42	0.78	0	0.08411	4.7	0.72210	5.5	0.06228	2.8	0.86	521	23	552	23	683	60	76
A-210	305348	907	128	0.25	2	0.13180	4.9	2.75400	5.4	0.15160	2.2	0.91	798	37	1343	40	2364	38	34
A-211	966	697	3	0.08	2	0.00488	6.0	0.04247	16.6	0.06310	15.0	0.36	31	2	42	7	711	329	4
A-212	93759	333	101	0.52	0	0.29580	4.6	4.19300	5.0	0.10280	2.0	0.92	1671	67	1673	41	1675	37	100
A-213	6557	71	13	0.35	0	0.18680	4.4	1.90600	6.0	0.07405	4.0	0.74	1104	45	1083	40	1042	81	106
A-214	19244	402	33	0.32	1	0.08312	5.4	0.72060	6.1	0.06290	2.8	0.89	515	27	551	26	704	59	73
A-215	143930	636	153	0.16	0	0.23560	4.8	3.26500	5.1	0.10060	1.7	0.94	1364	59	1473	40	1634	32	83

spot number	$^{207}Pb^a$ (cps)	U^b (ppm)	Pb^b (ppm)	$\frac{Th^b}{U}$	$\frac{^{206}Pb^c}{^{204}Pb}$	$\frac{^{206}Pb^c}{^{238}U}$	2 s %	$\frac{^{207}Pb^c}{^{235}U}$	2 s %	$\frac{^{207}Pb^c}{^{206}Pb}$	2 s %	rho^d	$\frac{^{206}Pb}{^{238}U}$	2 s (Ma)	$\frac{^{207}Pb}{^{235}U}$	2 s (Ma)	$\frac{^{207}Pb}{^{206}Pb}$	2 s (Ma)	conc %[e]
A-221	16139	391	31	2.00	0	0.08092	4.7	0.65020	5.4	0.05829	2.6	0.87	502	23	509	22	540	57	93
A-222	9105	558	-9	0.08	1	-0.01562	18.4	-0.16180	20.1	0.07517	8.1	0.91	-101	-19	-179	-39	1072	163	-9
A-223	1745	141	5	0.98	0	0.03493	4.9	0.24540	9.9	0.05096	8.6	0.50	221	11	223	20	238	199	93
A-224	4114	281	8	0.67	0	0.02798	4.6	0.20860	6.9	0.05409	5.2	0.66	178	8	192	12	374	117	48
A-225	2991	39	6	0.13	1	0.15220	5.6	1.65800	8.8	0.07899	6.8	0.64	914	47	992	56	1171	134	78
A-226	22131	109	27	0.63	0	0.24160	4.3	3.01400	5.2	0.09051	2.8	0.84	1395	54	1411	39	1436	54	97
A-227	7999	226	15	0.64	0	0.06943	4.6	0.53400	5.5	0.05579	3.1	0.83	433	19	434	19	444	68	98
A-228	44805	2071	65	0.64	6	0.03071	6.3	0.37720	8.9	0.08910	6.3	0.71	195	12	325	25	1406	121	14
A-229	5057	33	7	0.80	0	0.21700	4.9	2.39200	6.7	0.07998	4.6	0.73	1266	57	1240	48	1196	91	106
A-230	22258	98	40	0.88	0	0.39040	4.3	6.89100	4.9	0.12800	2.2	0.89	2125	78	2097	43	2071	39	103
A-231	10886	213	18	1.21	0	0.08340	4.4	0.71490	4.9	0.06219	2.2	0.90	516	22	548	21	680	46	76
A-232	48576	434	81	0.10	0	0.18700	4.7	1.89700	5.2	0.07360	2.3	0.90	1105	48	1080	35	1030	47	107
A-233	20442	155	29	0.39	0	0.18710	4.8	1.99800	5.4	0.07746	2.4	0.89	1106	49	1115	36	1133	48	98
A-234	9118	394	23	0.42	0	0.05842	4.8	0.43680	6.4	0.05424	4.2	0.75	366	17	368	20	380	95	96
A-235	34220	115	37	0.95	0	0.31140	4.5	4.62900	5.1	0.10790	2.3	0.89	1747	69	1755	42	1763	42	99
A-236	114981	620	158	0.24	1	0.24940	4.5	3.21100	4.9	0.09339	2.1	0.91	1436	57	1460	38	1495	39	96
A-237	29300	318	73	0.35	0	0.22840	4.2	2.71000	4.8	0.08610	2.3	0.88	1326	51	1331	36	1340	45	99
A-238	64334	245	70	0.10	0	0.27880	4.3	3.96700	4.8	0.10320	2.1	0.90	1585	60	1627	39	1683	39	94

TR2

spot number	$^{207}Pb^a$ (cps)	U^b (ppm)	Pb^b (ppm)	$\frac{Th^b}{U}$	$\frac{^{206}Pb^c}{^{204}Pb}$	$\frac{^{206}Pb^c}{^{238}U}$	2 s %	$\frac{^{207}Pb^c}{^{235}U}$	2 s %	$\frac{^{207}Pb^c}{^{206}Pb}$	2 s %	rho^d	$\frac{^{206}Pb}{^{238}U}$	2 s (Ma)	$\frac{^{207}Pb}{^{235}U}$	2 s (Ma)	$\frac{^{207}Pb}{^{206}Pb}$	2 s (Ma)	conc %[e]
a1	5872	108	18	0.4	8016	0.15856	2.2	1.59923	4.1	0.07315	3.5	0.53	949	19	970	26	1018	70	**93**
a2	7962	266	27	0.06	12705	0.1099	2.3	0.95105	5.4	0.06277	4.9	0.42	672	15	679	27	700	105	**96**
a3	4943	62	14	0.83	5773	0.20578	2.6	2.34742	4.6	0.08274	3.8	0.57	1206	29	1227	33	1263	74	**96**
a4	3321	422	15	0.45	6561	0.03456	2.1	0.24206	4.9	0.05079	4.4	0.44	219	5	220	10	231	102	**95**
a5	5606	243	28	1.55	9609	0.09341	1.9	0.75453	3.1	0.05859	2.5	0.6	576	10	571	14	552	54	**104**
a6	14002	199	40	0.7	15431	0.18069	1.8	1.93033	2.3	0.07748	1.5	0.78	1071	18	1092	16	1134	29	**94**
a7	36250	296	85	0.43	38032	0.27568	2	3.63285	2.5	0.09558	1.5	0.79	1570	27	1557	20	1539	28	**102**

204

spot number	$^{207}Pb^a$ (cps)	U^b (ppm)	Pb^b (ppm)	Th^b / U	$^{206}Pb^c$ / ^{204}Pb	$^{206}Pb^c$ / ^{238}U	2 s %	$^{207}Pb^c$ / ^{235}U	2 s %	$^{207}Pb^c$ / ^{206}Pb	2 s %	rho^d	^{206}Pb / ^{238}U	2 s (Ma)	^{207}Pb / ^{235}U	2 s (Ma)	^{207}Pb / ^{206}Pb	2 s (Ma)	conc $\%^e$
a8	33485	529	28	0.13	2603	0.0477	4.9	0.73211	10.7	0.11132	9.5	0.46	300	14	558	47	1821	172	16
a9	19634	142	45	0.56	19420	0.29483	2.4	4.12696	2.9	0.10152	1.7	0.81	1666	35	1660	24	1652	32	101
a10	1636	200	7	0.96	1542	0.03291	2.4	0.26071	6.3	0.05746	5.9	0.37	209	5	235	13	509	129	41
a11	1678	30	4	0.8	2265	0.11075	4.1	1.1345	7.4	0.07429	6.2	0.56	677	26	770	41	1049	124	65
a12	4989	161	15	0.5	2290	0.08882	2.3	0.76307	3.7	0.06231	2.9	0.63	549	12	576	16	685	62	80
a13	26911	186	66	0.97	8368	0.30647	2.1	4.31776	2.5	0.10218	1.4	0.83	1723	31	1697	21	1664	26	104
a14	6071	355	26	0.66	3958	0.06869	2.2	0.53355	3.3	0.05634	2.4	0.67	428	9	434	12	466	53	92
a15	22896	286	66	0.63	9513	0.21577	1.9	2.54111	2.6	0.08541	1.8	0.74	1260	22	1284	19	1325	34	95
a16	6824	215	18	0.26	9554	0.07791	2.3	0.72616	3.6	0.0676	2.8	0.63	484	11	554	16	856	58	56
a17	16974	197	46	0.52	20308	0.21851	1.8	2.52002	2.6	0.08364	1.8	0.71	1274	21	1278	19	1284	36	99
a18	2249	140	10	0.6	2709	0.06873	2.6	0.5227	4.7	0.05516	3.9	0.56	428	11	427	16	419	86	102
a19	3136	207	15	0.46	5618	0.07129	2.8	0.54938	4.6	0.05589	3.6	0.61	444	12	445	17	448	81	99
a20	2933	428	15	1.25	5586	0.02865	2.2	0.2074	5.4	0.05249	4.9	0.41	182	4	191	9	307	111	59
a21	4617	184	21	1.32	3662	0.09388	2.2	0.77892	3.9	0.06018	3.3	0.56	578	12	585	18	610	71	95
a22	4088	146	15	0.35	2953	0.10446	1.8	0.8956	3.4	0.06218	2.8	0.55	640	11	649	16	680	60	94
a23	7863	121	24	0.48	10213	0.19408	2.3	2.06227	3.1	0.07707	2.1	0.73	1143	24	1136	21	1123	42	102
a24	4541	78	16	0.83	3388	0.18509	2.2	1.8947	4.5	0.07424	3.9	0.49	1095	22	1079	30	1048	79	104
a25	18343	310	47	0.09	5154	0.15931	2.3	1.73458	3.1	0.07897	2	0.75	953	20	1021	20	1171	40	81
a26	3410	201	16	0.62	5998	0.07409	2.7	0.58532	4.5	0.0573	3.6	0.6	461	12	468	17	503	79	92
a27	3868	320	18	0.69	2999	0.05289	2.1	0.38788	3.9	0.05319	3.2	0.55	332	7	333	11	337	73	99
a28	13501	36	20	0.4	7570	0.5117	2.6	12.60592	3.3	0.17867	2.1	0.78	2664	56	2651	32	2641	34	101
a29	1417	209	7	0.93	2526	0.02983	3.4	0.23344	7.7	0.05676	6.9	0.44	189	6	213	15	482	152	39
a30	26084	217	66	0.58	8859	0.28475	2.1	3.99047	2.7	0.10164	1.7	0.77	1615	30	1632	22	1654	32	98
a31	3372	132	15	1.19	5590	0.09693	2.3	0.80942	4	0.06056	3.3	0.56	596	13	602	18	624	72	96
a32	10374	175	34	0.4	13502	0.1903	2.3	2.02049	3.2	0.077	2.3	0.7	1123	23	1122	22	1121	46	100
a33	10465	105	29	0.56	5278	0.25318	2.1	3.10863	3.6	0.08905	2.9	0.59	1455	28	1435	28	1405	55	104

spot number	$^{207}Pb^a$ (cps)	U^b (ppm)	Pb^b (ppm)	$\underline{Th^b}$ U	$\underline{^{206}Pb^c}$ ^{204}Pb	$\underline{^{206}Pb^c}$ ^{238}U	2 s %	$\underline{^{207}Pb^c}$ ^{235}U	2 s %	$\underline{^{207}Pb^c}$ ^{206}Pb	2 s %	rho^d	$\underline{^{206}Pb}$ ^{238}U (Ma)	2 s (Ma)	$\underline{^{207}Pb}$ ^{235}U (Ma)	2 s (Ma)	$\underline{^{207}Pb}$ ^{206}Pb (Ma)	2 s (Ma)	conc %^{e}
a35	3046	138	14	0.98	5072	0.08949	2.1	0.74248	4.3	0.06017	3.8	0.49	553	11	564	19	610	82	91
a36	1293	100	8	1.61	2412	0.05978	3.2	0.44254	8.3	0.05369	7.7	0.39	374	12	372	26	358	173	105
a37	11675	157	38	0.75	13914	0.21764	1.9	2.51987	2.5	0.08397	1.6	0.77	1269	22	1278	18	1292	31	98
a38	5539	334	27	1	9625	0.06923	1.9	0.55005	3.2	0.05762	2.7	0.57	432	8	445	12	515	58	84
a39	3890	70	14	0.87	5167	0.18333	2.3	1.90638	3.1	0.07542	2.1	0.73	1085	23	1083	21	1080	43	101
a40	31548	375	90	1.03	6397	0.20062	2.6	2.88902	3	0.10444	1.4	0.88	1179	28	1379	23	1705	26	69
a41	701	117	4	1.02	1008	0.03048	2.3	0.21151	8.1	0.05033	7.8	0.28	194	4	195	15	210	181	92
a42	5595	116	20	0.42	7682	0.16988	1.8	1.70911	3.3	0.07297	2.7	0.55	1011	17	1012	21	1013	55	100
a43	11972	799	52	0.78	977	0.0569	1.9	0.5673	3.8	0.07232	3.3	0.49	357	6	456	14	995	67	36
a44	12380	167	27	0.45	4295	0.15158	5.3	1.76708	5.7	0.08455	2	0.94	910	45	1033	38	1305	39	70
a45	35009	391	94	0.3	38897	0.23786	2	2.9588	2.4	0.09022	1.4	0.83	1376	25	1397	19	1430	26	96
a46	7843	165	29	0.48	10756	0.16579	1.9	1.67116	3.3	0.07311	2.7	0.56	989	17	998	21	1017	56	97
a47	80855	151	101	0.65	8783	0.54648	2.3	17.09699	2.6	0.22691	1.1	0.9	2811	53	2940	25	3030	18	93
a48	7116	469	38	1.05	5953	0.07011	2.1	0.54795	3.2	0.05668	2.4	0.66	437	9	444	11	479	53	91
a49	4724	277	30	1.54	6286	0.089	2.2	0.73613	3.5	0.05999	2.7	0.63	550	12	560	15	603	59	91
a50	49749	452	97	0.38	1517	0.19804	3.1	2.82573	3.4	0.10348	1.4	0.91	1165	33	1362	26	1688	26	69
a51	4506	277	17	0.79	949	0.05219	2	0.56291	5.3	0.07823	4.9	0.38	328	7	453	20	1153	98	28
a52	4549	546	22	0.33	3388	0.04073	1.9	0.30712	4.3	0.05468	3.9	0.43	257	5	272	10	399	87	64
a53	27840	458	97	0.51	35297	0.20169	1.7	2.19896	2.1	0.07907	1.3	0.8	1184	18	1181	15	1174	25	101
a54	28494	601	107	0.28	39706	0.1798	1.7	1.78412	2.3	0.07197	1.6	0.74	1066	17	1040	15	985	32	108
a55	6214	309	32	0.88	4879	0.09355	2.3	0.78024	3.1	0.06049	2.2	0.73	576	13	586	14	621	46	93
a56	1449	111	9	1.25	1979	0.06613	2.9	0.49829	6.9	0.05465	6.3	0.42	413	11	411	24	398	140	104
a57	37028	399	105	0.45	24882	0.24912	2.1	3.12835	2.6	0.09108	1.6	0.8	1434	27	1440	20	1448	30	99
a58	8066	166	35	0.91	10983	0.18271	1.8	1.85168	2.9	0.0735	2.3	0.63	1082	18	1064	19	1028	46	105
a59	2712	203	14	0.68	4946	0.06306	1.8	0.47856	7.3	0.05504	7.1	0.24	394	7	397	24	414	159	95
a60	10996	3293	37	0.07	1238	0.01135	6.7	0.10682	7.4	0.06823	3.2	0.9	73	5	103	7	876	66	8
a61	6814	123	25	0.3	8898	0.20123	1.9	2.12496	2.6	0.07659	1.8	0.72	1182	21	1157	18	1110	36	106

spot number	$^{207}Pb^a$ (cps)	U^b (ppm)	Pb^b (ppm)	$\frac{Th^b}{U}$	$\frac{^{206}Pb^c}{^{204}Pb}$	$\frac{^{206}Pb^c}{^{238}U}$	2 s %	$\frac{^{207}Pb^c}{^{235}U}$	2 s %	$\frac{^{207}Pb^c}{^{206}Pb}$	2 s %	rho^d	$\frac{^{206}Pb}{^{238}U}$	2 s (Ma)	$\frac{^{207}Pb}{^{235}U}$	2 s (Ma)	$\frac{^{207}Pb}{^{206}Pb}$	2 s (Ma)	conc %[e]
a62	1180	222	9	3.53	1380	0.02261	2.9	0.19698	9.9	0.06318	9.5	0.29	144	4	183	17	714	201	20
a63	6075	296	32	0.87	4317	0.0978	1.9	0.8148	3	0.06043	2.4	0.62	601	11	605	14	619	51	97
a64	13536	106	40	1.57	916	0.30295	2.6	4.722	4.2	0.11305	3.2	0.63	1706	40	1771	36	1849	58	92
a65	13222	630	65	0.61	21685	0.09867	2.3	0.83198	2.9	0.06115	1.7	0.81	607	14	615	13	645	36	94
a66	11822	986	63	0.53	12996	0.06068	1.8	0.49824	3.6	0.05955	3.1	0.5	380	7	411	12	587	68	65
a67	1199	75	6	0.48	1983	0.07602	2.2	0.6321	7.1	0.0603	6.8	0.31	472	10	497	28	614	146	77
a68	52828	472	150	0.52	51575	0.29499	2	4.17516	2.2	0.10265	1	0.9	1666	29	1669	18	1673	18	100
a69	11659	1102	57	0.56	20300	0.047	3.1	0.37405	4	0.05772	2.5	0.78	296	9	323	11	519	55	57
a70	4466	358	28	1.3	8136	0.06334	2	0.47924	3.4	0.05488	2.8	0.58	396	8	398	11	407	62	97
a71	28394	495	122	1.33	14353	0.19577	2.3	2.22428	2.6	0.0824	1.3	0.86	1153	24	1189	19	1255	26	92
a72	6880	102	25	0.66	2056	0.23061	2.3	2.83515	3.5	0.08916	2.6	0.67	1338	28	1365	26	1408	50	95
a73	3537	181	21	1.34	4393	0.09697	2.3	0.81605	4.7	0.06103	4.1	0.5	597	13	606	22	640	88	93
a74	2536	1156	17	0.69	5201	0.01384	2.2	0.09158	8.9	0.04799	8.6	0.25	89	2	89	8	99	204	90
a75	14528	185	52	0.74	16125	0.25436	2.3	3.1659	2.9	0.09027	1.8	0.79	1461	30	1449	23	1431	34	102
a76	4421	199	24	0.98	7292	0.10306	2.1	0.86355	4.1	0.06077	3.5	0.52	632	13	632	19	631	75	100
a77	14769	237	54	0.52	17370	0.21314	2.4	2.51937	3.1	0.08573	2	0.78	1246	27	1278	23	1332	38	94
a78	111519	1322	349	0.1	86431	0.27175	1.8	3.66888	2	0.09792	0.9	0.9	1550	25	1565	16	1585	16	98
a79	3883	351	22	0.37	7092	0.0613	2.1	0.46397	4.3	0.0549	3.8	0.48	384	8	387	14	408	85	94
a80	17906	297	66	0.26	21586	0.22178	1.8	2.53888	2.5	0.08303	1.6	0.75	1291	22	1283	18	1270	31	102
a81	21170	2154	65	0.1	3431	0.03118	2.6	0.25909	3.1	0.06027	1.6	0.85	198	5	234	6	613	36	32
a82	4819	60	18	0.79	2051	0.26797	2.9	3.52214	4.2	0.09533	2.9	0.71	1530	40	1532	33	1535	55	100
a83	7884	695	40	0.7	881	0.05299	5.6	0.3916	6.8	0.0536	3.9	0.83	333	18	336	20	354	87	94
a84	8882	1386	56	0.66	17108	0.03773	1.9	0.27076	2.8	0.05205	2.1	0.68	239	5	243	6	288	48	83
a85	35593	310	107	0.41	21699	0.32752	2.5	5.46424	2.8	0.121	1.3	0.88	1826	39	1895	24	1971	24	93
a86	3574	162	19	0.55	1394	0.11157	2.6	0.97465	5.1	0.06336	4.4	0.5	682	17	691	26	720	94	95
a87	22255	557	98	0.4	30481	0.1713	1.8	1.72706	2.4	0.07312	1.6	0.75	1019	17	1019	16	1017	33	100

spot number	$^{207}Pb^a$ (cps)	U^b (ppm)	Pb^b (ppm)	Th^b/U	$\frac{^{206}Pb^c}{^{204}Pb}$	$\frac{^{206}Pb^c}{^{238}U}$	2 s %	$\frac{^{207}Pb^c}{^{235}U}$	2 s %	$\frac{^{207}Pb^c}{^{206}Pb}$	2 s %	rho^d	$\frac{^{206}Pb}{^{238}U}$ (Ma)	2 s (Ma)	$\frac{^{207}Pb}{^{235}U}$ (Ma)	2 s (Ma)	$\frac{^{207}Pb}{^{206}Pb}$ (Ma)	2 s (Ma)	conc %^{e}
a89	8168	1130	51	0.37	14717	0.0445	2	0.3411	3.9	0.05559	3.3	0.52	281	6	298	10	436	74	64
a90	5310	2112	39	0.18	996	0.0151	2.2	0.11939	4.1	0.05733	3.5	0.54	97	2	115	4	504	76	19
a91	18848	448	78	0.03	12253	0.18476	1.8	1.93199	2.4	0.07584	1.6	0.73	1093	18	1092	16	1091	33	**100**
a92	15990	1017	71	0.94	314	0.04864	2.7	0.74692	4.7	0.11137	3.9	0.57	306	8	566	21	1822	70	17
a93	35108	391	109	0.37	4354	0.26911	2.5	3.75099	2.8	0.10109	1.1	0.91	1536	35	1582	23	1644	21	**93**
a94	5725	86	22	0.65	1911	0.23377	1.9	3.02245	2.7	0.09377	2	0.69	1354	23	1413	21	1504	37	**90**
a95	2573	345	27	3.06	1520	0.04674	2	0.35322	3.6	0.05481	3.1	0.54	294	6	307	10	404	68	73
a96	18519	196	66	1.1	1462	0.26766	2.8	4.50489	3.2	0.12207	1.6	0.87	1529	38	1732	27	1987	29	77
a97	4005	378	25	0.43	3660	0.06392	2.4	0.48995	3.9	0.0556	3.1	0.63	399	9	405	13	436	68	**92**
a98	3663	104	20	0.66	3206	0.17569	2.4	1.82264	5.3	0.07524	4.7	0.46	1043	24	1054	36	1075	95	**97**
a99	4021	992	36	1.26	2320	0.02669	2.1	0.21047	4.9	0.05718	4.4	0.43	170	4	194	9	499	98	34
a100	17045	1315	79	0.15	12610	0.06259	2.8	0.49103	3	0.0569	1	0.94	391	11	406	10	488	23	80
a101	21626	569	113	1.36	4099	0.17068	2.2	1.78445	2.8	0.07583	1.7	0.79	1016	21	1040	18	1091	34	**93**
a102	284	472	2	0.06	522	0.00409	2.9	0.031	9.5	0.055	9.1	0.31	26	1	31	3	412	203	6
a103	7141	1156	62	1.34	4449	0.04198	2	0.31394	2.5	0.05424	1.6	0.78	265	5	277	6	381	36	70

TR4 (1st & 2nd mount)

spot number	$^{207}Pb^a$ (cps)	U^b (ppm)	Pb^b (ppm)	Th^b/U	$\frac{^{206}Pb^c}{^{204}Pb}$	$\frac{^{206}Pb^c}{^{238}U}$	2 s %	$\frac{^{207}Pb^c}{^{235}U}$	2 s %	$\frac{^{207}Pb^c}{^{206}Pb}$	2 s %	rho^d	$\frac{^{206}Pb}{^{238}U}$ (Ma)	2 s (Ma)	$\frac{^{207}Pb}{^{235}U}$ (Ma)	2 s (Ma)	$\frac{^{207}Pb}{^{206}Pb}$ (Ma)	2 s (Ma)	conc %^{e}
a104	29580	1711	122	0.42	3883	0.06001	5.5	0.64873	5.6	0.07841	1.3	0.97	376	20	508	23	1157	26	32
a105	71248	791	145	0.26	371	0.14181	2.6	3.52637	3	0.18035	1.4	0.89	855	21	1533	24	2656	23	32
a106	4507	133	25	1.43	6445	0.14697	2.6	1.41848	4	0.07	3.1	0.64	884	21	897	24	928	63	**95**
a107	8623	732	50	0.48	2815	0.0646	2.9	0.55382	4.5	0.06218	3.5	0.64	404	11	447	17	680	74	59
a108	4178	475	36	1.63	7384	0.05868	2.7	0.4583	4.2	0.05664	3.3	0.63	368	10	383	14	478	72	77
a109	4790	436	28	0.04	4804	0.06756	2	0.52283	3.1	0.05613	2.3	0.65	421	8	427	11	457	52	**92**
a110	3805	155	19	0.53	468	0.11538	5.4	2.45375	13.3	0.15423	12.1	0.41	704	36	1259	100	2393	206	29
a111	931	1903	7	0.19	2008	0.00402	2.7	0.02576	6.7	0.04653	6.1	0.4	26	1	26	2	25	148	**103**
a112	43609	1671	232	0.14	10405	0.14456	2.4	1.41308	2.6	0.0709	1.1	0.91	870	20	894	16	954	22	**91**
a113	1203	157	10	1.28	2190	0.05344	2.2	0.40456	5.4	0.0549	5	0.4	336	7	345	16	408	111	82

spot number	$^{207}Pb^a$ (cps)	U^b (ppm)	Pb^b (ppm)	$\frac{Th^b}{U}$	$\frac{^{206}Pb^c}{^{204}Pb}$	$\frac{^{206}Pb^c}{^{238}U}$	2 s %	$\frac{^{207}Pb^c}{^{235}U}$	2 s %	$\frac{^{207}Pb^c}{^{206}Pb}$	2 s %	rho^d	$\frac{^{206}Pb}{^{238}U}$	2 s (Ma)	$\frac{^{207}Pb}{^{235}U}$	2 s (Ma)	$\frac{^{207}Pb}{^{206}Pb}$	2 s (Ma)	conc %^{e}
a114	4732	502	42	1.46	8590	0.06784	2.1	0.51672	3.1	0.05524	2.3	0.67	423	8	423	11	422	51	**100**
a115	10169	185	53	0.75	1296	0.25599	2.1	3.09357	3.4	0.08765	2.7	0.61	1469	27	1431	27	1375	52	**107**
a116	23087	746	123	0.24	31840	0.16714	1.8	1.67395	2.3	0.07264	1.5	0.77	996	16	999	15	1004	29	**99**
a117	23512	293	101	0.73	22660	0.30928	1.9	4.4339	2.2	0.10398	1	0.88	1737	29	1719	18	1696	19	**102**
a118	19109	338	78	0.57	21832	0.21016	1.9	2.49785	2.2	0.0862	1.1	0.87	1230	22	1271	16	1343	21	**92**
a119	7470	430	51	0.86	12198	0.10587	2.3	0.89606	3	0.06139	1.9	0.76	649	14	650	14	653	41	**99**
a120	5416	461	42	1.31	3610	0.07331	2	0.57366	3.3	0.05675	2.7	0.59	456	9	460	12	482	59	**95**
a121	8773	529	69	1.58	4350	0.10424	1.9	0.90183	3.1	0.06275	2.5	0.62	639	12	653	15	700	52	**91**
a122	5816	58	24	1.41	5255	0.32171	1.9	4.91827	3.4	0.11088	2.8	0.56	1798	30	1805	29	1814	51	**99**
a123	35691	684	138	0.45	4394	0.18724	2.4	2.29805	2.6	0.08901	1.2	0.9	1106	24	1212	19	1404	22	79
a124	16534	427	79	0.45	3646	0.17792	2.4	2.04351	3.3	0.0833	2.3	0.72	1056	23	1130	23	1276	45	83
a125	5860	517	44	0.92	3036	0.07479	1.7	0.58618	2.7	0.05684	2.1	0.64	465	8	468	10	485	46	**96**
a126	3382	88	17	0.39	4354	0.19006	2.8	2.0361	4.4	0.0777	3.4	0.64	1122	29	1128	31	1139	68	**98**
a127	3524	310	26	1.11	4309	0.06987	2.1	0.63335	5.5	0.06575	5.1	0.39	435	9	498	22	798	107	55
a128	11953	598	76	1.28	18631	0.0999	2.4	0.88521	4.1	0.06427	3.2	0.6	614	14	644	20	751	69	82
a129	26638	517	135	0.88	29211	0.23274	2.1	2.93124	2.4	0.09134	1.1	0.89	1349	26	1390	18	1454	21	**93**
a130	2385	1027	20	0.78	4970	0.0176	2.5	0.11661	4.1	0.04806	3.3	0.61	112	3	112	4	102	78	**110**
a131	43258	527	169	0.37	40879	0.30936	1.9	4.50253	2.2	0.10556	1.1	0.87	1738	29	1731	18	1724	20	**101**
a132	25959	477	123	0.75	28744	0.23143	2.3	2.88531	2.6	0.09042	1.3	0.87	1342	27	1378	20	1434	24	**94**
a133	15655	188	66	0.78	3479	0.31153	1.9	4.63922	2.6	0.10801	1.8	0.73	1748	30	1756	22	1766	33	**99**
a134	5415	539	42	0.8	9585	0.07083	2.1	0.55306	3.9	0.05663	3.3	0.53	441	9	447	14	477	74	**92**
a135	11286	268	64	0.92	12101	0.21015	1.8	2.39245	2.6	0.08257	1.8	0.71	1230	20	1240	18	1259	35	**98**
a136	2429	184	20	1	1760	0.09219	1.9	0.76419	3.9	0.06012	3.4	0.48	569	10	576	17	608	74	**94**
a137	14666	2997	115	0.57	1819	0.03565	8	0.27624	8.6	0.0562	3.1	0.93	226	18	248	19	460	69	49
a138	23339	518	123	0.41	6788	0.23153	2.1	2.68645	2.6	0.08415	1.6	0.8	1342	26	1325	20	1296	31	**104**
a139	9232	704	79	1.46	2163	0.08941	1.9	0.76221	2.9	0.06183	2.2	0.66	552	10	575	13	668	47	83
a140	9740	948	91	0.7	1055	0.08926	2	0.83012	4.4	0.06818	3.9	0.45	551	10	619	20	874	81	63

spot number	^{207}Pb[a] (cps)	U[b] (ppm)	Pb[b] (ppm)	Th[b]/U	^{206}Pb[c]/^{204}Pb	^{206}Pb[c]/^{238}U	2 s %	^{207}Pb[c]/^{235}U	2 s %	^{207}Pb[c]/^{206}Pb	2 s %	rho[d]	^{206}Pb/^{238}U (Ma)	2 s (Ma)	^{207}Pb/^{235}U (Ma)	2 s (Ma)	^{207}Pb/^{206}Pb (Ma)	2 s (Ma)	conc %[e]
a141	3442	1780	27	0.58	1970	0.01388	2.2	0.09164	12	0.04788	11.8	0.19	89	2	89	10	93	279	95
a142	12402	290	70	0.63	14874	0.22483	1.9	2.58937	2.8	0.08353	2.1	0.68	1307	23	1298	21	1282	40	102
a143	6907	409	38	0.62	442	0.06806	2.9	0.9633	9.2	0.10264	8.7	0.32	424	12	685	47	1672	162	25
a144	3312	51	21	2.03	1220	0.30244	2	4.29559	4.1	0.10301	3.6	0.49	1703	30	1693	35	1679	66	101
a145	8264	1411	115	0.42	7206	0.05102	2.2	0.37042	3.8	0.05265	3.1	0.57	321	7	320	10	314	70	102
a146	2946	913	31	1.62	3782	0.02594	2.5	0.17671	11.8	0.04941	11.5	0.21	165	4	165	18	167	269	99
a147	1735	277	15	0.75	3097	0.04803	2.7	0.37177	5.5	0.05614	4.8	0.49	302	8	321	15	458	107	66
a148	143350	2389	588	0.33	1052	0.23433	2.8	4.67921	3.1	0.14482	1.1	0.93	1357	35	1764	26	2286	19	59
a149	14140	1087	91	0.34	2056	0.08164	3	0.74072	3.6	0.06581	2	0.83	506	14	563	16	800	42	63
a150	9864	741	78	1.04	6238	0.08708	2	0.78147	3.3	0.06508	2.6	0.62	538	10	586	15	777	54	69
a151	5081	524	42	0.78	1214	0.07238	2.3	0.66223	4.2	0.06636	3.5	0.55	450	10	516	17	818	74	55
a152	6102	1036	46	0.27	4542	0.04474	3.2	0.35901	4.3	0.0582	3	0.73	282	9	311	12	537	65	53
a153	13110	239	67	1.29	12558	0.218	3.4	3.14035	4.2	0.10448	2.5	0.8	1271	39	1443	33	1705	46	75
a154	2977	378	26	0.43	5048	0.06694	2.1	0.54318	4.1	0.05885	3.5	0.52	418	9	441	15	562	76	74
a155	23042	222	99	0.46	17902	0.4172	1.9	7.41602	2.2	0.12892	1.1	0.87	2248	37	2163	20	2083	19	108
a156	7258	567	53	0.86	676	0.07887	3.1	0.90331	3.7	0.08306	1.9	0.85	489	15	653	18	1271	37	39
a157	67429	811	300	0.71	14892	0.329	2.1	5.40765	2.3	0.11921	1	0.9	1834	33	1886	20	1944	18	94
a158	19969	369	142	1.24	8741	0.32023	2	4.86636	2.6	0.11022	1.6	0.78	1791	31	1796	22	1803	29	99
a159	2846	484	31	1.62	965	0.04955	1.9	0.41452	4.9	0.06068	4.5	0.39	312	6	352	15	628	97	50
a160	67116	2058	410	0.78	6397	0.14747	3.2	1.78995	3.3	0.08803	0.8	0.97	887	26	1042	21	1383	16	64
a161	7855	716	76	0.92	12977	0.09461	2.1	0.7892	2.8	0.0605	1.9	0.74	583	11	591	12	622	40	94
a162	4151	151	31	0.75	1740	0.18505	2.1	2.06121	3.8	0.08079	3.1	0.56	1094	21	1136	26	1216	62	90
a163	2267	222	20	0.28	3622	0.08848	2.1	0.7613	5.2	0.0624	4.7	0.4	547	11	575	23	688	101	79
a164	7658	596	78	1.03	12392	0.11353	2.4	0.96871	3.3	0.06188	2.3	0.72	693	16	688	17	670	50	103
a165	1492	205	15	0.75	831	0.06589	2.1	0.52685	6.7	0.05799	6.3	0.32	411	9	430	24	530	138	78
a166	32395	1109	169	0.41	3789	0.14507	3	1.72095	3.3	0.08604	1.4	0.91	873	24	1016	21	1339	26	65
a167	5676	2442	72	0.97	5217	0.02583	1.9	0.17551	3.3	0.04928	2.7	0.56	164	3	164	5	161	64	102

spot number	$^{207}Pb^a$ (cps)	U^b (ppm)	Pb^b (ppm)	$\frac{Th^b}{U}$	$\frac{^{206}Pb^c}{^{204}Pb}$	$\frac{^{206}Pb^c}{^{238}U}$	2 s %	$\frac{^{207}Pb^c}{^{235}U}$	2 s %	$\frac{^{207}Pb^c}{^{206}Pb}$	2 s %	rho^d	$\frac{^{206}Pb}{^{238}U}$	2 s (Ma)	$\frac{^{207}Pb}{^{235}U}$	2 s (Ma)	$\frac{^{207}Pb}{^{206}Pb}$	2 s (Ma)	conc %^{e}
a168	7756	241	69	0.58	9933	0.27499	2.1	2.97162	2.7	0.07838	1.8	0.75	1566	29	1400	21	1156	36	135
a169	29731	1749	218	0.68	2627	0.11148	2.6	1.24061	3.1	0.08071	1.6	0.86	681	17	819	17	1214	31	56
a170	14486	809	109	0.06	1508	0.14167	1.8	1.53095	2.5	0.07838	1.7	0.72	854	15	943	16	1156	35	74
a171	8060	1142	82	0.34	14268	0.07251	1.8	0.5648	3.1	0.05649	2.5	0.59	451	8	455	12	472	56	96
a172	4754	420	50	0.95	644	0.10462	2.3	0.87369	4.2	0.06057	3.6	0.53	641	14	638	20	624	78	103
a173	2250	236	23	1.32	1362	0.07736	2.2	0.76142	6.3	0.07138	5.9	0.34	480	10	575	28	968	121	50
a174	1165	124	14	1.58	1091	0.09099	2.4	0.75281	5.8	0.06001	5.2	0.42	561	13	570	25	604	113	93
a175	56923	780	302	0.24	40440	0.38	2.3	6.78793	2.6	0.12956	1.1	0.9	2076	41	2084	23	2092	20	99
a176	23025	921	185	0.45	29717	0.19418	2	2.07817	2.6	0.07762	1.6	0.78	1144	21	1142	18	1137	32	101
a177	11731	256	81	0.92	11848	0.26914	2.4	3.68466	3.5	0.09929	2.5	0.69	1536	33	1568	28	1611	47	95
a178	5427	198	44	0.79	2005	0.19874	2.2	2.15644	3.2	0.07869	2.3	0.7	1169	24	1167	22	1164	45	100
a179	29205	19182	207	0.06	1127	0.01064	4.4	0.11647	5	0.0794	2.5	0.87	68	3	112	5	1182	48	6
a180	13872	350	99	0.53	14715	0.26626	1.9	3.46663	2.6	0.09443	1.7	0.75	1522	26	1520	20	1517	32	100
a181	2533	1295	39	1.13	5131	0.02588	2	0.17656	4.6	0.04949	4.1	0.44	165	3	165	7	171	96	96
a182	5710	835	62	0.74	768	0.06602	1.9	0.62405	4	0.06855	3.5	0.47	412	8	492	16	885	73	47
a183	3994	687	44	0.3	4058	0.06593	2.2	0.5067	3.7	0.05574	3	0.59	412	9	416	13	442	67	93
a184	9286	1912	111	0.54	1547	0.05507	2.1	0.48787	3.1	0.06425	2.2	0.7	346	7	403	10	750	47	46
a185	12835	264	88	0.55	12235	0.30891	2.1	4.47954	2.6	0.10517	1.4	0.82	1735	32	1727	21	1717	27	101
a186	3707	213	38	0.6	5089	0.16993	2.1	1.71186	4.3	0.07306	3.7	0.49	1012	20	1013	28	1016	76	100
a187	16072	709	156	1.27	2640	0.19306	2.1	2.45081	2.9	0.09207	2	0.72	1138	22	1258	21	1469	37	77
a188	9812	517	84	0.22	1418	0.16002	2.6	1.84605	3.6	0.08367	2.5	0.73	957	23	1062	24	1285	48	74
a189	10910	2075	146	0.77	19652	0.06367	2.1	0.48588	2.9	0.05535	1.9	0.73	398	8	402	10	426	43	93
a190	14915	2858	200	0.66	4151	0.06539	2.1	0.51766	2.7	0.05741	1.7	0.77	408	8	424	9	507	38	80
a191	5342	743	75	0.83	8885	0.09226	1.8	0.76598	3.3	0.06022	2.8	0.54	569	10	577	15	611	60	93
a192	4128	584	76	2.08	7084	0.09412	2	0.75713	2.9	0.05834	2.1	0.68	580	11	572	13	543	47	107
a193	2450	1392	40	0.76	4776	0.02645	2.1	0.18653	4.5	0.05114	4	0.46	168	4	174	7	247	92	68

spot number	$^{207}Pb^a$ (cps)	U^b (ppm)	Pb^b (ppm)	$\underline{Th^b}$ U	$\underline{^{206}Pb^c}$ ^{204}Pb	$\underline{^{206}Pb^c}$ ^{238}U	2 s %	$\underline{^{207}Pb^c}$ ^{235}U	2 s %	$\underline{^{207}Pb^c}$ ^{206}Pb	2 s %	rho^d	$\underline{^{206}Pb}$ ^{238}U	2 s (Ma)	$\underline{^{207}Pb}$ ^{235}U	2 s (Ma)	$\underline{^{207}Pb}$ ^{206}Pb	2 s (Ma)	conc %^{e}
a195	8105	1559	139	1.35	5505	0.07256	1.9	0.5546	2.9	0.05543	2.2	0.65	452	8	448	11	430	49	**105**
a196	50409	1348	438	0.55	26751	0.30274	1.9	4.20679	2.2	0.10078	1.1	0.87	1705	28	1675	18	1639	20	**104**
A-5	9274	352	36	0.88	0	0.1034	2.8	0.8775	4.2	0.06157	3.2	0.66	634	17	640	20	659	68	**96**
A-6	65711	362	133	0.26	0	0.3516	2.5	6.114	2.9	0.1262	1.4	0.87	1942	42	1992	25	2045	25	**95**
A-7	26995	289	68	0.29	0	0.2315	3.1	2.944	3.7	0.09226	1.9	0.85	1342	38	1393	28	1472	36	**91**
A-8	19489	289	58	1.05	0	0.1981	2.7	2.136	3.7	0.07822	2.5	0.74	1165	29	1161	25	1152	49	**101**
A-9	36171	237	76	0.26	0	0.3128	2.5	4.678	3	0.1085	1.7	0.83	1754	39	1763	25	1774	31	**99**
A-10	915	163	4	1.41	1	0.02355	3.8	0.1629	13.9	0.05016	13	0.27	150	6	153	20	202	311	74
A-11	86182	722	204	0.34	0	0.2755	2.3	3.861	2.8	0.1017	1.6	0.83	1569	32	1606	23	1654	29	**95**
A-12	8873	289	32	0.27	0	0.1117	3.1	0.9686	4.3	0.0629	2.9	0.73	683	20	688	21	704	62	**97**
A-13	19893	393	66	0.17	0	0.1689	2.4	1.7	3.1	0.07299	2	0.76	1006	22	1008	20	1013	41	**99**
A-14	33502	633	109	0.19	0	0.1718	2.4	1.736	3.2	0.07329	2.2	0.74	1022	22	1022	21	1021	44	**100**
A-15	3611	140	15	1.07	0	0.1052	2.6	0.8786	6.7	0.06061	6.1	0.39	645	16	640	32	625	132	**103**
A-16	88788	641	203	0.69	0	0.3085	2.6	4.48	3	0.1054	1.5	0.86	1733	39	1727	25	1720	28	**101**
A-17	1115	69	3	0.82	4	0.03988	3.6	0.4878	14	0.08873	14	0.26	252	9	403	47	1398	260	18
A-18	36340	1052	131	0.09	1	0.1255	4.8	1.147	5.1	0.06633	1.9	0.93	762	34	776	28	816	40	**93**
A-19	20390	306	58	0.28	0	0.1906	3.2	1.996	4	0.07595	2.5	0.79	1125	33	1114	27	1093	49	**103**
A-20	36631	700	110	0.21	0	0.1565	2.7	1.573	3.3	0.07291	1.9	0.81	937	23	959	20	1011	39	**93**
A-21	10577	185	33	0.43	0	0.1775	2.1	1.842	3.9	0.07528	3.2	0.54	1053	20	1060	25	1075	65	**98**
A-22	2786	205	13	0.27	0	0.06283	2.7	0.4733	5.1	0.05464	4.3	0.53	393	10	393	16	397	96	**99**
A-23	11226	85	26	0.43	0	0.3021	2.7	4.447	4.3	0.1068	3.3	0.65	1702	41	1721	35	1745	60	**98**
A-24	2918	582	14	1.03	0	0.02484	2.6	0.1677	4.6	0.049	3.8	0.56	158	4	157	7	147	89	**108**
A-29	1949	373	9	1.13	1	0.02522	2.9	0.1671	9.4	0.04806	8.9	0.31	161	5	157	14	101	211	158
A-30	11898	569	33	0.53	2	0.05832	3.5	0.4722	7	0.05873	6.1	0.5	365	12	393	23	557	132	66
A-31	6406	484	27	0.43	0	0.05705	2.8	0.4192	4.5	0.0533	3.5	0.63	358	10	355	13	341	79	**105**
A-32	16925	310	55	0.83	0	0.1753	2.3	1.804	3.3	0.07463	2.4	0.68	1041	22	1047	22	1058	49	**98**
A-33	1304	23	4	0.67	1	0.1704	3.5	1.746	10.4	0.07436	9.8	0.33	1014	33	1026	67	1051	198	**97**

spot number	$^{207}Pb^a$ (cps)	U^b (ppm)	Pb^b (ppm)	Th^b/U	$^{206}Pb^c$/^{204}Pb	$^{206}Pb^c$/^{238}U	2 s %	$^{207}Pb^c$/^{235}U	2 s %	$^{207}Pb^c$/^{206}Pb	2 s %	rho^d	^{206}Pb/^{238}U (Ma)	2 s (Ma)	^{207}Pb/^{235}U (Ma)	2 s (Ma)	^{207}Pb/^{206}Pb (Ma)	2 s (Ma)	conc %^{e}
A-34	25985	543	71	0.17	0	0.1308	4.2	1.359	5	0.07534	2.8	0.83	793	31	871	29	1077	57	74
A-35	8269	170	25	0.37	0	0.1495	2.4	1.436	5.3	0.06966	4.7	0.45	898	20	904	32	918	97	98
A-36	4300	268	17	0.57	0	0.06551	2.5	0.4936	5.6	0.05467	5	0.44	409	10	407	19	398	112	103
A-37	7386	489	30	0.19	0	0.0621	3.6	0.4638	5.6	0.05418	4.3	0.64	388	13	387	18	378	97	103
A-38	437	55	2	0.47	2	0.03492	5.2	0.2333	20.1	0.04847	19	0.26	221	11	213	39	122	458	182
A-39	16943	602	62	0.44	0	0.1034	3.1	0.8587	4.2	0.06022	2.8	0.73	635	19	629	20	611	61	104
A-40	43609	349	104	0.41	0	0.2896	2.6	4.073	3.2	0.102	1.9	0.81	1640	38	1649	26	1661	35	99
A-41	3247	117	12	1.94	0	0.1025	3	0.8733	6.4	0.06179	5.7	0.47	629	18	637	30	666	122	94
A-42	3650	195	14	0.7	0	0.07191	3.6	0.563	5.9	0.0568	4.6	0.62	448	16	454	21	483	101	93
A-43	25688	347	64	0.18	0	0.185	3.2	1.903	4.8	0.07461	3.6	0.66	1094	32	1082	32	1057	72	103
A-44	8397	541	34	0.5	0	0.0638	2.7	0.489	4.2	0.0556	3.3	0.63	399	10	404	14	436	73	91
A-45	1314	155	5	0.82	2	0.03294	2.8	0.2581	16.1	0.05684	16	0.18	209	6	233	33	485	349	43
A-46	2226	77	8	1.55	0	0.1084	3.3	0.9282	6.3	0.06211	5.4	0.53	664	21	667	31	677	115	98
A-47	2071	696	10	0.73	1	0.01427	3.2	0.09203	7.5	0.04677	6.7	0.43	91	3	89	6	37	161	247
A-48	40538	295	92	0.57	0	0.305	2.4	4.302	3	0.1023	1.9	0.79	1716	36	1694	25	1666	34	103
A-53	36261	321	89	0.48	0	0.2702	3.4	3.792	4.1	0.1018	2.3	0.82	1542	46	1591	33	1657	43	93
A-54	7153	536	20	0.58	0	0.03717	5.2	0.2731	7.1	0.05331	4.9	0.73	235	12	245	16	341	111	69
A-55	9047	106	25	0.76	0	0.235	2.5	2.709	3.7	0.08363	2.7	0.68	1361	31	1331	27	1283	53	106
A-56	7592	265	27	0.3	0	0.1013	2.9	0.8579	4.4	0.06146	3.3	0.65	622	17	629	21	655	72	95
A-57	4534	63	12	0.55	0	0.1975	3.2	2.113	6.3	0.07761	5.4	0.51	1162	34	1153	43	1136	107	102
A-58	1257	202	5	0.83	1	0.02711	3.2	0.1773	9.6	0.04746	9.1	0.33	172	5	166	15	72	217	241
A-59	27108	202	63	1.26	0	0.3043	3.1	4.021	4.2	0.09588	2.7	0.76	1712	47	1639	34	1545	51	111
A-60	14465	288	34	0.07	0	0.1186	3.5	1.386	4.6	0.08481	3	0.76	722	24	883	27	1311	57	55
A-61	2896	133	12	1.13	0	0.09386	2.3	0.7736	5.9	0.05979	5.4	0.39	578	13	582	26	596	118	97
A-62	5783	225	24	1.64	0	0.107	2.7	0.8739	4.3	0.05923	3.3	0.63	656	17	638	20	575	72	114
A-63	25563	257	57	0.21	0	0.2194	3.1	2.919	3.9	0.0965	2.4	0.78	1279	36	1387	30	1557	46	82

spot number	$^{207}Pb^a$ (cps)	U^b (ppm)	Pb^b (ppm)	$\frac{Th^b}{U}$	$\frac{^{206}Pb^c}{^{204}Pb}$	$\frac{^{206}Pb^c}{^{238}U}$	2 s %	$\frac{^{207}Pb^c}{^{235}U}$	2 s %	$\frac{^{207}Pb^c}{^{206}Pb}$	2 s %	rho^d	$\frac{^{206}Pb}{^{238}U}$	2 s (Ma)	$\frac{^{207}Pb}{^{235}U}$	2 s (Ma)	$\frac{^{207}Pb}{^{206}Pb}$	2 s (Ma)	conc %^{e}
A-65	16177	1096	50	0.03	3	0.04634	4.6	0.3768	5.8	0.05898	3.5	0.8	292	13	325	16	566	76	52
A-66	3878	71	13	1.24	0	0.1809	3.1	1.838	5	0.07371	4	0.61	1072	30	1059	33	1033	81	**104**
A-67	26001	73	39	0.47	0	0.4939	2.8	11.29	3.5	0.1658	2.1	0.8	2588	60	2547	33	2515	35	**103**
A-68	133266	854	310	0.19	0	0.348	2.8	5.881	3.4	0.1226	2	0.81	1925	46	1958	29	1994	35	**97**
A-69	3290	213	14	0.5	0	0.06684	2.9	0.5048	6	0.05479	5.3	0.48	417	12	415	21	403	119	**103**
A-70	2813	374	12	1.91	0	0.03228	2.4	0.2348	5.4	0.05277	4.8	0.45	205	5	214	10	318	109	64
A-71	6715	113	21	0.37	0	0.1846	2.7	1.945	6.3	0.07646	5.7	0.43	1092	28	1097	42	1106	114	**99**
A-72	5104	226	21	0.14	0	0.09525	2.3	0.7475	4.9	0.05694	4.3	0.48	587	13	567	21	488	95	120
A-77	9627	386	38	0.89	0	0.099	2.3	0.826	3.6	0.06053	2.8	0.63	609	13	611	16	622	60	**98**
A-78	4956	412	22	0.73	0	0.05537	2.5	0.3832	4.6	0.05021	3.8	0.55	347	8	329	13	204	89	170
A-79	1130	191	6	1.15	1	0.03149	3.1	0.1966	10.3	0.04529	9.9	0.3	200	6	182	17	-41	239	-493
A-80	60406	149	84	0.69	0	0.5133	2.3	12.78	2.8	0.1807	1.6	0.81	2671	49	2664	26	2659	27	**100**
A-81	9446	617	39	0.59	0	0.06425	2.6	0.484	4.5	0.05465	3.7	0.58	401	10	401	15	397	83	**101**
A-82	64585	883	193	0.84	0	0.2155	3.6	2.541	4.2	0.08553	2.1	0.86	1258	41	1284	30	1327	41	**95**
A-83	17322	111	36	0.62	0	0.3139	2.8	4.75	3.6	0.1098	2.3	0.78	1760	43	1776	30	1795	41	**98**
A-84	13578	263	43	0.32	0	0.1647	2.9	1.613	4.2	0.07104	3	0.69	983	26	975	26	958	62	**103**
A-85	8896	152	27	0.29	0	0.1775	2.3	1.813	4	0.07409	3.3	0.57	1054	22	1050	26	1043	66	**101**
A-86	6603	470	22	0.71	1	0.04767	3	0.3769	7.5	0.05736	6.8	0.41	300	9	325	21	505	150	59
A-87	6351	240	24	0.55	0	0.1022	2.7	0.871	3.8	0.06182	2.7	0.71	627	16	636	18	667	58	**94**
A-88	26780	640	95	0.43	2	0.1491	6	1.405	7.1	0.06836	3.8	0.84	896	50	891	42	879	79	**102**
A-89	4915	395	21	0.01	0	0.05514	2.5	0.4076	3.9	0.05363	3	0.64	346	8	347	12	355	68	**97**
A-90	29099	506	77	0.04	0	0.152	2.8	1.632	3.4	0.07793	1.8	0.85	912	24	983	21	1144	36	80
A-91	37844	2766	86	1.29	6	0.03171	3.6	0.2409	18.1	0.0551	18	0.2	201	7	219	36	416	396	48
A-92	9920	173	32	0.45	0	0.1832	2.2	1.862	3.3	0.07372	2.5	0.66	1085	22	1068	22	1033	50	**105**
A-93	5656	666	26	1.15	0	0.03967	3.2	0.2836	5.2	0.05186	4.1	0.61	251	8	254	12	279	95	90
A-94	23922	1702	52	0.12	4	0.03056	4.1	0.2716	6.2	0.06447	4.6	0.67	194	8	244	13	756	96	26
A-95	2409	269	10	0.73	1	0.03868	3.2	0.2718	7.8	0.05097	7.2	0.4	245	8	244	17	239	165	**102**

TR5

spot number	$^{207}Pb^a$ (cps)	U^b (ppm)	Pb^b (ppm)	$\underline{Th^b}$ U	$\underline{^{206}Pb^c}$ ^{204}Pb	$\underline{^{206}Pb^c}$ ^{238}U	2 s %	$\underline{^{207}Pb^c}$ ^{235}U	2 s %	$\underline{^{207}Pb^c}$ ^{206}Pb	2 s %	rho^d	$\underline{^{206}Pb}$ ^{238}U	2 s (Ma)	$\underline{^{207}Pb}$ ^{235}U	2 s (Ma)	$\underline{^{207}Pb}$ ^{206}Pb	2 s (Ma)	conc %[e]
a1	319	380	1	0.02	633	0.00392	2.1	0.02762	11.0	0.05104	10.8	0.19	25	1	28	3	243	249	10
a2	32796	499	109	0.29	10522	0.22359	2.1	2.56761	2.2	0.08329	0.8	0.94	1301	24	1291	16	1276	15	102
a3	7704	554	35	0.18	13883	0.06540	1.9	0.50215	3.2	0.05569	2.6	0.58	408	7	413	11	440	58	93
a4	80900	979	224	0.46	2675	0.21983	4.6	2.75219	6.0	0.09080	3.8	0.78	1281	54	1343	46	1442	72	89
a5	38934	913	125	0.41	20737	0.13468	2.0	1.42270	2.4	0.07661	1.3	0.84	815	16	898	15	1111	26	73
a6	22759	424	83	0.70	30664	0.17946	1.8	1.83577	2.3	0.07419	1.4	0.79	1064	18	1058	16	1047	29	102
a7	45638	307	108	0.46	9046	0.33122	2.0	4.98371	2.8	0.10913	1.9	0.73	1844	33	1817	24	1785	35	103
a8	1986	374	12	1.31	2933	0.02566	2.0	0.17507	5.0	0.04949	4.5	0.41	163	3	164	8	171	106	95
a9	7758	66	30	2.76	7052	0.28984	2.0	3.90126	3.2	0.09762	2.5	0.63	1641	29	1614	26	1579	47	104
a10	4901	327	27	1.16	3255	0.07041	2.1	0.53813	4.2	0.05543	3.6	0.51	439	9	437	15	430	81	102
a11	6477	681	29	0.58	3746	0.03906	2.1	0.30608	4.0	0.05684	3.4	0.52	247	5	271	10	485	76	51
a12	24239	208	68	0.83	24070	0.28735	2.0	3.99077	2.4	0.10073	1.3	0.84	1628	29	1632	20	1638	24	99
a13	40996	2404	-95	0.20	1395	-0.04811	3.2	-0.52628	3.9	0.07933	2.1	0.84	-318	-11	-759	-43	1180	42	-27
a14	15787	1009	80	1.23	2661	0.06609	2.1	0.57651	3.7	0.06326	3.1	0.57	413	9	462	14	717	65	58
a15	34592	83	59	1.25	18077	0.55277	1.9	14.58552	2.1	0.19137	1.1	0.86	2837	43	2789	21	2754	18	103
a16	10111	410	47	0.95	14592	0.10019	1.9	0.84742	2.7	0.06134	1.9	0.70	616	11	623	13	651	41	95
a17	59259	380	147	0.67	51658	0.34819	1.8	5.51572	2.0	0.11489	0.7	0.93	1926	31	1903	17	1878	13	103
a18	20532	397	82	0.97	11382	0.17857	1.9	1.81896	2.5	0.07388	1.6	0.75	1059	18	1052	17	1038	33	102
a19	10103	62	24	0.57	8524	0.35122	2.3	5.72140	3.4	0.11815	2.4	0.68	1940	39	1935	29	1928	44	101
a20	79807	529	175	0.18	72346	0.32856	1.8	4.99935	2.0	0.11036	0.8	0.92	1831	29	1819	17	1805	14	101
a21	9155	283	32	0.33	14239	0.11142	2.1	0.98833	3.2	0.06434	2.4	0.65	681	13	698	16	753	51	90
a22	3773	270	19	0.81	5479	0.06216	2.3	0.48269	4.8	0.05632	4.2	0.48	389	9	400	16	465	94	84
a23	22435	934	89	0.07	13667	0.10250	1.8	0.85097	2.3	0.06021	1.3	0.81	629	11	625	11	611	29	103

spot number	$^{207}Pb^a$ (cps)	U^b (ppm)	Pb^b (ppm)	$\underline{Th^b}$ U	$\underline{^{206}Pb^c}$ ^{204}Pb	$\underline{^{206}Pb^c}$ ^{238}U	2 s %	$\underline{^{207}Pb^c}$ ^{235}U	2 s %	$\underline{^{207}Pb^c}$ ^{206}Pb	2 s %	rho^d	$\underline{^{206}Pb}$ ^{238}U	2 s (Ma)	$\underline{^{207}Pb}$ ^{235}U	2 s (Ma)	$\underline{^{207}Pb}$ ^{206}Pb	2 s (Ma)	conc %^{e}
a25	14054	645	61	0.39	18440	0.09431	2.0	0.77680	2.6	0.05974	1.7	0.77	581	11	584	12	594	36	98
a26	45583	947	151	0.64	2604	0.14783	2.1	1.62070	2.7	0.07951	1.6	0.80	889	18	978	17	1185	32	75
a27	42501	726	149	0.42	54983	0.20052	1.8	2.13804	2.0	0.07733	0.9	0.89	1178	19	1161	14	1130	18	104
a28	18340	294	63	0.67	23605	0.19905	2.0	2.13272	2.5	0.07771	1.6	0.78	1170	21	1159	18	1139	31	103
a29	23609	250	76	0.92	25657	0.26601	2.0	3.37362	2.3	0.09198	1.1	0.88	1521	27	1498	18	1467	20	104
a30	24642	325	89	1.09	28853	0.23227	1.8	2.73471	2.4	0.08539	1.5	0.77	1346	22	1338	18	1324	29	102
a31	2231	81	9	0.77	3572	0.10366	2.5	0.89353	4.4	0.06251	3.6	0.56	636	15	648	21	692	78	92
a32	17756	189	55	0.77	16895	0.26054	1.9	3.34473	2.6	0.09311	1.8	0.72	1493	25	1492	21	1490	34	100
a33	37446	584	119	0.35	32254	0.19914	1.8	2.32816	2.3	0.08479	1.3	0.81	1171	20	1221	16	1311	25	89
a34	15444	178	46	0.56	17510	0.24298	2.0	2.95601	2.3	0.08824	1.2	0.85	1402	25	1396	18	1388	23	101
a35	3424	264	17	0.70	6202	0.06037	1.9	0.45971	3.7	0.05522	3.1	0.52	378	7	384	12	421	70	90
a36	5959	113	21	0.78	6869	0.17019	2.1	1.72757	2.9	0.07362	2.0	0.71	1013	19	1019	19	1031	41	98
a37	2449	413	14	0.98	4786	0.03075	2.2	0.21799	5.6	0.05141	5.2	0.39	195	4	200	10	259	119	75
a38	65633	503	204	1.67	61374	0.31301	1.9	4.61528	2.0	0.10694	0.6	0.94	1755	29	1752	17	1748	12	100
a39	90458	261	135	0.36	48530	0.47069	2.0	12.10203	2.1	0.18647	0.9	0.92	2487	41	2612	20	2711	14	92
a40	98928	625	248	0.84	83289	0.34790	2.0	5.69457	2.1	0.11872	0.7	0.94	1925	33	1931	18	1937	13	99
a41	9432	817	46	0.88	13982	0.05005	2.3	0.37513	5.5	0.05436	5.0	0.41	315	7	323	15	386	113	82
a42	7815	804	44	0.88	12804	0.04975	1.9	0.36381	2.9	0.05304	2.2	0.65	313	6	315	8	330	50	95
a43	63528	2750	166	0.26	336	0.05408	2.9	0.82998	3.1	0.11131	1.1	0.94	340	9	614	14	1821	19	19
a44	42864	637	131	0.78	1712	0.18268	1.9	2.39902	2.2	0.09524	1.0	0.88	1082	19	1242	16	1533	19	71
a45	156192	1131	348	0.08	149145	0.31632	2.0	4.57080	2.1	0.10480	0.6	0.96	1772	31	1744	18	1711	11	104
a46	395483	2747	970	0.08	368266	0.36254	1.9	5.38136	2.3	0.10766	1.2	0.84	1994	33	1882	20	1760	23	113
a47	39781	341	125	1.43	39575	0.28997	1.9	4.01298	2.2	0.10037	1.1	0.86	1641	28	1637	18	1631	21	101
a48	1956	634	11	0.99	3789	0.01526	3.4	0.10913	7.9	0.05188	7.1	0.43	98	3	105	8	280	163	35
a49	3274	669	21	1.27	3693	0.02578	2.0	0.17699	4.2	0.04979	3.6	0.49	164	3	165	6	185	85	89
a50	8856	360	39	0.82	14661	0.09503	2.0	0.79097	2.8	0.06037	1.9	0.72	585	11	592	13	617	42	95

spot number	$^{207}Pb^a$ (cps)	U^b (ppm)	Pb^b (ppm)	$\underline{Th^b}$ U	$\underline{^{206}Pb^c}$ ^{204}Pb	$\underline{^{206}Pb^c}$ ^{238}U	2 s %	$\underline{^{207}Pb^c}$ ^{235}U	2 s %	$\underline{^{207}Pb^c}$ ^{206}Pb	2 s %	rho^d	$\underline{^{206}Pb}$ ^{238}U	2 s (Ma)	$\underline{^{207}Pb}$ ^{235}U	2 s (Ma)	$\underline{^{207}Pb}$ ^{206}Pb	2 s (Ma)	conc %[e]
a51	2105	153	11	0.63	1781	0.06642	2.2	0.50869	6.1	0.05554	5.7	0.36	415	9	418	21	434	127	95
a52	10562	171	37	0.50	13377	0.20747	1.9	2.25828	2.6	0.07895	1.7	0.75	1215	22	1199	18	1171	34	104
a53	4118	573	19	0.18	4058	0.03483	2.5	0.25177	3.8	0.05243	2.9	0.65	221	5	228	8	304	67	73
a54	8670	105	27	0.51	9774	0.24035	2.2	2.94319	3.3	0.08881	2.5	0.66	1388	27	1393	25	1400	47	99
a55	7747	509	38	1.09	13685	0.06262	2.0	0.48880	2.7	0.05661	1.9	0.73	392	7	404	9	476	41	82
a56	12534	356	40	0.42	16437	0.10183	2.5	1.06985	3.1	0.07620	1.8	0.82	625	15	739	16	1100	35	57
a57	5578	243	30	1.24	8951	0.10159	2.1	0.87386	3.1	0.06238	2.3	0.68	624	12	638	15	687	49	91
a58	2081	31	8	0.97	2544	0.20945	2.3	2.35708	3.9	0.08162	3.2	0.59	1226	26	1230	28	1236	62	99
a59	13481	150	44	0.96	14775	0.24915	2.1	3.13001	2.8	0.09111	1.9	0.75	1434	27	1440	22	1449	36	99
a60	72276	1018	236	0.44	35928	0.22353	1.9	2.56954	2.0	0.08337	0.7	0.94	1301	22	1292	15	1278	14	102
a61	8263	524	41	0.79	14934	0.07126	1.9	0.54422	2.8	0.05539	2.0	0.68	444	8	441	10	428	45	104
a62	22222	115	24	0.59	17124	0.16158	3.9	2.89169	4.3	0.12980	1.6	0.92	966	35	1380	33	2095	29	46
a63	427	142	3	0.51	945	0.01833	2.5	0.11528	8.2	0.04561	7.8	0.31	117	3	111	9	-23	190	-514
a64	13922	198	43	0.34	16710	0.21222	2.0	2.44816	2.7	0.08367	1.8	0.75	1241	23	1257	20	1285	35	97
a65	11712	821	55	0.45	21357	0.06620	1.9	0.50138	2.7	0.05493	1.9	0.71	413	8	413	9	409	42	101
a66	7048	471	29	0.13	12878	0.06592	2.0	0.49784	3.1	0.05477	2.3	0.64	412	8	410	10	403	52	102
a67	21599	275	22	0.70	1512	0.06049	4.7	0.84493	5.5	0.10130	3.0	0.84	379	17	622	26	1648	55	23
a68	14094	571	62	0.66	9265	0.10163	1.9	0.84207	2.3	0.06009	1.3	0.84	624	11	620	11	607	27	103
a69	2632	514	14	0.68	3848	0.02608	2.1	0.17907	4.3	0.04980	3.7	0.50	166	4	167	7	185	86	89
a70	2679	96	5	0.47	169	0.03406	3.1	0.89537	13.7	0.19067	13.3	0.23	216	7	649	68	2748	219	8
a71	54246	1036	185	0.28	6340	0.17758	2.2	1.85648	2.4	0.07582	0.9	0.92	1054	21	1066	16	1090	19	97
a72	85728	1180	276	0.64	1649	0.22503	2.7	3.04693	3.0	0.09820	1.1	0.92	1308	32	1419	23	1590	21	82
a73	9853	201	40	0.94	13432	0.17578	1.9	1.78126	2.7	0.07349	1.9	0.71	1044	19	1039	18	1028	38	102
a74	2145	107	11	0.90	1957	0.08740	2.1	0.70010	4.7	0.05810	4.2	0.45	540	11	539	20	533	93	101
a75	37370	604	123	0.53	31872	0.19421	1.8	2.14413	2.2	0.08007	1.1	0.85	1144	19	1163	15	1199	22	95

spot number	$^{207}Pb^a$ (cps)	U^b (ppm)	Pb^b (ppm)	$\frac{Th^b}{U}$	$\frac{^{206}Pb^c}{^{204}Pb}$	$\frac{^{206}Pb^c}{^{238}U}$	2 s %	$\frac{^{207}Pb^c}{^{235}U}$	2 s %	$\frac{^{207}Pb^c}{^{206}Pb}$	2 s %	rho^d	$\frac{^{206}Pb}{^{238}U}$	2 s (Ma)	$\frac{^{207}Pb}{^{235}U}$	2 s (Ma)	$\frac{^{207}Pb}{^{206}Pb}$	2 s (Ma)	conc %^{e}
a77	49532	1009	173	0.28	19168	0.17226	1.8	1.72277	2.1	0.07253	1.0	0.88	1025	17	1017	13	1001	20	102
a78	170	149	1	0.08	286	0.00400	9.2	0.03293	29.1	0.05977	27.6	0.32	26	2	33	9	595	599	4
a79	85329	536	185	1.08	31371	0.25981	2.3	4.14794	2.5	0.11579	1.1	0.89	1489	30	1664	21	1892	20	79
a80	37560	718	93	0.58	3351	0.11157	2.3	1.17933	2.8	0.07666	1.5	0.84	682	15	791	15	1112	30	61
a81	8446	627	51	1.70	15397	0.06396	1.8	0.48436	2.5	0.05492	1.8	0.71	400	7	401	8	409	40	98
a82	6776	1566	37	0.48	6032	0.02299	2.0	0.15693	3.3	0.04951	2.7	0.60	147	3	148	5	172	62	85
a83	4282	81	17	1.00	5731	0.18437	2.1	1.90246	3.3	0.07484	2.6	0.62	1091	21	1082	22	1064	53	102
a84	13342	567	58	1.34	2931	0.07692	2.8	0.64440	4.4	0.06076	3.4	0.63	478	13	505	17	631	73	76
a85	3951	879	21	0.51	1153	0.02367	2.0	0.15703	4.3	0.04813	3.8	0.47	151	3	148	6	105	90	143
a86	16291	3353	65	0.35	907	0.01652	3.3	0.16994	3.7	0.07460	1.6	0.90	106	3	159	5	1058	33	10
a87	4208	291	19	0.32	2339	0.06487	2.1	0.56417	3.7	0.06308	3.1	0.56	405	8	454	14	711	66	57
a88	12089	946	53	0.31	2655	0.05600	2.0	0.46188	5.6	0.05982	5.2	0.36	351	7	386	18	597	113	59
a89	31360	526	109	0.43	40319	0.20092	1.9	2.15778	2.3	0.07789	1.2	0.84	1180	21	1168	16	1144	25	103
a90	14843	1109	74	0.64	9158	0.06280	2.2	0.47964	3.1	0.05539	2.1	0.72	393	8	398	10	428	48	92
a91	1976	152	10	0.61	3502	0.06033	2.1	0.46740	4.7	0.05619	4.2	0.44	378	8	389	15	460	94	82
a92	4882	210	23	0.83	4955	0.09841	2.0	0.87028	3.7	0.06414	3.1	0.55	605	12	636	18	746	66	81
a93	3579	367	23	1.59	2724	0.04890	2.0	0.35379	4.3	0.05248	3.8	0.47	308	6	308	11	306	86	100
a94	2369	101	12	1.03	2509	0.10246	2.3	0.84698	5.7	0.05996	5.2	0.41	629	14	623	27	602	113	104
a95	8192	174	35	1.17	11262	0.16699	2.1	1.67602	3.4	0.07279	2.7	0.62	996	19	999	22	1008	54	99
a96	6502	455	29	0.19	5028	0.06612	1.9	0.53167	2.7	0.05832	1.9	0.71	413	8	433	10	542	42	76
a97	56678	470	160	0.99	13531	0.28707	1.9	4.14165	2.1	0.10464	1.0	0.89	1627	27	1663	17	1708	18	95
a98	1294	173	7	0.46	2399	0.03691	2.1	0.27481	5.3	0.05400	4.9	0.39	234	5	247	12	371	110	63
a99	13822	1268	61	0.56	1134	0.04460	2.0	0.41821	2.9	0.06800	2.2	0.67	281	5	355	9	869	45	32
a100	9152	353	73	1.17	180	0.05320	2.4	1.06734	9.1	0.14552	8.8	0.27	334	8	737	49	2294	151	15
a101	6039	581	33	0.93	11172	0.04993	1.9	0.37254	3.4	0.05411	2.9	0.55	314	6	322	10	376	65	84
a102	3879	170	20	0.99	6165	0.10574	2.1	0.89899	3.9	0.06166	3.3	0.53	648	13	651	19	662	70	98

spot number	$^{207}Pb^a$ (cps)	U^b (ppm)	Pb^b (ppm)	$\underline{Th^b}$ U	$\underline{^{206}Pb^c}$ ^{204}Pb	$\underline{^{206}Pb^c}$ ^{238}U	2 s %	$\underline{^{207}Pb^c}$ ^{235}U	2 s %	$\underline{^{207}Pb^c}$ ^{206}Pb	2 s %	rho^d	$\underline{^{206}Pb}$ ^{238}U	2 s (Ma)	$\underline{^{207}Pb}$ ^{235}U	2 s (Ma)	$\underline{^{207}Pb}$ ^{206}Pb	2 s (Ma)	conc %^{e}
a103	61992	1502	230	0.28	10603	0.15528	1.8	1.62276	2.7	0.07579	2.0	0.67	931	16	979	17	1090	41	85
a104	8728	131	30	0.53	10714	0.21644	2.0	2.43609	3.2	0.08163	2.5	0.62	1263	23	1253	23	1237	49	102
a105	48674	593	160	0.54	16884	0.25404	2.0	3.10694	2.2	0.08870	0.9	0.91	1459	26	1434	17	1398	17	104
a106	7923	687	50	0.31	743	0.03630	2.9	0.34427	7.9	0.06879	7.4	0.36	230	6	300	21	892	153	26
a107	4088	320	21	0.31	7776	0.06537	1.9	0.47433	3.0	0.05263	2.3	0.63	408	7	394	10	313	52	131
a108	11219	137	39	0.80	12552	0.25364	2.3	3.12719	3.4	0.08942	2.4	0.70	1457	31	1439	26	1413	46	103
a109	2975	149	14	0.34	5103	0.09771	2.2	0.78684	4.1	0.05841	3.5	0.53	601	12	589	18	545	76	110
a110	84489	649	200	0.32	19502	0.29901	1.8	4.33866	2.2	0.10524	1.2	0.83	1686	27	1701	18	1718	22	98
a111	2057	137	12	0.88	1921	0.07756	2.1	0.58722	4.4	0.05491	3.9	0.47	482	10	469	17	409	86	118
a112	3299	617	21	1.48	6581	0.02720	1.9	0.18846	3.7	0.05024	3.1	0.52	173	3	175	6	206	73	84
a113	8744	608	42	0.41	15669	0.06882	1.9	0.53127	2.9	0.05599	2.1	0.67	429	8	433	10	452	48	95
a114	84177	486	194	0.07	69366	0.40457	1.9	6.76788	2.2	0.12133	1.0	0.89	2190	36	2082	19	1976	17	111
a115	7282	519	32	0.12	13037	0.06534	2.0	0.50295	2.9	0.05583	2.1	0.69	408	8	414	10	446	46	92
a116	9439	677	46	0.51	6969	0.06608	2.0	0.51799	2.9	0.05685	2.1	0.68	412	8	424	10	486	47	85
a117	10522	505	51	1.13	17941	0.08223	2.0	0.66450	3.2	0.05861	2.5	0.62	509	10	517	13	553	55	92
a118	7623	149	26	0.39	10314	0.17089	1.9	1.74072	3.0	0.07388	2.3	0.65	1017	18	1024	20	1038	47	98
a119	19829	378	69	0.40	27151	0.17853	1.8	1.79893	2.3	0.07308	1.4	0.79	1059	18	1045	15	1016	29	104
a120	3669	712	22	1.18	1951	0.02539	2.0	0.17431	3.7	0.04980	3.1	0.54	162	3	163	6	186	73	87
a121	6478	447	32	0.60	4751	0.06863	2.0	0.52731	3.0	0.05573	2.2	0.67	428	8	430	11	441	50	97
a122	8045	647	46	1.31	5859	0.05891	2.1	0.44326	3.4	0.05457	2.7	0.61	369	8	373	11	395	61	93
a123	53777	472	150	0.69	13337	0.28495	1.9	4.07284	2.2	0.10366	1.1	0.88	1616	28	1649	18	1691	20	96
a124	2823	570	17	1.03	5666	0.02691	2.1	0.18563	4.0	0.05003	3.5	0.52	171	4	173	6	196	80	87
a125	10140	622	53	1.09	2124	0.07362	2.0	0.62491	3.1	0.06157	2.4	0.64	458	9	493	12	659	51	69
a126	52296	417	138	0.35	49574	0.32030	1.9	4.66282	2.1	0.10558	0.8	0.91	1791	30	1761	17	1724	16	104
a127	18244	243	57	0.42	17755	0.22825	1.9	2.69324	2.6	0.08558	1.8	0.73	1325	23	1327	19	1329	34	100

spot number	$^{207}Pb^a$ (cps)	U^b (ppm)	Pb^b (ppm)	$\underline{Th^b}$ U	$\underline{^{206}Pb^c}$ ^{204}Pb	$\underline{^{206}Pb^c}$ ^{238}U	2 s %	$\underline{^{207}Pb^c}$ ^{235}U	2 s %	$\underline{^{207}Pb^c}$ ^{206}Pb	2 s %	rho^d	$\underline{^{206}Pb}$ ^{238}U	2 s (Ma)	$\underline{^{207}Pb}$ ^{235}U	2 s (Ma)	$\underline{^{207}Pb}$ ^{206}Pb	2 s (Ma)	conc $\%^e$
a129	4524	207	22	0.89	7533	0.09200	2.2	0.75887	4.3	0.05983	3.7	0.51	567	12	573	19	597	81	95
a130	9068	161	32	0.41	11436	0.19002	2.0	2.07749	2.4	0.07929	1.3	0.84	1121	21	1141	16	1179	25	95
a131	3292	155	15	0.62	5509	0.09326	2.1	0.76788	3.6	0.05971	2.9	0.58	575	11	579	16	593	63	97
a132	2120	348	11	0.46	4133	0.02980	2.1	0.21095	4.4	0.05135	3.9	0.47	189	4	194	8	256	90	74
a133	63495	599	167	0.34	64513	0.27331	2.0	3.70067	2.2	0.09820	1.0	0.89	1558	28	1572	18	1590	19	98
a134	6018	106	23	0.94	7703	0.18682	2.1	2.01381	3.5	0.07818	2.8	0.59	1104	21	1120	24	1151	56	96
a135	39587	319	108	0.67	38811	0.30772	1.9	4.32999	2.1	0.10205	0.9	0.91	1729	29	1699	17	1662	16	104
a136	2653	161	14	0.64	4737	0.08014	2.0	0.61826	4.4	0.05595	3.9	0.45	497	9	489	17	450	88	110
a137	2655	216	13	0.62	2591	0.05795	2.1	0.43468	4.1	0.05440	3.5	0.51	363	7	366	13	388	78	94
a138	29957	152	62	0.48	16043	0.37228	2.2	6.62304	2.4	0.12903	0.9	0.92	2040	39	2062	21	2085	16	98
a139	156973	1798	409	0.54	23870	0.21192	2.8	2.87625	3.3	0.09844	1.7	0.86	1239	32	1376	25	1595	31	78
a140	7135	518	33	1.04	3057	0.05401	5.2	0.46839	6.1	0.06289	3.2	0.85	339	17	390	20	705	68	48
a141	8384	222	31	0.61	12506	0.12951	3.3	1.19525	5.0	0.06693	3.7	0.67	785	25	798	28	836	77	94
a142	26321	483	84	0.28	35369	0.17328	2.6	1.77985	3.2	0.07450	2.0	0.79	1030	25	1038	21	1055	40	98
a143	15581	361	63	0.56	22086	0.16562	1.8	1.61361	2.6	0.07066	1.8	0.70	988	17	976	16	948	38	104
a144	2420	50	9	0.17	3426	0.18122	2.2	1.77408	5.7	0.07100	5.3	0.38	1074	22	1036	38	957	108	112
a145	11278	857	50	0.90	1116	0.04930	1.9	0.47381	3.5	0.06971	3.0	0.53	310	6	394	12	920	61	34
a146	1883	100	11	0.87	1330	0.09947	2.3	0.76318	4.4	0.05565	3.7	0.53	611	14	576	20	438	83	139
a147	54611	465	158	0.57	52269	0.31403	1.9	4.51833	2.3	0.10435	1.2	0.85	1761	30	1734	19	1703	22	103
a148	5885	335	25	0.32	3063	0.07346	2.4	0.60665	5.2	0.05990	4.6	0.47	457	11	481	20	600	99	76
a149	6653	1355	44	0.16	2396	0.02242	2.3	0.15130	4.1	0.04896	3.4	0.56	143	3	143	6	146	80	98
a150	8213	616	44	0.79	13090	0.06584	1.9	0.49045	2.6	0.05402	1.8	0.73	411	8	405	9	372	40	110
a151	7139	1282	41	1.07	3290	0.02755	1.9	0.19031	5.2	0.05010	4.9	0.37	175	3	177	9	200	113	88
a152	30089	514	109	0.64	38481	0.19616	1.9	2.11569	2.2	0.07823	1.1	0.88	1155	21	1154	15	1153	21	100
a153	36761	394	105	0.50	12483	0.25026	2.0	3.12259	2.2	0.09049	1.1	0.88	1440	25	1438	17	1436	20	100
a154	19899	448	78	0.19	27965	0.18038	2.1	1.77031	3.2	0.07118	2.5	0.64	1069	20	1035	21	963	51	111

spot number	$^{207}Pb^a$ (cps)	U^b (ppm)	Pb^b (ppm)	$\underline{Th^b}$ U	$\underline{^{206}Pb^c}$ ^{204}Pb	$\underline{^{206}Pb^c}$ ^{238}U	2 s %	$\underline{^{207}Pb^c}$ ^{235}U	2 s %	$\underline{^{207}Pb^c}$ ^{206}Pb	2 s %	rho^d	$\underline{^{206}Pb}$ ^{238}U	2 s (Ma)	$\underline{^{207}Pb}$ ^{235}U	2 s (Ma)	$\underline{^{207}Pb}$ ^{206}Pb	2 s (Ma)	conc %^{e}
a155	41772	1338	126	0.25	304	0.08659	2.2	1.54498	7.2	0.12940	6.8	0.30	535	11	948	45	2090	120	26
a156	24101	323	78	0.50	28042	0.22759	1.9	2.69934	2.5	0.08602	1.5	0.78	1322	23	1328	18	1339	30	99
a157	26555	235	97	2.19	7753	0.28728	2.0	4.04208	2.6	0.10204	1.7	0.77	1628	29	1643	21	1662	31	98
a158	746	160	5	1.90	1366	0.02623	3.9	0.20007	12.9	0.05533	12.3	0.30	167	6	185	22	425	274	39
a159	29942	399	89	0.22	15364	0.22427	2.0	2.69750	2.4	0.08724	1.4	0.83	1304	24	1328	18	1366	26	96
a160	16270	728	73	0.62	21155	0.09479	1.9	0.77656	2.2	0.05942	1.1	0.87	584	10	584	10	582	23	100
a161	13138	224	47	0.50	11167	0.20195	1.8	2.22088	2.6	0.07976	1.9	0.69	1186	19	1188	18	1191	37	100
a162	3191	236	17	0.54	5870	0.07077	2.2	0.53089	8.1	0.05441	7.8	0.28	441	10	432	29	388	175	114
a163	8926	602	42	0.26	16000	0.07154	1.9	0.55088	2.6	0.05585	1.7	0.74	445	8	446	9	446	39	100
a164	5529	404	37	1.56	10345	0.07299	2.0	0.53830	3.5	0.05349	2.9	0.56	454	9	437	13	349	67	130
a165	1664	161	10	0.76	3135	0.05475	2.0	0.40055	4.5	0.05306	4.0	0.44	344	7	342	13	331	91	104
a166	7347	519	40	0.91	13140	0.06888	1.9	0.53232	3.2	0.05605	2.6	0.59	429	8	433	11	454	57	95
a167	7217	738	32	0.63	5192	0.03946	2.3	0.28788	3.6	0.05291	2.9	0.62	249	6	257	8	325	65	77
a168	1759	284	11	1.54	3379	0.03139	3.0	0.22725	11.0	0.05250	10.6	0.27	199	6	208	21	307	241	65
a169	14103	267	53	0.73	15132	0.18436	2.0	1.89077	2.8	0.07438	2.0	0.69	1091	20	1078	19	1052	41	104
a170	4342	815	24	0.67	2846	0.02732	2.1	0.17850	7.3	0.04738	7.0	0.29	174	4	167	11	69	166	253

Paleolake deposits

SD

A-5	1256	5	0.14	0.78	0.95	0.026687	2.2	0.16805	5.9	0.04568	5.4	0.38	170	3.7				
A-6	101	181	0.60	0.98	2.14	0.003409	4.6	0.01849	21.9	0.03935	21.4	0.21	22	1.0				
A-7	212	265	1.26	0.74	1.54	0.004870	3.1	0.02851	15.5	0.04247	15.2	0.20	31	1.0				
A-8	941	136	5.13	0.82	1.19	0.038786	2.4	0.23654	8.0	0.04424	7.6	0.30	245	5.8				
A-9	553	251	3.19	0.42	0.86	0.012974	2.8	0.08556	7.5	0.04784	6.9	0.38	83	2.3	83	5.98	91	2.3	92

spot number	^{207}Pb[a] (cps)	U[b] (ppm)	Pb[b] (ppm)	Th[b]/U	^{206}Pb[c]/^{204}Pb	^{206}Pb[c]/^{238}U	2 s %	^{207}Pb[c]/^{235}U	2 s %	^{207}Pb[c]/^{206}Pb	2 s %	rho[d]	^{206}Pb/^{238}U (Ma)	2 s (Ma)	^{207}Pb/^{235}U (Ma)	2 s (Ma)	^{207}Pb/^{206}Pb (Ma)	2 s (Ma)	conc %[e]
A-11	10989	99	32.43	0.35	0.85	0.317658	2.1	4.53277	2.8	0.10352	1.9	0.74	1778	32.4	1737	23.54	1688	32.4	105
A-12	1292	3069	8.19	1.10	1.62	0.002755	2.4	0.01374	11.4	0.03619	11.2	0.21	18	0.4					
A-13	265	335	1.56	0.50	1.03	0.004777	2.9	0.03069	13.3	0.04660	12.9	0.22	31	0.9	31	4.01			
A-14	9913	517	51.61	0.78	1.19	0.100853	2.2	0.83667	5.4	0.06018	4.9	0.41	619	13.2	617	24.99	610	13.2	102
A-15	10948	143	35.95	0.74	0.57	0.247263	2.2	3.00981	3.8	0.08831	3.1	0.57	1424	28.0	1410	29.22	1389	28.0	103
A-16	43049	139	90.93	0.27	0.50	0.590671	2.2	15.88751	5.2	0.19514	4.7	0.42	2992	52.7	2870	49.99	2785	52.7	107
A-17	17118	214	50.80	0.58	0.32	0.234123	2.2	2.93452	3.1	0.09093	2.1	0.72	1356	27.0	1391	23.11	1445	27.0	94
A-18	185	171	0.61	0.80	4.08	0.003624	4.8	0.02837	32.0	0.05680	31.7	0.15	23	1.1	28	8.97			
A-19	260	320	1.21	0.67	1.55	0.003847	3.4	0.02923	12.7	0.05511	12.3	0.26	25	0.8	29	3.67			
A-20	1273	115	6.77	0.71	1.22	0.060075	2.5	0.44638	5.8	0.05391	5.2	0.44	376	9.2	375	18.04	366	9.2	103
A-21	78094	407	140.30	0.32	0.15	0.320902	2.9	7.00459	3.3	0.15836	1.6	0.87	1794	45.1	2112	29.31	2438	45.1	74
A-22	79	144	0.48	1.47	2.39	0.003451	4.9	0.02211	19.7	0.04647	19.1	0.25	22	1.1	22	4.33			
A-23	8917	118	31.19	0.55	0.36	0.260011	2.1	3.19834	3.1	0.08924	2.2	0.69	1490	28.4	1457	23.93	1409	28.4	106
A-24	26833	201	73.62	0.41	0.24	0.353846	2.1	5.61887	2.5	0.11520	1.4	0.82	1953	34.7	1919	21.60	1883	34.7	104
A-29	775	511	2.36	1.18	5.19	0.004674	2.8	0.03823	16.9	0.05934	16.7	0.16	30	0.8	38	6.33			
A-30	1265	1311	6.29	1.16	3.98	0.004999	2.3	0.01813	19.7	0.02631	19.5	0.12	32	0.7					
A-31	5207	81	18.93	1.12	0.27	0.230603	2.2	2.70259	4.6	0.08502	4.0	0.48	1338	26.2	1329	33.83	1316	26.2	102
A-32	7563	182	31.62	0.15	0.18	0.174115	2.1	1.75280	2.8	0.07303	1.8	0.77	1035	20.5	1028	18.07	1014	20.5	102
A-33	1179	53	5.13	1.89	1.89	0.098549	3.2	0.82771	10.3	0.06093	9.8	0.31	606	18.4	612	47.51	636	18.4	95
A-34	17010	534	63.59	0.57	0.03	0.116753	2.4	1.53649	2.9	0.09547	1.7	0.82	712	16.1	945	17.91	1537	16.1	46
A-35	307	468	1.94	0.49	1.01	0.004249	2.7	0.02726	7.4	0.04654	6.9	0.36	27	0.7	27	1.99			
A-36	19884	184	59.63	0.46	0.19	0.314077	2.1	4.61762	2.6	0.10666	1.5	0.81	1761	32.0	1752	21.39	1743	32.0	101
A-37	11948	7	0.70	0.14	2.61	0.099253	3.2	0.81019	5.4	0.05922	4.4	0.59	610	18.7	603	24.57	575	18.7	106
A-38	178	333	1.10	1.00	2.17	0.003416	3.2	0.01774	19.7	0.03766	19.4	0.17	22	0.7					
A-39	12464	446	55.51	0.64	0.18	0.125072	2.4	1.14259	3.2	0.06628	2.1	0.76	760	17.3	774	17.22	814	17.3	93
A-40	5371	630	30.56	0.60	0.74	0.049386	2.1	0.36006	5.6	0.05289	5.2	0.38	311	6.5	312	15.18	324	6.5	96

spot number	$^{207}Pb^a$ (cps)	U^b (ppm)	Pb^b (ppm)	$\underline{Th^b}$ U	$\underline{^{206}Pb^c}$ ^{204}Pb	$\underline{^{206}Pb^c}$ ^{238}U	2 s %	$\underline{^{207}Pb^c}$ ^{235}U	2 s %	$\underline{^{207}Pb^c}$ ^{206}Pb	2 s %	rho^d	$\underline{^{206}Pb}$ ^{238}U	2 s (Ma)	$\underline{^{207}Pb}$ ^{235}U	2 s (Ma)	$\underline{^{207}Pb}$ ^{206}Pb	2 s (Ma)	conc %^{e}
A-41	874	748	2.24	1.09	8.43	0.003061	2.8	0.01999	18.3	0.04737	18.1	0.15	20	0.5	20	3.65			
A-42	10701	164	36.62	0.69	0.74	0.220672	2.2	2.54550	3.3	0.08369	2.5	0.66	1285	25.4	1285	23.99	1285	25.4	**100**
A-43	232	328	1.49	0.76	1.01	0.004637	2.7	0.02979	12.3	0.04662	12.0	0.22	30	0.8	30	3.62			
A-44	43880	341	107.99	0.29	0.15	0.306943	2.3	4.66553	3.0	0.11027	1.9	0.77	1726	35.0	1761	25.16	1803	35.0	**96**
A-45	173	344	1.05	1.08	1.00	0.003115	3.8	0.01995	12.2	0.04646	11.6	0.31	20	0.8	20	2.43			
A-46	3598	192	8.65	0.28	1.34	0.045849	3.9	0.34132	9.7	0.05401	8.8	0.41	289	11.1	298	24.94	371	11.1	78
A-47	944	2150	5.77	0.87	0.98	0.002755	2.4	0.01635	8.1	0.04307	7.7	0.29	18	0.4					
A-48	383	875	2.41	0.74	1.56	0.002834	2.7	0.01503	12.7	0.03848	12.4	0.21	18	0.5					
A-53	311	150	1.71	2.41	4.29	0.011821	2.9	0.04586	25.0	0.02814	24.8	0.11	76	2.2					
A-54	177	259	0.88	0.93	2.65	0.003462	3.6	0.02217	12.5	0.04645	11.9	0.29	22	0.8	22	2.75			
A-55	16243	146	46.92	0.32	0.24	0.311557	2.1	4.61090	2.7	0.10737	1.7	0.77	1748	31.7	1751	22.46	1755	31.7	**100**
A-56	207	336	1.16	0.90	1.79	0.003540	3.0	0.02271	13.5	0.04654	13.1	0.22	23	0.7	23	3.04			
A-57	5258	416	28.65	0.38	0.54	0.069887	2.1	0.53545	3.3	0.05558	2.6	0.62	435	8.7	435	11.69	435	8.7	**100**
A-58	199	206	0.70	0.88	3.02	0.003415	3.8	0.03106	14.3	0.06597	13.8	0.27	22	0.8	31	4.37			
A-59	3367	171	16.49	0.75	0.69	0.097196	2.2	0.80373	4.5	0.05999	4.0	0.48	598	12.4	599	20.48	603	12.4	**99**
A-60	30674	259	98.21	0.65	0.00	0.365114	2.9	5.85937	3.3	0.11643	1.5	0.89	2006	49.9	1955	28.26	1902	49.9	**106**
A-61	19554	1131	83.99	0.33	0.41	0.075133	2.2	0.60995	3.1	0.05890	2.2	0.72	467	10.0	484	11.94	563	10.0	83
A-62	4063	46	13.22	0.93	0.98	0.281834	2.1	3.67003	4.1	0.09447	3.5	0.52	1601	30.5	1565	32.66	1517	30.5	**106**
A-63	9210	115	30.34	0.52	0.54	0.260581	2.2	3.22123	3.5	0.08968	2.7	0.63	1493	29.0	1462	26.98	1418	29.0	**105**
A-64	152	187	0.98	0.93	2.04	0.005372	3.5	0.03051	17.7	0.04120	17.4	0.20	35	1.2					
A-65	372	495	2.03	0.85	1.32	0.004230	2.9	0.02234	17.0	0.03832	16.8	0.17	27	0.8					
A-66	463	154	2.67	0.76	2.34	0.017875	2.6	0.08852	22.0	0.03593	21.9	0.12	114	2.9					
A-67	397	575	2.31	1.08	1.01	0.004135	2.8	0.02289	11.8	0.04016	11.4	0.24	27	0.8					
A-68	4078	107	17.31	0.91	1.19	0.162418	3.1	1.44656	19.0	0.06461	18.8	0.16	970	27.7	908	114.14	761	27.7	127
A-69	11637	912	59.24	0.88	0.76	0.065949	2.2	0.50014	4.3	0.05502	3.8	0.50	412	8.6	412	14.71	412	8.6	**100**

spot number	$^{207}Pb^a$ (cps)	U^b (ppm)	Pb^b (ppm)	$\frac{Th^b}{U}$	$\frac{^{206}Pb^c}{^{204}Pb}$	$\frac{^{206}Pb^c}{^{238}U}$	2 s %	$\frac{^{207}Pb^c}{^{235}U}$	2 s %	$\frac{^{207}Pb^c}{^{206}Pb}$	2 s %	rho^d	$\frac{^{206}Pb}{^{238}U}$	2 s (Ma)	$\frac{^{207}Pb}{^{235}U}$	2 s (Ma)	$\frac{^{207}Pb}{^{206}Pb}$	2 s (Ma)	conc %e
A-71	6076	368	28.34	0.69	1.10	0.078150	2.2	0.61045	4.1	0.05667	3.4	0.53	485	10.2	484	15.67	478	10.2	**101**
A-72	3649	303	19.47	0.43	0.48	0.065169	2.2	0.49254	3.8	0.05483	3.1	0.59	407	8.8	407	12.73	405	8.8	**101**
A-77	374	559	2.17	0.50	1.50	0.004014	2.8	0.01994	17.2	0.03604	17.0	0.16	26	0.7					
A-78	33585	555	131.60	0.10	0.10	0.234662	2.4	2.75249	2.6	0.08510	0.9	0.94	1359	29.7	1343	19.27	1317	29.7	**103**
A-79	78	134	0.48	0.51	2.14	0.003685	5.0	0.02362	21.6	0.04650	21.0	0.23	24	1.2	24	5.06			
A-80	765	1446	4.92	0.93	0.77	0.003481	2.5	0.02232	7.0	0.04653	6.6	0.36	22	0.6	22	1.56			
A-81	37092	367	118.52	0.57	0.32	0.315128	2.1	4.43649	2.5	0.10214	1.4	0.84	1766	33.0	1719	21.07	1663	33.0	**106**
A-82	154	301	1.06	0.97	1.29	0.003619	3.0	0.02058	17.3	0.04126	17.0	0.18	23	0.7					
A-83	126	151	0.78	0.72	2.16	0.005341	3.4	0.03059	19.0	0.04155	18.7	0.18	34	1.1					
A-84	186	405	1.15	0.67	1.36	0.002921	3.3	0.01595	16.9	0.03962	16.6	0.20	19	0.6					
A-85	9116	134	32.25	0.84	0.53	0.237638	2.2	2.79875	3.5	0.08544	2.7	0.63	1374	27.3	1355	26.10	1325	27.3	**104**
A-86	7247	176	30.31	0.20	0.19	0.171890	2.1	1.74672	3.1	0.07372	2.3	0.67	1023	19.9	1026	20.15	1033	19.9	**99**
A-87	2311	244	13.86	1.05	1.08	0.058214	2.6	0.36715	13.0	0.04576	12.7	0.20	365	9.3					
A-88	31758	278	94.09	0.36	0.16	0.327741	2.1	4.97114	2.3	0.11004	1.0	0.90	1827	33.4	1814	19.71	1800	33.4	**102**
A-89	252	351	1.37	0.67	0.53	0.003983	3.5	0.02571	16.2	0.04682	15.8	0.22	26	0.9	26	4.13			
A-90	103	205	0.69	0.79	1.68	0.003500	3.4	0.01744	19.2	0.03615	18.9	0.18	23	0.8					
A-91	27774	301	84.47	0.25	0.23	0.273697	2.1	3.82576	2.6	0.10141	1.5	0.82	1560	29.2	1598	20.77	1650	29.2	**95**
A-92	604	916	3.46	0.72	1.59	0.003867	2.8	0.02480	7.5	0.04652	7.0	0.37	25	0.7	25	1.85			
A-93	412	901	2.79	0.64	1.17	0.003197	2.9	0.01558	20.3	0.03534	20.1	0.14	21	0.6					
A-94	31258	302	80.28	0.14	0.04	0.258883	2.2	3.74950	2.6	0.10507	1.3	0.87	1484	29.6	1582	20.71	1715	29.6	87
A-95	966	2032	5.54	0.59	0.93	0.002792	2.6	0.01786	8.3	0.04641	7.9	0.31	18	0.5	18	1.48			
A-96	210	371	1.36	0.87	1.71	0.003763	3.1	0.02002	14.3	0.03860	13.9	0.22	24	0.7					
A-101	5452	12272	35.70	0.67	0.95	0.002975	2.4	0.01937	12.5	0.04725	12.2	0.19	19	0.5	19	2.41			
A-102	335	644	2.04	0.75	0.74	0.003237	2.7	0.02072	9.3	0.04644	8.9	0.29	21	0.6	21	1.91			
A-103	18149	327	53.10	1.72	0.00	0.158870	2.5	2.13570	5.3	0.09753	4.7	0.46	950	21.7	1160	36.66	1577	21.7	60
A-104	515	701	2.68	0.48	1.90	0.003915	2.7	0.02513	13.8	0.04657	13.5	0.20	25	0.7	25	3.43			

spot number	^{207}Pb[a] (cps)	U[b] (ppm)	Pb[b] (ppm)	Th[b] U	$\frac{^{206}\text{Pb}^c}{^{204}\text{Pb}}$	$\frac{^{206}\text{Pb}^c}{^{238}\text{U}}$	2 s %	$\frac{^{207}\text{Pb}^c}{^{235}\text{U}}$	2 s %	$\frac{^{207}\text{Pb}^c}{^{206}\text{Pb}}$	2 s %	rho[d]	$\frac{^{206}\text{Pb}}{^{238}\text{U}}$	2 s (Ma)	$\frac{^{207}\text{Pb}}{^{235}\text{U}}$	2 s (Ma)	$\frac{^{207}\text{Pb}}{^{206}\text{Pb}}$	2 s (Ma)	conc %[e]
A-105	1657	131	5.95	0.47	0.00	0.045910	2.8	0.38973	10.0	0.06159	9.6	0.28	289	7.8	334	28.42	659	7.8	44
A-106	247	2	0.01	0.88	12.68	0.003033	6.4	0.03133	34.3	0.07494	33.7	0.19	20	1.3	31	10.58			
A-107	794	328	5.04	0.62	0.90	0.015693	2.6	0.10390	8.4	0.04803	8.0	0.31	100	2.6	100	8.06	100	2.6	100
A-108	223	448	1.53	0.93	1.33	0.003506	3.3	0.01895	12.8	0.03920	12.4	0.25	23	0.7					
A-109	6730	86	24.12	1.43	1.02	0.276534	2.2	3.60673	2.8	0.09462	1.7	0.78	1574	30.5	1551	22.24	1520	30.5	104
A-110	1023	1521	5.71	1.15	2.60	0.003837	2.5	0.02481	13.6	0.04691	13.4	0.18	25	0.6	25	3.34			
A-111	736	1607	4.32	0.86	1.22	0.002757	2.7	0.01645	9.6	0.04327	9.2	0.28	18	0.5					
A-112	522	851	3.46	0.55	1.00	0.004182	2.5	0.02393	8.7	0.04151	8.4	0.28	27	0.7					
A-113	321	349	1.18	0.98	4.49	0.003448	3.1	0.02208	10.7	0.04647	10.3	0.29	22	0.7	22	2.35			
A-114	32	125	0.22	0.96	3.35	0.001733	7.6	0.01656	30.2	0.06931	29.2	0.25	11	0.8	17	4.99			
A-115	6034	88	22.08	1.16	0.57	0.247354	2.1	2.89989	3.8	0.08505	3.1	0.56	1425	27.0	1382	28.38	1316	27.0	108
A-116	27352	780	61.40	1.11	0.00	0.076212	5.0	1.16687	5.2	0.11108	1.4	0.96	473	22.8	785	28.42	1817	22.8	26
A-117	3346	177	17.31	0.40	0.64	0.098944	2.3	0.81945	4.4	0.06008	3.8	0.51	608	13.1	608	20.10	606	13.1	100
A-118	148957	1557	483.30	0.07	0.02	0.302503	2.2	4.28980	2.4	0.10288	1.0	0.92	1704	33.6	1691	20.16	1676	33.6	102
A-119	112	172	0.67	1.16	2.86	0.003992	3.6	0.02561	16.4	0.04653	16.0	0.22	26	0.9	26	4.15			
A-120	1342	107	7.53	0.57	0.61	0.071664	2.3	0.55693	4.2	0.05638	3.5	0.54	446	9.8	450	15.14	467	9.8	96
A-125	15575	376	66.30	0.33	0.34	0.176497	2.1	1.75925	3.0	0.07231	2.1	0.71	1048	20.3	1031	19.25	994	20.3	105
A-126	2297	353	12.66	0.33	1.39	0.036732	2.5	0.22155	7.0	0.04376	6.6	0.36	233	5.8					
A-127	174	255	1.06	1.23	2.04	0.004279	3.6	0.02392	20.1	0.04056	19.8	0.18	28	1.0					
A-128	2805	307	15.75	0.53	0.68	0.052272	2.2	0.38093	5.0	0.05287	4.5	0.44	328	7.1	328	14.11	323	7.1	102
A-129	811	293	1.10	0.62	17.02	0.003799	3.6	0.03108	20.6	0.05935	20.3	0.17	24	0.9	31	6.32			
A-130	3643	651	20.86	0.40	0.48	0.032676	2.3	0.22681	3.9	0.05036	3.2	0.58	207	4.7	208	7.39	211	4.7	98
A-131	32524	301	102.64	0.85	0.06	0.330645	2.2	4.97862	2.9	0.10924	1.8	0.78	1842	36.0	1816	24.26	1786	36.0	103
A-132	334	444	2.04	1.01	1.54	0.004696	2.8	0.03018	12.0	0.04663	11.7	0.23	30	0.8	30	3.58			
A-133	11450	157	38.43	0.68	0.11	0.241016	2.2	3.08781	2.7	0.09295	1.5	0.82	1392	27.6	1430	20.59	1486	27.6	94

spot number	$^{207}Pb^a$ (cps)	U^b (ppm)	Pb^b (ppm)	$\frac{Th^b}{U}$	$\frac{^{206}Pb^c}{^{204}Pb}$	$\frac{^{206}Pb^c}{^{238}U}$	2 s %	$\frac{^{207}Pb^c}{^{235}U}$	2 s %	$\frac{^{207}Pb^c}{^{206}Pb}$	2 s %	rho^d	$\frac{^{206}Pb}{^{238}U}$	2 s (Ma)	$\frac{^{207}Pb}{^{235}U}$	2 s (Ma)	$\frac{^{207}Pb}{^{206}Pb}$	2 s (Ma)	conc %[e]
A-135	28932	288	89.41	0.72	0.73	0.303357	2.1	4.13026	3.0	0.09878	2.1	0.71	1708	31.6	1660	24.23	1601	31.6	107
A-136	90	215	0.62	0.90	2.01	0.002929	3.6	0.01875	14.8	0.04645	14.4	0.24	19	0.7	19	2.77			
A-137	3665	46	13.33	1.49	0.63	0.282175	2.3	3.66703	4.9	0.09428	4.3	0.46	1602	32.1	1564	39.05	1513	32.1	106
A-138	266	422	1.69	0.46	1.27	0.004090	3.1	0.02628	9.2	0.04661	8.7	0.34	26	0.8	26	2.40			
A-139	251	267	1.30	0.87	3.62	0.005002	3.2	0.03007	18.6	0.04361	18.3	0.17	32	1.0					
A-140	996	2109	6.41	0.60	0.57	0.003110	2.2	0.01992	5.5	0.04647	5.0	0.40	20	0.4	20	1.09			
A-141	4337	377	24.55	0.53	0.35	0.066142	2.2	0.49918	3.5	0.05475	2.7	0.63	413	8.7	411	11.81	401	8.7	103
A-142	447	915	2.82	0.80	1.13	0.003185	2.7	0.01596	13.6	0.03636	13.3	0.20	20	0.5					
A-143	281	382	1.73	0.48	0.89	0.004630	3.1	0.02976	10.2	0.04664	9.7	0.30	30	0.9	30	2.99			
A-144	23678	248	49.70	0.51	0.04	0.193981	4.0	2.92636	4.2	0.10944	1.3	0.95	1143	41.7	1389	31.65	1790	41.7	64
A-149	6700	562	34.12	0.02	0.08	0.061508	2.3	0.48546	3.0	0.05726	1.9	0.78	385	8.7	402	9.99	501	8.7	77
A-150	36894	6	1.90	0.36	0.54	0.302422	2.2	4.28421	2.8	0.10277	1.8	0.77	1703	32.7	1690	23.37	1674	32.7	102
A-151	140	252	0.85	0.88	1.17	0.003465	3.1	0.02221	17.0	0.04651	16.8	0.18	22	0.7	22	3.76			
A-152	4438	373	24.98	0.32	0.47	0.067961	2.1	0.51926	3.4	0.05543	2.6	0.63	424	8.6	425	11.65	429	8.6	99
A-153	19791	182	49.32	0.54	0.62	0.265024	2.2	3.67188	3.0	0.10051	2.1	0.72	1515	29.7	1565	24.30	1633	29.7	93
A-154	17903	240	64.36	0.64	0.59	0.264523	2.2	3.22196	2.9	0.08837	1.9	0.75	1513	29.4	1462	22.38	1390	29.4	109
A-155	14486	482	58.81	0.02	0.32	0.122405	2.2	1.16888	3.0	0.06928	2.1	0.73	744	15.4	786	16.45	906	15.4	82
A-156	386	439	2.35	2.48	1.01	0.005486	3.6	0.03533	12.1	0.04673	11.5	0.30	35	1.3	35	4.18			
A-157	239	305	1.56	1.12	1.15	0.005271	2.9	0.02880	12.9	0.03964	12.6	0.22	34	1.0					
A-158	166	205	1.03	0.66	1.66	0.005160	3.8	0.02855	18.6	0.04014	18.2	0.21	33	1.3					
A-159	5121	439	28.82	0.08	0.20	0.066652	2.3	0.50436	3.2	0.05490	2.2	0.72	416	9.2	415	10.83	407	9.2	102
A-160	250	308	1.62	0.71	1.34	0.005404	3.1	0.03054	12.3	0.04100	11.9	0.25	35	1.1					
A-161	1072	273	6.65	0.58	0.83	0.024918	2.3	0.16919	6.1	0.04926	5.6	0.39	159	3.7	159	8.93	160	3.7	99
A-162	319	433	1.97	0.62	1.19	0.004669	3.1	0.02838	7.2	0.04410	6.5	0.43	30	0.9					
A-163	372	93	0.51	1.26	14.61	0.005515	6.1	0.04581	46.2	0.06026	45.8	0.13	35	2.1	45	20.57			
A-164	268	497	1.41	1.27	2.50	0.002906	2.9	0.01707	19.6	0.04260	19.3	0.15	19	0.6					

spot number	$^{207}Pb^a$ (cps)	U^b (ppm)	Pb^b (ppm)	$\underline{Th^b}$ U	$\underline{^{206}Pb^c}$ ^{204}Pb	$\underline{^{206}Pb^c}$ ^{238}U	2 s %	$\underline{^{207}Pb^c}$ ^{235}U	2 s %	$\underline{^{207}Pb^c}$ ^{206}Pb	2 s %	rho^d	$\underline{^{206}Pb}$ ^{238}U	2 s (Ma)	$\underline{^{207}Pb}$ ^{235}U	2 s (Ma)	$\underline{^{207}Pb}$ ^{206}Pb	2 s (Ma)	conc %^{e}
A-165	35561	647	82.00	0.14	0.03	0.124686	2.6	1.56046	2.9	0.09079	1.3	0.90	757	18.8	955	18.04	1442	18.8	53
A-166	3534	98	12.45	0.53	0.80	0.127610	3.1	1.14750	7.5	0.06524	6.9	0.41	774	22.3	776	40.88	781	22.3	99
A-167	77	165	0.42	1.21	3.27	0.002593	5.3	0.01328	40.8	0.03716	40.4	0.13	17	0.9					
A-168	22939	290	60.92	0.70	0.00	0.204093	4.7	3.00950	4.9	0.10698	1.3	0.96	1197	51.3	1410	37.13	1748	51.3	68
A-173	1510	364	9.38	0.43	0.43	0.026301	2.3	0.17885	5.4	0.04933	4.8	0.43	167	3.8	167	8.25	163	3.8	103
A-174	2419	200	14.05	0.45	0.34	0.071174	2.5	0.54772	3.6	0.05583	2.6	0.69	443	10.5	444	12.82	445	10.5	100
A-175	6033	87	22.23	1.29	0.29	0.252232	2.2	3.13290	3.3	0.09011	2.5	0.68	1450	29.1	1441	25.56	1427	29.1	102
A-176	46378	657	158.24	0.20	0.23	0.236796	2.5	3.00373	2.8	0.09203	1.2	0.90	1370	30.8	1409	21.11	1467	30.8	93
A-177	171	212	1.07	0.94	1.09	0.005188	3.1	0.03341	15.2	0.04672	14.9	0.21	33	1.0	33	5.01			
A-178	317	255	1.42	0.60	2.92	0.005725	3.3	0.03692	18.3	0.04678	18.0	0.18	37	1.2	37	6.61			
A-179	2573	58	11.10	0.33	0.47	0.191480	2.2	1.98871	3.7	0.07535	2.9	0.61	1129	23.3	1112	24.93	1077	23.3	105
A-180	278	272	1.32	0.57	4.20	0.004959	3.4	0.03193	18.3	0.04672	18.0	0.19	32	1.1	32	5.75			
A-181	11	757	0.08	0.49	4.84	0.000106	9.2	0.00084	64.6	0.05719	64.0	0.14	1	0.1					
A-182	8745	135	31.61	0.93	0.85	0.232820	2.1	2.63945	3.8	0.08225	3.2	0.55	1349	25.3	1312	27.88	1251	25.3	108
A-183	4822	250	24.42	1.24	0.87	0.098673	2.2	0.81802	4.2	0.06014	3.6	0.52	607	12.6	607	19.04	608	12.6	100
A-184	281	414	1.45	1.00	2.78	0.003595	3.2	0.02305	15.1	0.04651	14.7	0.21	23	0.7	23	3.45			
A-185	9374	126	34.47	0.27	0.33	0.270246	2.1	3.37750	2.9	0.09067	2.0	0.73	1542	29.1	1499	22.74	1439	29.1	107
A-186	951	1815	4.91	0.82	1.99	0.002767	2.6	0.01768	8.6	0.04637	8.2	0.30	18	0.5	18	1.53			
A-187	15703	408	68.26	0.29	0.31	0.167265	2.2	1.66174	2.6	0.07207	1.4	0.85	997	20.5	994	16.57	987	20.5	101
A-188	261	349	1.78	0.39	0.80	0.005224	3.2	0.03145	12.7	0.04368	12.2	0.25	34	1.1					
A-189	168	338	1.21	1.23	2.64	0.003719	3.6	0.01731	27.7	0.03377	27.5	0.13	24	0.9					
A-190	12072	118	38.00	0.41	0.39	0.313663	2.1	4.45735	2.8	0.10310	1.9	0.74	1759	32.2	1723	23.39	1680	32.2	105
A-191	118882	266	185.63	0.11	0.00	0.608196	2.3	20.28937	2.4	0.24202	0.7	0.96	3063	55.0	3105	22.83	3133	55.0	98
A-192	48381	475	155.68	0.33	0.26	0.319139	2.1	4.53737	2.5	0.10315	1.2	0.86	1786	33.4	1738	20.61	1681	33.4	106
B-5	8865	458	43.91	0.48	0.28	0.097037	2.2	0.78120	3.8	0.05840	3.1	0.58	597	12.6	586	17.09	544	12.6	110

spot number	$^{207}Pb^a$ (cps)	U^b (ppm)	Pb^b (ppm)	$\frac{Th^b}{U}$	$\frac{^{206}Pb^c}{^{204}Pb}$	$\frac{^{206}Pb^c}{^{238}U}$	2 s %	$\frac{^{207}Pb^c}{^{235}U}$	2 s %	$\frac{^{207}Pb^c}{^{206}Pb}$	2 s %	rho^d	$\frac{^{206}Pb}{^{238}U}$ (Ma)	2 s (Ma)	$\frac{^{207}Pb}{^{235}U}$ (Ma)	2 s (Ma)	$\frac{^{207}Pb}{^{206}Pb}$ (Ma)	2 s (Ma)	conc %^{e}
B-7	29681	296	93.73	0.07	0.00	0.307686	2.1	4.42038	2.6	0.10423	1.5	0.81	1729	31.4	1716	21.23	1700	31.4	**102**
B-8	317	424	2.00	0.62	0.75	0.004839	3.8	0.03120	10.6	0.04678	9.9	0.36	31	1.2	31	3.25			
B-9	756	1202	4.30	1.48	2.00	0.003656	2.6	0.02351	15.8	0.04665	15.6	0.17	24	0.6	24	3.69			
B-10	132	296	0.88	0.96	2.03	0.003100	4.1	0.01053	31.4	0.02464	31.1	0.13	20	0.8					
B-11	141	352	0.96	0.83	1.74	0.002797	3.8	0.01790	18.5	0.04642	18.1	0.20	18	0.7	18	3.31			
B-12	89	158	0.50	0.77	0.32	0.003218	5.2	0.02116	21.0	0.04771	20.4	0.25	21	1.1	21	4.43			
B-13	102	101	0.58	0.76	1.50	0.005849	4.7	0.03851	24.2	0.04777	23.7	0.19	38	1.8	38	9.11			
B-14	273	407	1.79	0.49	0.28	0.004510	3.2	0.02637	12.1	0.04242	11.6	0.27	29	0.9					
B-15	28589	290	91.64	0.19	0.12	0.307985	2.2	4.35943	2.6	0.10269	1.4	0.84	1731	33.2	1705	21.53	1673	33.2	**103**
B-16	18174	161	54.22	0.68	0.33	0.326910	2.2	4.76589	2.9	0.10576	2.0	0.73	1823	34.2	1779	24.72	1727	34.2	**106**
B-17	2133	3727	14.09	0.29	0.23	0.003870	2.7	0.02439	4.6	0.04572	3.7	0.58	25	0.7					
B-18	7177	125	27.12	0.65	0.26	0.215524	2.2	2.39631	3.1	0.08066	2.3	0.69	1258	24.9	1241	22.54	1213	24.9	**104**
B-19	44117	1399	143.97	0.97	0.00	0.100781	2.9	1.35066	3.1	0.09723	1.2	0.92	619	17.0	868	18.20	1571	17.0	39
B-20	230	379	1.43	0.52	1.27	0.003888	3.3	0.01910	23.3	0.03563	23.1	0.14	25	0.8					
B-21	337	313	1.42	1.03	1.42	0.004510	2.4	0.04813	20.9	0.07741	20.8	0.11	29	0.7	48	9.76			
B-22	1949	52	8.22	0.60	0.74	0.159346	2.4	1.50893	4.2	0.06870	3.5	0.56	953	20.9	934	25.90	889	20.9	**107**
B-23	616	1158	3.87	1.00	0.94	0.003418	2.5	0.02193	10.3	0.04656	10.0	0.24	22	0.5	22	2.24			
B-24	19346	183	58.44	0.30	0.17	0.310977	2.1	4.51739	2.4	0.10539	1.1	0.89	1746	32.2	1734	19.72	1721	32.2	**101**
B-29	362	619	2.11	0.99	0.69	0.003500	2.8	0.02121	11.5	0.04396	11.1	0.25	23	0.6					
B-30	4823	267	24.80	1.12	0.84	0.094036	2.1	0.75929	4.0	0.05858	3.4	0.54	579	11.8	574	17.44	551	11.8	**105**
B-31	28009	428	74.04	1.36	1.11	0.171096	3.2	2.00337	5.1	0.08495	3.9	0.64	1018	30.3	1117	34.27	1314	30.3	77
B-32	844	1615	5.57	0.09	0.38	0.003531	2.5	0.02262	5.3	0.04649	4.7	0.47	23	0.6	23	1.19			
B-33	470	1127	3.18	0.64	1.28	0.002927	2.3	0.01320	16.0	0.03271	15.8	0.14	19	0.4					
B-34	10846	109	35.05	0.58	0.25	0.313196	2.3	4.35631	2.6	0.10091	1.3	0.86	1756	34.7	1704	21.57	1640	34.7	107
B-35	227	356	1.46	0.67	0.40	0.004222	3.2	0.02581	12.7	0.04435	12.3	0.25	27	0.9					
B-36	991	244	5.54	0.48	0.28	0.023119	3.1	0.16386	7.1	0.05142	6.5	0.43	147	4.4	154	10.21	259	4.4	57

spot number	$^{207}Pb^a$ (cps)	U^b (ppm)	Pb^b (ppm)	$\frac{Th^b}{U}$	$\frac{^{206}Pb^c}{^{204}Pb}$	$\frac{^{206}Pb^c}{^{238}U}$	2 s %	$\frac{^{207}Pb^c}{^{235}U}$	2 s %	$\frac{^{207}Pb^c}{^{206}Pb}$	2 s %	rho^d	$\frac{^{206}Pb}{^{238}U}$ (Ma)	2 s (Ma)	$\frac{^{207}Pb}{^{235}U}$ (Ma)	2 s (Ma)	$\frac{^{207}Pb}{^{206}Pb}$ (Ma)	2 s (Ma)	conc %^{e}
B-37	332	580	1.77	1.68	1.73	0.003116	3.0	0.02024	12.8	0.04711	12.5	0.24	20	0.6	20	2.59			
B-38	6462	38	16.50	1.29	0.69	0.416162	2.2	7.31011	3.5	0.12743	2.8	0.61	2243	41.3	2150	31.69	2062	41.3	109
B-39	5960	326	31.13	0.43	0.33	0.096526	2.1	0.78087	3.1	0.05869	2.2	0.69	594	12.1	586	13.76	555	12.1	107
B-40	14108	261	55.32	0.59	0.43	0.210875	2.1	2.26627	3.3	0.07797	2.5	0.64	1233	23.5	1202	23.15	1145	23.5	108
B-41	8866	809	30.76	0.45	5.81	0.038397	3.3	0.32301	6.4	0.06103	5.5	0.51	243	7.8	284	15.94	640	7.8	38
B-42	14209	150	26.43	0.30	0.84	0.172674	3.8	2.37355	5.3	0.09972	3.7	0.71	1027	36.1	1235	38.05	1618	36.1	63
B-43	272	306	1.55	0.45	1.15	0.005188	2.8	0.03336	12.1	0.04665	11.8	0.23	33	0.9	33	3.97			
B-44	17744	185	57.17	0.75	0.21	0.300918	2.2	4.30266	2.9	0.10373	2.0	0.74	1696	32.2	1694	24.09	1691	32.2	100
B-45	95	87	0.55	0.83	3.85	0.006386	3.6	0.04686	29.5	0.05324	29.3	0.12	41	1.5	47	13.40			
B-46	5893	61	18.70	0.82	0.46	0.299386	2.1	4.10728	3.4	0.09953	2.7	0.62	1688	31.6	1656	28.10	1615	31.6	105
B-47	1131	1290	7.09	0.93	1.08	0.005620	2.5	0.03619	8.8	0.04672	8.5	0.28	36	0.9	36	3.13			
B-48	7760	425	18.30	0.31	6.92	0.044568	2.7	0.21232	6.5	0.03456	5.9	0.41	281	7.3					
B-53	65161	643	165.50	0.74	0.80	0.252199	2.2	3.31979	2.8	0.09550	1.8	0.78	1450	28.8	1486	22.22	1537	28.8	94
B-54	22565	283	76.64	0.30	0.10	0.265861	2.4	3.43756	2.9	0.09380	1.6	0.84	1520	32.8	1513	22.79	1504	32.8	101
B-55	434	502	2.60	0.61	0.51	0.005286	3.0	0.03405	7.6	0.04674	7.0	0.39	34	1.0	34	2.55			
B-56	561	165	2.96	0.39	1.13	0.018232	2.8	0.12495	8.7	0.04972	8.2	0.32	116	3.2	120	9.76	181	3.2	64
B-57	6483	334	28.49	0.42	1.36	0.086353	2.3	0.67790	5.5	0.05695	5.0	0.41	534	11.7	526	22.70	489	11.7	109
B-58	19188	185	60.77	0.44	0.28	0.319350	2.1	4.52441	2.8	0.10278	1.9	0.74	1787	32.9	1735	23.57	1674	32.9	107
B-59	28770	1515	148.67	4.06	0.00	0.095890	2.5	1.31510	2.8	0.09950	1.1	0.91	590	14.3	852	15.99	1614	14.3	37
B-60	228	286	1.40	0.55	1.34	0.005014	3.4	0.03230	12.6	0.04674	12.1	0.27	32	1.1	32	3.99			
B-61	298	459	1.99	1.11	1.33	0.004424	3.0	0.02843	14.4	0.04662	14.1	0.21	28	0.8	28	4.06			
B-62	7023	90	24.40	-0.17	-0.07	0.266328	2.6	3.63750	6.9	0.09909	6.4	0.37	1522	34.9	1558	55.24	1606	34.9	95
B-63	22974	326	68.75	0.03	0.06	0.206807	2.5	2.66426	3.0	0.09346	1.6	0.84	1212	27.6	1319	21.80	1497	27.6	81
B-64	1266	82	0.67	0.87	56.61	0.008165	6.8	0.09432	65.0	0.08381	64.7	0.10	52	3.6	92	56.93			
B-65	22181	314	79.36	0.18	0.13	0.248962	2.1	3.07063	2.4	0.08948	1.3	0.85	1433	26.4	1425	18.46	1414	26.4	101

spot number	$^{207}Pb^a$ (cps)	U^b (ppm)	Pb^b (ppm)	Th^b U	$^{206}Pb^c$ ^{204}Pb	$^{206}Pb^c$ ^{238}U	2 s %	$^{207}Pb^c$ ^{235}U	2 s %	$^{207}Pb^c$ ^{206}Pb	2 s %	rho^d	^{206}Pb ^{238}U (Ma)	2 s (Ma)	^{207}Pb ^{235}U (Ma)	2 s (Ma)	^{207}Pb ^{206}Pb (Ma)	2 s (Ma)	conc %^{e}
B-67	971	2161	5.93	0.81	0.55	0.002817	2.4	0.01672	6.7	0.04307	6.2	0.36	18	0.4					
B-68	3335	78	14.09	1.07	0.06	0.179085	2.2	2.02318	3.8	0.08196	3.1	0.57	1062	21.6	1123	26.11	1244	21.6	85
B-69	486	537	1.90	0.69	4.21	0.003620	2.9	0.02323	14.8	0.04654	14.6	0.19	23	0.7	23	3.42			
B-70	24690	174	58.02	1.08	1.31	0.323320	2.2	4.80133	3.1	0.10773	2.2	0.71	1806	34.4	1785	25.96	1761	34.4	103
B-71	3759	651	19.89	0.43	0.68	0.031154	2.5	0.21521	6.1	0.05012	5.6	0.40	198	4.8	198	11.02	200	4.8	99
B-72	31149	208	53.74	3.30	2.69	0.247596	3.3	4.20261	3.9	0.12314	2.0	0.86	1426	42.8	1675	32.04	2002	42.8	71
B-77	6385	59	20.04	0.27	0.28	0.328939	2.4	4.75504	3.1	0.10487	2.0	0.76	1833	37.8	1777	26.07	1712	37.8	107
B-78	1552	15	0.14	2.52	83.49														
B-79	167	355	1.06	0.63	0.21	0.003048	4.2	0.01985	14.4	0.04724	13.8	0.29	20	0.8	20	2.85			
B-80	163	280	0.92	0.93	3.58	0.003431	4.2	0.01126	44.9	0.02382	44.8	0.09	22	0.9					
B-81	5394	296	21.40	0.15	0.57	0.072997	2.6	0.60430	3.5	0.06006	2.3	0.74	454	11.3	480	13.21	605	11.3	75
B-82	395	423	2.25	0.70	1.34	0.005452	2.7	0.03517	11.7	0.04680	11.3	0.24	35	1.0	35	4.02			
B-83	307	216	1.17	0.66	5.22	0.005615	3.1	0.02472	30.1	0.03194	30.0	0.10	36	1.1					
B-84	508	880	3.86	2.44	-14.15	0.004093	8.2	0.08681	28.3	0.15386	27.1	0.29	26	2.1	85	22.96			
B-85	292	394	1.86	0.67	1.95	0.004821	2.5	0.03101	9.9	0.04667	9.6	0.25	31	0.8	31	3.04			
B-86	16063	219	56.20	0.30	0.33	0.253081	2.1	3.09648	2.5	0.08876	1.4	0.83	1454	27.5	1432	19.51	1399	27.5	104
B-87	8487	124	28.93	0.52	0.61	0.230772	2.2	2.67679	4.2	0.08415	3.6	0.53	1339	26.9	1322	31.03	1295	26.9	103
B-88	303	447	1.71	0.55	1.83	0.003919	2.5	0.02231	16.3	0.04129	16.1	0.15	25	0.6					
B-89	24628	242	70.21	0.27	0.22	0.282556	2.2	3.95421	2.8	0.10153	1.8	0.78	1604	30.6	1625	22.50	1652	30.6	97
B-90	1867	41	7.68	0.44	0.60	0.187424	2.3	1.91780	6.2	0.07423	5.8	0.37	1107	23.3	1087	41.38	1047	23.3	106
B-91	3357	91	15.32	0.47	0.53	0.168490	2.2	1.63832	4.5	0.07054	4.0	0.49	1004	20.7	985	28.64	944	20.7	106
B-92	878	1836	6.12	0.94	0.47	0.003407	2.3	0.02262	10.4	0.04817	10.1	0.22	22	0.5	23	2.34			
B-93	4997	441	27.67	0.54	0.49	0.063768	2.1	0.47975	3.4	0.05458	2.6	0.63	398	8.2	398	11.12	394	8.2	101
B-94	1265	548	7.86	0.40	0.88	0.014759	2.3	0.08372	10.1	0.04115	9.8	0.23	94	2.1					
B-95	2871	263	15.50	0.27	0.34	0.059928	2.2	0.44264	3.8	0.05359	3.1	0.57	375	7.9	372	11.75	353	7.9	106
B-96	40371	393	123.01	0.32	0.20	0.304395	2.1	4.40848	2.5	0.10507	1.2	0.87	1713	32.0	1714	20.30	1715	32.0	100

spot number	$^{207}Pb^a$ (cps)	U^b (ppm)	Pb^b (ppm)	$\underline{Th^b}$ U	$\underline{^{206}Pb^c}$ ^{204}Pb	$\underline{^{206}Pb^c}$ ^{238}U	2 s %	$\underline{^{207}Pb^c}$ ^{235}U	2 s %	$\underline{^{207}Pb^c}$ ^{206}Pb	2 s %	rho^d	$\underline{^{206}Pb}$ ^{238}U	2 s (Ma)	$\underline{^{207}Pb}$ ^{235}U	2 s (Ma)	$\underline{^{207}Pb}$ ^{206}Pb	2 s (Ma)	conc %^{e}
B-101	153	314	1.18	0.77	3.14	0.003926	3.0	0.01049	67.1	0.01939	67.0	0.04	25	0.8					
B-102	1448	648	9.05	0.45	0.97	0.014351	2.2	0.08226	8.3	0.04158	8.0	0.26	92	2.0					
B-103	319	645	2.27	0.55	0.70	0.003601	2.6	0.02328	7.7	0.04690	7.3	0.34	23	0.6	23	1.79			
B-104	275	340	1.76	0.78	0.85	0.005322	3.5	0.03044	13.4	0.04150	12.9	0.26	34	1.2					
B-105	334	452	2.30	0.62	0.96	0.005195	2.7	0.03345	10.1	0.04670	9.8	0.26	33	0.9	33	3.33			
B-106	3418	282	17.85	0.43	0.90	0.064256	2.8	0.47956	5.1	0.05414	4.3	0.55	401	11.0	398	16.90	376	11.0	107
B-107	1046	858	4.87	0.34	2.22	0.005804	3.2	0.03747	8.7	0.04684	8.1	0.37	37	1.2	37	3.20			
B-108	380	424	2.27	0.87	1.22	0.005481	3.2	0.03533	12.8	0.04676	12.4	0.25	35	1.1	35	4.44			
B-109	196	182	0.86	0.62	1.64	0.004778	4.3	0.04202	18.0	0.06381	17.4	0.24	31	1.3	42	7.35			
B-110	411	714	2.96	0.64	0.91	0.004270	2.6	0.02188	10.2	0.03717	9.9	0.26	27	0.7					
B-111	27187	607	99.74	0.52	2.94	0.162616	2.9	1.88968	4.9	0.08430	3.9	0.60	971	26.6	1077	32.49	1299	26.6	75
B-112	969	2536	5.78	2.10	2.48	0.002373	3.0	0.00880	32.0	0.02692	31.9	0.09	15	0.5					
B-113	1638	2780	8.04	0.28	1.53	0.002958	2.8	0.01893	8.8	0.04643	8.3	0.32	19	0.5	19	1.66			
B-114	9386	734	42.03	0.17	0.33	0.057925	3.1	0.47179	4.0	0.05909	2.6	0.77	363	11.0	392	13.16	570	11.0	64
B-115	1704	2279	6.80	2.67	6.38	0.003109	2.8	0.01122	55.7	0.02618	55.6	0.05	20	0.6					
B-116	924	1476	6.37	0.92	0.47	0.004432	2.4	0.02621	7.4	0.04291	7.0	0.32	29	0.7					
B-117	95	183	0.36	1.46	1.22	0.001975	5.5	0.01742	18.6	0.06401	17.8	0.30	13	0.7	18	3.23			
B-118	19433	195	49.83	0.46	0.44	0.248250	3.0	3.55904	4.0	0.10401	2.6	0.76	1429	38.5	1540	31.38	1696	38.5	84
B-119	10383	212	41.68	0.32	0.17	0.195533	2.2	2.12240	2.5	0.07875	1.3	0.86	1151	23.1	1156	17.53	1165	23.1	99
B-120	256	359	1.58	0.50	1.15	0.004510	3.1	0.02898	17.1	0.04661	16.9	0.18	29	0.9	29	4.90			
B-125	32886	531	100.38	0.81	0.15	0.184928	2.2	2.50645	3.1	0.09833	2.2	0.70	1094	21.9	1274	22.41	1592	21.9	69
B-126	190	209	0.80	1.02	2.52	0.003924	4.0	0.02519	18.6	0.04657	18.1	0.22	25	1.0	25	4.63			
B-127	154	182	0.97	0.71	0.98	0.005457	3.3	0.03227	19.4	0.04290	19.1	0.17	35	1.2					
B-128	742	688	3.24	0.67	2.46	0.004821	2.7	0.03100	7.4	0.04665	6.9	0.37	31	0.8	31	2.26			
B-129	23526	306	74.11	0.59	0.38	0.238411	2.2	3.00645	3.2	0.09149	2.3	0.69	1378	27.0	1409	24.09	1456	27.0	95

spot number	$^{207}Pb^a$ (cps)	U^b (ppm)	Pb^b (ppm)	$\frac{Th^b}{U}$	$\frac{^{206}Pb^c}{^{204}Pb}$	$\frac{^{206}Pb^c}{^{238}U}$	2 s %	$\frac{^{207}Pb^c}{^{235}U}$	2 s %	$\frac{^{207}Pb^c}{^{206}Pb}$	2 s %	rho^d	$\frac{^{206}Pb}{^{238}U}$	2 s (Ma)	$\frac{^{207}Pb}{^{235}U}$	2 s (Ma)	$\frac{^{207}Pb}{^{206}Pb}$	2 s (Ma)	conc $\%^e$
B-131	16095	100	43.51	0.67	0.48	0.416908	2.4	7.40432	3.3	0.12885	2.2	0.74	2246	46.0	2161	29.32	2082	46.0	108
B-132	44691	486	139.26	0.35	0.03	0.278824	2.1	3.97777	2.2	0.10350	0.7	0.95	1585	29.3	1630	17.84	1687	29.3	94
B-133	32199	458	108.95	0.26	0.19	0.234753	2.2	2.87499	2.6	0.08885	1.4	0.84	1359	26.4	1375	19.44	1400	26.4	97
B-134	167	192	1.14	0.88	0.63	0.006104	3.1	0.03854	12.7	0.04581	12.3	0.25	39	1.2					
B-135	314	394	2.12	0.71	0.79	0.005535	3.0	0.03164	14.7	0.04147	14.4	0.21	36	1.1					
B-136	3732	625	22.97	0.48	0.58	0.037437	2.1	0.26129	3.2	0.05064	2.5	0.64	237	4.8	236	6.77	224	4.8	106
B-137	2550	196	14.62	0.43	0.31	0.075673	2.2	0.58219	3.8	0.05581	3.2	0.56	470	9.8	466	14.34	444	9.8	106
B-138	290	343	1.54	0.82	3.17	0.004650	3.0	0.01968	27.8	0.03071	27.7	0.11	30	0.9					
B-139	1329	93	7.01	0.63	0.29	0.076718	2.2	0.60146	5.7	0.05688	5.3	0.38	477	10.0	478	21.81	486	10.0	98
B-140	32078	649	113.03	0.09	0.87	0.173426	2.2	1.85198	2.9	0.07747	1.8	0.78	1031	21.4	1064	19.03	1133	21.4	91
B-141	157	298	1.04	0.55	1.31	0.003622	4.3	0.01669	29.2	0.03344	28.9	0.15	23	1.0					
B-142	162	238	1.21	0.74	0.18	0.005249	3.7	0.03021	16.7	0.04176	16.3	0.22	34	1.3					
B-143	2469	241	14.26	0.39	0.46	0.060192	2.2	0.44409	3.6	0.05353	2.9	0.59	377	7.9	373	11.35	350	7.9	108
B-144	9557	497	35.72	0.67	4.64	0.073582	2.3	0.46438	9.5	0.04579	9.2	0.25	458	10.4					
B-149	3101	592	15.79	1.35	2.41	0.027224	2.3	0.18505	8.5	0.04931	8.2	0.27	173	3.8	172	13.44	162	3.8	107
B-150	240	276	1.46	0.80	0.36	0.005386	3.1	0.03770	16.3	0.05078	16.0	0.19	35	1.1	38	6.02			
B-151	1148	287	7.68	0.55	0.74	0.027439	2.3	0.16010	8.4	0.04233	8.1	0.27	175	3.9					
B-152	6	77		0.55	-41.49	-0.000222	67.7	-0.00924	108.0	0.30222	84.1	0.63	-1	-1.0					
B-153	150	213	0.76	0.96	3.36	0.003708	4.5	0.01837	42.7	0.03593	42.5	0.10	24	1.1					
B-154	247	172	1.12	0.81	5.18	0.006703	3.8	0.03486	53.8	0.03772	53.7	0.07	43	1.6					
B-155	632	902	4.07	0.77	0.72	0.004615	3.0	0.02970	12.1	0.04668	11.7	0.25	30	0.9	30	3.54			
B-156	981	419	6.76	0.37	0.56	0.016499	2.5	0.10975	6.3	0.04826	5.8	0.39	105	2.6	106	6.33	111	2.6	95
B-157	3870	99	17.79	0.52	0.14	0.179741	2.2	1.88279	3.7	0.07599	3.0	0.60	1066	22.0	1075	24.84	1094	22.0	97
B-158	30611	319	104.31	0.51	0.01	0.318518	2.1	4.60234	2.4	0.10483	1.0	0.91	1782	33.3	1750	19.61	1711	33.3	104
B-159	8687	412	46.89	0.29	0.36	0.114889	2.1	0.97134	2.9	0.06134	2.1	0.70	701	13.8	689	14.74	650	13.8	108
B-160	22777	221	72.54	0.29	0.10	0.319011	2.1	4.70517	2.5	0.10700	1.3	0.86	1785	33.4	1768	20.92	1748	33.4	102

spot number	$^{207}Pb^a$ (cps)	U^b (ppm)	Pb^b (ppm)	Th^b/U	$^{206}Pb^c$/^{204}Pb	$^{206}Pb^c$/^{238}U	2 s %	$^{207}Pb^c$/^{235}U	2 s %	$^{207}Pb^c$/^{206}Pb	2 s %	rho^d	^{206}Pb/^{238}U (Ma)	2 s (Ma)	^{207}Pb/^{235}U (Ma)	2 s (Ma)	^{207}Pb/^{206}Pb (Ma)	2 s (Ma)	conc %^{e}
B-161	1429	2117	9.85	1.13	0.23	0.004760	2.4	0.03072	5.9	0.04682	5.4	0.40	31	0.7	31	1.79			
B-162	21371	222	70.36	0.03	0.00	0.306286	2.1	4.84291	2.7	0.11471	1.8	0.77	1722	31.7	1792	23.05	1875	31.7	92
B-163	561	392	1.45	0.87	6.62	0.003718	3.9	0.03307	23.8	0.06454	23.5	0.16	24	0.9	33	7.74			
B-164	211	11	1.04	2.13	2.73	0.100624	4.1	0.62993	22.1	0.04542	21.7	0.19	618	24.3					
B-165	17574	339	65.81	0.99	0.00	0.190952	2.4	2.44567	3.1	0.09292	1.9	0.78	1127	24.7	1256	22.18	1486	24.7	76
B-166	13287	93	35.79	1.10	0.41	0.369308	2.3	6.02323	3.2	0.11832	2.3	0.70	2026	39.6	1979	28.28	1931	39.6	105
B-167	24464	377	87.99	0.34	0.15	0.229800	2.2	2.84096	2.5	0.08969	1.3	0.85	1333	25.9	1366	19.06	1418	25.9	94
B-168	8384	266	38.62	1.32	2.95	0.148184	2.1	0.98876	14.2	0.04841	14.1	0.15	891	17.9	698	71.82	119	17.9	751
B-173	9237	299	11.07	0.17	0.18	0.037086	4.1	0.36330	4.8	0.07107	2.5	0.86	235	9.6	315	13.07	959	9.6	24
B-174	267	420	1.42	0.60	2.02	0.003463	3.2	0.02219	14.5	0.04649	14.1	0.22	22	0.7	22	3.19			
B-175	219	390	1.28	0.91	1.69	0.003370	2.7	0.02160	7.6	0.04650	7.1	0.36	22	0.6	22	1.64			
B-176	2249	220	13.55	0.05	0.07	0.062557	2.2	0.47350	4.3	0.05491	3.7	0.51	391	8.4	394	14.04	408	8.4	96
B-177	36497	290	76.70	0.84	0.00	0.244723	2.3	5.59127	2.9	0.16575	1.7	0.81	1411	29.3	1915	24.69	2515	29.3	56
B-178	18710	278	60.01	0.64	0.64	0.213172	2.3	2.55260	3.6	0.08687	2.8	0.64	1246	26.2	1287	26.25	1357	26.2	92
B-179	149	108	0.41	0.84	10.25	0.003885	5.3	0.02181	66.1	0.04073	65.9	0.08	25	1.3					

RPD2

spot number	$^{207}Pb^a$ (cps)	U^b (ppm)	Pb^b (ppm)	Th^b/U	$^{206}Pb^c$/^{204}Pb	$^{206}Pb^c$/^{238}U	2 s %	$^{207}Pb^c$/^{235}U	2 s %	$^{207}Pb^c$/^{206}Pb	2 s %	rho^d	^{206}Pb/^{238}U (Ma)	2 s (Ma)	^{207}Pb/^{235}U (Ma)	2 s (Ma)	^{207}Pb/^{206}Pb (Ma)	2 s (Ma)	conc %^{e}
A-5	420	157	2.25	0.56	0.93	0.014656	2.7	0.08777	11.7	0.04345	11.4	0.23	94	2.5					
A-6	972	1052	5.72	0.36	0.54	0.005568	2.3	0.03590	6.0	0.04678	5.6	0.38	36	0.8	36	2.13			
A-7	346	350	1.83	0.48	0.56	0.005343	2.6	0.03440	10.4	0.04671	10.0	0.25	34	0.9	34	3.50			
A-8	1758	280	9.65	0.30	0.53	0.035191	2.1	0.24657	4.3	0.05083	3.7	0.49	223	4.6	224	8.57	233	4.6	96
A-9	1050	391	5.92	0.36	0.36	0.015471	2.3	0.10262	4.8	0.04812	4.3	0.47	99	2.2	99	4.57	104	2.2	95
A-10	76	209	0.46	0.83	1.58	0.002254	3.5	0.01073	31.8	0.03453	31.6	0.11	15	0.5					
A-11	287	364	1.67	0.37	0.60	0.004702	2.9	0.03026	10.0	0.04669	9.6	0.29	30	0.9	30	2.98			
A-12	851	334	5.20	0.35	0.28	0.015919	2.2	0.10530	4.8	0.04799	4.2	0.46	102	2.3	102	4.64	98	2.3	104

spot	$^{207}Pb^a$	U^b	Pb^b	$\frac{Th^b}{U}$	$\frac{^{206}Pb^c}{^{204}Pb}$	$\frac{^{206}Pb^c}{^{238}U}$	2 s	$\frac{^{207}Pb^c}{^{235}U}$	2 s	$\frac{^{207}Pb^c}{^{206}Pb}$	2 s		$\frac{^{206}Pb}{^{238}U}$	2 s	$\frac{^{207}Pb}{^{235}U}$	2 s	$\frac{^{207}Pb}{^{206}Pb}$	2 s	
number	(cps)	(ppm)	(ppm)	U			%		%		%	rho^d		(Ma)		(Ma)		(Ma)	conc %[e]
A-14	1368	1162	6.52	0.27	1.94	0.005761	2.3	0.03413	7.3	0.04298	6.9	0.31	37	0.8					
A-15	2431	971	12.88	0.54	0.91	0.013619	2.2	0.08055	5.8	0.04291	5.3	0.38	87	1.9					
A-16	1099	436	6.30	0.29	0.26	0.014742	2.2	0.09722	5.9	0.04784	5.5	0.38	94	2.1	94	5.32	91	2.1	104
A-17	1222	246	6.76	0.58	0.77	0.028022	2.3	0.19271	7.4	0.04989	7.0	0.31	178	4.0	179	12.14	189	4.0	94
A-18	984	227	3.52	0.36	5.40	0.015874	2.5	0.10483	11.2	0.04791	10.9	0.22	102	2.5	101	10.79	94	2.5	108
A-19	2046	433	10.53	0.53	0.48	0.024836	2.4	0.16938	6.6	0.04948	6.2	0.36	158	3.8	159	9.77	170	3.8	93
A-20	330	112	1.81	0.66	1.17	0.016634	3.2	0.10379	9.3	0.04527	8.7	0.35	106	3.4					
A-21	2206	821	12.41	1.06	0.16	0.015438	2.1	0.10179	8.7	0.04783	8.5	0.24	99	2.1	98	8.18	90	2.1	109
A-22	1393	513	10.15	0.74	0.83	0.020308	2.3	0.12350	10.7	0.04412	10.4	0.21	130	2.9					
A-23	957	320	6.31	0.64	0.49	0.020157	2.3	0.13468	8.6	0.04847	8.3	0.27	129	2.9	128	10.35	122	2.9	106
A-24	577	226	3.70	0.34	0.66	0.016804	2.4	0.09927	7.8	0.04286	7.4	0.30	107	2.5					
A-29	202	761	0.88	0.58	4.75	0.001181	4.6	0.00706	37.4	0.04337	37.1	0.12	8	0.4					
A-30	18571	223	51.50	0.14	0.14	0.226640	2.2	2.87575	2.6	0.09205	1.3	0.86	1317	26.7	1376	19.55	1468	26.7	90
A-31	1206	350	5.38	0.40	2.08	0.015698	2.3	0.10429	6.4	0.04819	6.0	0.36	100	2.3	101	6.13	108	2.3	93
A-32	450	370	2.37	0.79	1.01	0.006531	3.3	0.04649	15.4	0.05164	15.1	0.21	42	1.4	46	6.95			
A-33	5984	125	23.22	0.82	0.31	0.185731	2.1	1.88575	3.1	0.07366	2.2	0.69	1098	21.3	1076	20.24	1032	21.3	106
A-34	3770	1828	21.62	0.02	0.17	0.012085	2.1	0.08004	3.2	0.04805	2.4	0.65	77	1.6	78	2.41			
A-35	1408	582	8.29	0.39	0.64	0.014552	2.2	0.09627	5.1	0.04799	4.6	0.44	93	2.0	93	4.52	98	2.0	95
A-36	1021	423	5.95	0.30	0.41	0.014368	2.3	0.09492	5.9	0.04793	5.5	0.38	92	2.1	92	5.20	95	2.1	97
A-37	3669	522	19.44	0.47	0.54	0.037929	2.1	0.26569	3.5	0.05082	2.7	0.62	240	5.1	239	7.43	232	5.1	103
A-38	1190	399	5.91	0.33	1.34	0.015153	2.3	0.10029	6.0	0.04802	5.5	0.39	97	2.2	97	5.55	99	2.2	98
A-39	643	1768	3.99	0.34	0.51	0.002311	2.3	0.01477	5.5	0.04635	5.0	0.41	15	0.3	15	0.82			
A-40	727	520	2.56	0.31	4.64	0.005013	2.6	0.03621	7.4	0.05240	6.9	0.36	32	0.8	36	2.62			
A-41	283	155	0.25	0.55	23.82	0.001626	7.9	0.01254	58.5	0.05592	58.0	0.13	10	0.8	13	7.36			
A-42	6049	5880	31.18	0.29	1.22	0.005424	2.2	0.03497	4.0	0.04677	3.3	0.54	35	0.7	35	1.37			
A-43	504	180	2.72	0.34	0.92	0.015477	2.4	0.10238	8.4	0.04799	8.1	0.29	99	2.4	99	7.96	98	2.4	101
A-44	3187	1471	19.57	0.38	0.50	0.013625	2.1	0.08457	4.0	0.04503	3.4	0.52	87	1.8					

spot number	$^{207}Pb^a$ (cps)	U^b (ppm)	Pb^b (ppm)	Th^b/U	$\frac{^{206}Pb^c}{^{204}Pb}$	$\frac{^{206}Pb^c}{^{238}U}$	2 s %	$\frac{^{207}Pb^c}{^{235}U}$	2 s %	$\frac{^{207}Pb^c}{^{206}Pb}$	2 s %	rho^d	$\frac{^{206}Pb}{^{238}U}$	2 s (Ma)	$\frac{^{207}Pb}{^{235}U}$	2 s (Ma)	$\frac{^{207}Pb}{^{206}Pb}$	2 s (Ma)	conc %^{e}
A-45	1497	210	8.14	0.41	0.24	0.039443	2.3	0.27938	4.9	0.05139	4.4	0.46	249	5.6	250	10.95	258	5.6	97
A-46	45829	121	78.60	0.59	0.00	0.586210	2.1	15.92400	2.3	0.19707	1.1	0.89	2974	49.7	2872	22.36	2802	49.7	106
A-47	293	809	1.53	1.64	2.20	0.001935	2.8	0.01170	13.7	0.04388	13.5	0.20	12	0.3					
A-48	512	170	2.57	0.37	1.03	0.015375	2.7	0.10860	9.1	0.05124	8.7	0.29	98	2.6	105	9.04	251	2.6	39
A-53	2121	809	11.92	0.29	0.36	0.015068	2.3	0.09968	5.1	0.04799	4.5	0.45	96	2.2	96	4.69	98	2.2	98
A-54	2075	891	12.35	0.36	0.73	0.014193	2.2	0.08814	5.5	0.04505	5.0	0.41	91	2.0					
A-55	3781	790	18.06	0.74	0.95	0.023344	2.4	0.15800	4.5	0.04910	3.9	0.52	149	3.5	149	6.29	152	3.5	98
A-56	213	337	1.34	0.78	0.56	0.004088	2.8	0.02364	11.0	0.04194	10.6	0.25	26	0.7					
A-57	252	256	1.37	0.56	1.00	0.005474	2.6	0.03530	10.9	0.04678	10.6	0.24	35	0.9	35	3.78			
A-58	277	168	0.96	0.55	5.20	0.005760	3.4	0.04467	18.9	0.05627	18.6	0.18	37	1.3	44	8.22			
A-59	15749	2467	29.56	0.65	4.62	0.011675	2.7	0.16517	10.4	0.10264	10.0	0.26	75	2.0	155	14.92			
A-60	1196	499	6.73	0.37	0.83	0.013791	2.4	0.09078	6.8	0.04776	6.3	0.35	88	2.1	88	5.72	87	2.1	102
A-61	1419	505	8.54	0.45	0.58	0.017263	2.2	0.11493	4.1	0.04830	3.4	0.54	110	2.4	110	4.25	113	2.4	97
A-62	2268	821	12.63	0.37	0.65	0.015711	2.2	0.10433	4.8	0.04818	4.2	0.46	100	2.2	101	4.59	107	2.2	94
A-63	2195	2586	12.91	0.27	0.43	0.005106	2.2	0.03291	4.9	0.04675	4.3	0.46	33	0.7	33	1.58			
A-64	1	71	0.07	0.45	2.73	0.000998	7.9	-0.00110	332.8	-0.00801	332.7	0.02	6	0.5					
A-65	615	217	3.25	0.42	0.88	0.015315	2.6	0.10093	9.1	0.04781	8.7	0.29	98	2.5	98	8.43	89	2.5	110
A-66	5635	2423	34.00	0.14	0.26	0.014340	2.2	0.09487	3.1	0.04799	2.2	0.70	92	2.0	92	2.71	98	2.0	93
A-67	20988	5313	116.64	0.07	0.10	0.022415	2.1	0.15143	2.3	0.04901	1.0	0.89	143	2.9	143	3.11	148	2.9	97
A-68	3884	728	9.53	0.51	9.00	0.013422	2.2	0.07945	13.0	0.04295	12.8	0.17	86	1.9					
A-69	249	253	0.91	0.79	5.75	0.003708	3.6	0.01726	35.1	0.03376	35.0	0.10	24	0.9					
A-70	1510	589	8.48	0.63	0.83	0.014717	2.6	0.09699	4.9	0.04781	4.1	0.54	94	2.5	94	4.38	89	2.5	105
A-71	2398	935	13.53	0.57	1.02	0.014836	2.2	0.09010	5.8	0.04406	5.4	0.37	95	2.0					
A-72	876	226	3.36	0.35	2.51	0.015190	2.6	0.10072	6.6	0.04810	6.1	0.39	97	2.5	97	6.15	104	2.5	94
A-77	809	328	4.64	0.49	0.39	0.014438	2.3	0.09540	7.3	0.04794	6.9	0.32	92	2.2	93	6.48	96	2.2	97
A-78	396	223	1.42	0.69	4.57	0.006559	3.1	0.03650	20.7	0.04037	20.5	0.15	42	1.3					

spot number	$^{207}Pb^a$ (cps)	U^b (ppm)	Pb^b (ppm)	$\frac{Th^b}{U}$	$\frac{^{206}Pb^c}{^{204}Pb}$	$\frac{^{206}Pb^c}{^{238}U}$	2 s %	$\frac{^{207}Pb^c}{^{235}U}$	2 s %	$\frac{^{207}Pb^c}{^{206}Pb}$	2 s %	rho^d	$\frac{^{206}Pb}{^{238}U}$	2 s (Ma)	$\frac{^{207}Pb}{^{235}U}$ (Ma)	2 s (Ma)	$\frac{^{207}Pb}{^{206}Pb}$	2 s (Ma)	conc %^{e}
A-80	527	211	3.04	0.56	0.45	0.014726	2.5	0.09723	6.9	0.04790	6.4	0.37	94	2.4	94	6.21	94	2.4	101
A-81	7973	1204	45.25	0.49	0.47	0.038286	2.1	0.26935	3.0	0.05104	2.2	0.68	242	4.9	242	6.55	242	4.9	100
A-82	1169	1319	6.92	0.07	0.22	0.005369	2.5	0.03457	7.1	0.04671	6.6	0.36	35	0.9	35	2.42			
A-83	247	912	1.47	0.82	1.49	0.001647	2.6	0.01051	13.8	0.04626	13.6	0.19	11	0.3	11	1.46			
A-84	5574	2566	33.07	0.19	0.23	0.013171	2.1	0.08658	3.2	0.04769	2.4	0.67	84	1.8	84	2.56	83	1.8	101
A-85	393	467	1.97	0.43	1.11	0.004310	2.8	0.02787	10.2	0.04691	9.8	0.28	28	0.8	28	2.81			
A-86	181	225	1.15	0.58	0.79	0.005252	3.1	0.02972	14.9	0.04106	14.6	0.21	34	1.0					
A-87	726	296	4.47	0.28	0.40	0.015415	2.4	0.10215	5.5	0.04807	4.9	0.44	99	2.4	99	5.14	102	2.4	96
A-88	525	617	2.87	1.14	1.92	0.004760	2.4	0.03014	14.7	0.04593	14.5	0.16	31	0.7					
A-89	141	118	0.75	0.55	2.57	0.006449	3.2	0.04600	18.3	0.05175	18.0	0.17	41	1.3	46	8.18			
A-90	27	183	0.19	0.48	3.56	0.001075	6.4	0.00397	60.8	0.02678	60.5	0.10	7	0.4					
A-91	904	353	5.46	0.50	0.52	0.015800	2.4	0.10474	4.7	0.04809	4.1	0.50	101	2.4	101	4.56	103	2.4	98
A-92	557	218	3.45	0.89	0.76	0.016177	2.6	0.10749	6.7	0.04820	6.2	0.39	103	2.7	104	6.65	109	2.7	95
A-93	983	1315	4.90	0.56	1.66	0.003813	2.4	0.02394	9.0	0.04555	8.7	0.27	25	0.6					
A-94	596	654	3.35	0.39	0.78	0.005260	2.5	0.03149	9.5	0.04343	9.1	0.27	34	0.8					
A-95	932	349	5.39	0.67	0.76	0.015847	2.3	0.08992	7.3	0.04117	7.0	0.31	101	2.3					
A-96	2973	1167	16.83	0.51	0.83	0.014733	2.1	0.09745	4.6	0.04799	4.0	0.47	94	2.0	94	4.13	98	2.0	96
A-101	1779	782	10.30	0.28	0.47	0.013462	2.3	0.08850	8.5	0.04769	8.1	0.27	86	1.9	86	6.98	83	1.9	103
A-102	1630	631	8.43	0.31	0.70	0.013672	2.3	0.09000	5.5	0.04776	5.0	0.42	88	2.0	88	4.63	87	2.0	101
A-103	1007	417	6.57	0.60	0.57	0.016126	2.3	0.10004	6.6	0.04501	6.2	0.35	103	2.4					
A-104	2116	813	11.78	0.31	1.07	0.014821	2.2	0.09777	5.2	0.04785	4.7	0.42	95	2.1	95	4.74	91	2.1	104
A-105	1164	387	5.93	0.28	0.95	0.015658	2.5	0.10413	6.6	0.04825	6.1	0.39	100	2.5	101	6.28	111	2.5	90
A-106	1056	419	5.75	0.47	0.89	0.014017	2.4	0.09247	7.0	0.04786	6.6	0.34	90	2.1	90	6.02	92	2.1	98
A-107	181	490	1.04	0.33	0.83	0.002178	3.5	0.01256	13.9	0.04184	13.4	0.25	14	0.5					
A-108	1802	4116	10.83	0.67	0.65	0.002692	2.3	0.01722	4.5	0.04641	3.9	0.51	17	0.4	17	0.77			
A-109	8887	7889	36.53	0.11	2.72	0.004738	2.2	0.03047	3.7	0.04666	3.0	0.58	30	0.7	30	1.12			
A-110	179	140	0.79	0.69	2.81	0.005779	3.3	0.03424	18.9	0.04299	18.6	0.17	37	1.2					

spot number	$^{207}Pb^a$ (cps)	U^b (ppm)	Pb^b (ppm)	$\underline{Th^b}$ U	$\underline{^{206}Pb^c}$ ^{204}Pb	$\underline{^{206}Pb^c}$ ^{238}U	2 s %	$\underline{^{207}Pb^c}$ ^{235}U	2 s %	$\underline{^{207}Pb^c}$ ^{206}Pb	2 s %	rho^d	$\underline{^{206}Pb}$ ^{238}U	2 s (Ma)	$\underline{^{207}Pb}$ ^{235}U	2 s (Ma)	$\underline{^{207}Pb}$ ^{206}Pb	2 s (Ma)	conc %^{e}
A-111	3600	410	13.32	0.33	3.04	0.033152	2.3	0.23009	4.8	0.05035	4.2	0.48	210	4.7	210	9.06	211	4.7	100
A-112	4228	1540	20.04	0.21	1.61	0.013297	2.1	0.08740	4.4	0.04769	3.8	0.48	85	1.8	85	3.56	83	1.8	103
A-113	891	395	5.88	0.45	0.27	0.015225	2.2	0.09565	5.4	0.04558	4.9	0.41	97	2.2					
A-114	0	692	0.04	0.54	-6.14	0.000057	16.9	0.00036											
A-115	749	284	4.28	0.62	0.95	0.015379	2.4	0.10187	8.2	0.04806	7.9	0.30	98	2.4	99	7.72	101	2.4	97
A-116	1592	532	7.08	0.47	1.71	0.013620	2.2	0.08955	7.1	0.04770	6.8	0.31	87	1.9	87	5.94	84	1.9	104
A-117	2804	1000	14.65	0.36	1.27	0.014963	2.2	0.09877	5.8	0.04789	5.4	0.38	96	2.1	96	5.30	93	2.1	103
A-118	195	307	1.15	0.66	0.80	0.003833	3.3	0.02579	11.0	0.04881	10.5	0.29	25	0.8	26	2.82			
A-119	233	353	1.58	0.72	1.10	0.004608	3.2	0.02439	14.0	0.03840	13.6	0.23	30	1.0					
A-120	187	234	0.89	1.19	4.46	0.003950	4.8	0.01627	48.7	0.02988	48.4	0.10	25	1.2					
A-125	644	326	1.67	0.53	5.34	0.005216	3.6	0.03958	17.5	0.05506	17.2	0.21	34	1.2	39	6.78			
A-126	17980	402	17.88	0.31	0.19	0.044381	5.3	0.46012	6.0	0.07521	2.9	0.88	280	14.5	384	19.26	1074	14.5	26
A-127	1823	786	10.81	0.24	0.43	0.014049	2.2	0.09253	4.4	0.04778	3.8	0.50	90	2.0	90	3.81	88	2.0	102
A-128	588	250	3.75	0.60	0.83	0.015343	2.4	0.09735	7.3	0.04603	6.8	0.34	98	2.4					
A-129	2157	886	13.76	0.41	0.51	0.015871	2.2	0.10499	4.3	0.04799	3.7	0.50	102	2.2	101	4.14	98	2.2	103
A-130	1956	324	11.33	0.33	0.42	0.035632	2.2	0.24809	4.2	0.05051	3.6	0.53	226	5.0	225	8.54	218	5.0	104
A-131	1099	1147	6.50	0.85	0.81	0.005800	2.3	0.03736	7.7	0.04673	7.3	0.31	37	0.9	37	2.80			
A-132	433	173	2.58	0.27	0.56	0.015238	2.6	0.10103	8.8	0.04810	8.4	0.29	97	2.5	98	8.16	103	2.5	94
A-133	19816	190	60.20	0.62	0.18	0.308019	2.1	4.45706	2.5	0.10498	1.3	0.85	1731	32.6	1723	20.82	1713	32.6	101
A-134	2970	749	11.91	0.41	2.33	0.016128	2.6	0.12495	8.8	0.05621	8.4	0.30	103	2.7	120	9.97	460	2.7	22
A-135	45	90	0.21	0.77	4.58	0.002369	6.9	0.01227	64.4	0.03756	64.0	0.11	15	1.1					
A-136	1857	825	11.14	0.22	0.46	0.013798	2.3	0.09074	4.3	0.04771	3.6	0.53	88	2.0	88	3.61	84	2.0	105
A-137	1467	645	8.48	0.46	0.57	0.013499	2.7	0.07992	6.3	0.04295	5.7	0.43	86	2.4					
A-138	2961	1120	19.48	0.35	0.64	0.017822	2.6	0.11188	4.6	0.04554	3.8	0.57	114	3.0					
A-139	602	243	3.79	0.37	0.85	0.015919	2.8	0.10564	7.8	0.04814	7.3	0.36	102	2.9	102	7.58	106	2.9	96
A-140	11983	5505	70.91	0.42	0.84	0.013165	2.3	0.08673	5.9	0.04779	5.5	0.39	84	2.0	84	4.82	88	2.0	95

spot number	$^{207}Pb^a$ (cps)	U^b (ppm)	Pb^b (ppm)	Th^b/U	$^{206}Pb^c$/^{204}Pb	$^{206}Pb^c$/^{238}U	2 s %	$^{207}Pb^c$/^{235}U	2 s %	$^{207}Pb^c$/^{206}Pb	2 s %	rho^d	^{206}Pb/^{238}U	2 s (Ma)	^{207}Pb/^{235}U	2 s (Ma)	^{207}Pb/^{206}Pb	2 s (Ma)	conc %^{c}
A-142	258	93	1.50	0.62	1.24	0.016519	2.9	0.11015	10.1	0.04838	9.7	0.28	106	3.0	106	10.22	117	3.0	90
A-143	115	290	0.62	0.27	0.60	0.002198	3.6	0.01404	19.1	0.04632	18.7	0.19	14	0.5	14	2.68			
A-144	4086	1829	25.59	0.50	0.32	0.014301	2.2	0.09447	3.3	0.04792	2.5	0.66	92	2.0	92	2.89	95	2.0	97
A-149	41	804	0.05	0.39	24.37	0.000072	24.1	-0.00084	35.0	-0.08473	25.4	0.69	0	0.1					
A-150	4879	2250	29.68	0.13	0.23	0.013484	2.1	0.08889	3.2	0.04783	2.4	0.66	86	1.8	86	2.68	90	1.8	96
A-151	761	331	4.70	0.41	0.56	0.014523	2.3	0.09572	6.1	0.04782	5.6	0.38	93	2.2	93	5.41	90	2.2	104
A-152	2321	909	13.44	0.97	1.28	0.015187	2.2	0.08899	7.0	0.04251	6.6	0.32	97	2.2					
A-153	1686	745	10.89	0.40	0.42	0.014952	2.3	0.09889	4.6	0.04798	4.0	0.50	96	2.2	96	4.18	98	2.2	98
A-154	1918	796	11.30	0.29	0.47	0.014504	2.2	0.09569	3.9	0.04786	3.1	0.58	93	2.1	93	3.42	92	2.1	101
A-155	3641	967	13.38	0.53	2.18	0.014068	2.6	0.10441	6.4	0.05384	5.9	0.40	90	2.3	101	6.15	364	2.3	25
A-156	1579	376	9.90	0.42	0.46	0.026997	2.3	0.16898	4.8	0.04541	4.2	0.48	172	3.9					
A-157	2159	990	13.78	0.28	0.56	0.014228	2.3	0.09376	4.0	0.04781	3.2	0.58	91	2.1	91	3.46	89	2.1	102
A-158	959	402	6.15	0.37	0.51	0.015634	2.3	0.10317	5.5	0.04788	5.0	0.41	100	2.2	100	5.20	92	2.2	108
A-159	2278	1040	14.57	0.56	0.49	0.014312	2.2	0.09425	4.1	0.04778	3.4	0.54	92	2.0	91	3.55	88	2.0	105
A-160	13694	295	37.78	0.45	1.40	0.127706	2.4	1.35710	3.4	0.07709	2.5	0.69	775	17.3	871	19.92	1123	17.3	69
A-161	4436	1449	20.16	0.76	0.00	0.013950	2.2	0.13352	5.1	0.06944	4.6	0.43	89	1.9	127	6.07	911	1.9	10
A-162	1710	719	9.68	0.41	0.56	0.013746	2.2	0.09069	4.5	0.04787	3.9	0.49	88	1.9	88	3.82	92	1.9	96
A-163	1065	343	4.93	0.41	2.51	0.014677	2.3	0.09691	6.9	0.04790	6.5	0.34	94	2.2	94	6.17	94	2.2	100
A-164	b.d.																		
A-165	6464	7358	38.47	0.22	0.54	0.005328	2.1	0.03767	4.0	0.05130	3.5	0.51	34	0.7	38	1.49			
A-166	7590	1217	44.60	0.43	0.38	0.037350	2.1	0.26188	3.1	0.05087	2.3	0.68	236	4.8	236	6.50	234	4.8	101
A-167	14771	365	6.63	0.20	1.86	0.017950	54.3	0.21235	54.7	0.08583	6.2	0.99	115	61.8	196	97.26	1334	61.8	9
A-168	5107	2143	28.42	0.24	0.52	0.013556	2.5	0.08938	3.8	0.04783	2.9	0.66	87	2.2	87	3.16	90	2.2	96
A-173	1569	607	9.39	0.54	0.66	0.015803	2.1	0.10463	5.0	0.04803	4.6	0.43	101	2.2	101	4.84	100	2.2	101
A-174	1785	803	11.88	0.41	0.61	0.015161	2.8	0.09451	4.8	0.04522	3.9	0.58	97	2.7					
A-175	28	187	0.11	0.46	2.19	0.000597	9.2	0.00228	168.7	0.02765	168.4	0.05	4	0.4					
A-176	1047	386	5.58	0.40	0.48	0.014764	2.6	0.09737	7.5	0.04784	7.1	0.35	94	2.5	94	6.79	91	2.5	104

spot number	$^{207}Pb^{a}$ (cps)	U^{b} (ppm)	Pb^{b} (ppm)	$\underline{Th^{b}}$ U	$\underline{^{206}Pb^{c}}$ ^{204}Pb	$\underline{^{206}Pb^{c}}$ ^{238}U	2 s %	$\underline{^{207}Pb^{c}}$ ^{235}U	2 s %	$\underline{^{207}Pb^{c}}$ ^{206}Pb	2 s %	rho^{d}	$\underline{^{206}Pb}$ ^{238}U	2 s (Ma)	$\underline{^{207}Pb}$ ^{235}U	2 s (Ma)	$\underline{^{207}Pb}$ ^{206}Pb	2 s (Ma)	conc %^{e}
A-177	399	139	2.18	0.43	1.55	0.016016	2.8	0.10577	12.5	0.04791	12.1	0.22	102	2.8	102	12.10	94	2.8	109
A-178	886	359	5.07	0.45	0.66	0.014427	2.4	0.09521	8.1	0.04788	7.8	0.29	92	2.2	92	7.17	92	2.2	100
A-179	2272	656	9.74	0.51	2.85	0.015158	2.2	0.10049	6.8	0.04810	6.4	0.33	97	2.1	97	6.28	103	2.1	94
A-180	1182	610	3.90	0.39	5.13	0.006535	3.1	0.04319	15.5	0.04795	15.2	0.20	42	1.3	43	6.52			
A-181	203	906	1.19	0.62	1.06	0.001349	3.2	0.00860	13.7	0.04624	13.3	0.23	9	0.3	9	1.19			
A-182	1224	268	7.17	0.46	0.20	0.027339	2.6	0.18736	6.5	0.04972	6.0	0.40	174	4.5	174	10.48	181	4.5	96
A-183	726	231	3.30	0.51	2.48	0.014584	2.6	0.09825	9.3	0.04887	8.9	0.28	93	2.4	95	8.45	141	2.4	66

RPD3

spot number	$^{207}Pb^{a}$ (cps)	U^{b} (ppm)	Pb^{b} (ppm)	$\underline{Th^{b}}$ U	$\underline{^{206}Pb^{c}}$ ^{204}Pb	$\underline{^{206}Pb^{c}}$ ^{238}U	2 s %	$\underline{^{207}Pb^{c}}$ ^{235}U	2 s %	$\underline{^{207}Pb^{c}}$ ^{206}Pb	2 s %	rho^{d}	$\underline{^{206}Pb}$ ^{238}U	2 s (Ma)	$\underline{^{207}Pb}$ ^{235}U	2 s (Ma)	$\underline{^{207}Pb}$ ^{206}Pb	2 s (Ma)	conc %^{e}
A-5	723	5	0.07	0.49	0.51	0.016472	2.5	0.10941	6.6	0.04819	6.1	0.38	105	2.6	105	6.62	108	2.6	98
A-6	796	306	4.77	0.48	0.86	0.015922	2.3	0.10570	6.6	0.04816	6.1	0.35	102	2.3	102	6.36	107	2.3	96
A-7	4080	560	21.19	0.31	0.38	0.038540	2.2	0.27418	3.9	0.05161	3.2	0.57	244	5.3	246	8.55	268	5.3	91
A-8	1166	466	7.73	0.69	0.70	0.017052	2.2	0.09784	6.5	0.04162	6.1	0.35	109	2.4					
A-9	28	89	0.20	0.41	0.47	0.002368	4.9	0.01454	31.8	0.04454	31.4	0.15	15	0.7					
A-10	7440	1131	41.47	0.49	0.47	0.037370	2.1	0.26276	3.2	0.05101	2.5	0.64	237	4.8	237	6.83	241	4.8	98
A-11	553	218	3.12	0.56	1.03	0.014693	2.6	0.08486	10.6	0.04190	10.3	0.25	94	2.4					
A-12	317	179	0.44	0.50	18.75	0.002488	4.9	0.01750	43.8	0.05101	43.5	0.11	16	0.8	18	7.65			
A-13	902	416	1.59	0.39	12.56	0.003886	3.1	0.02783	15.1	0.05195	14.7	0.21	25	0.8	28	4.14			
A-14	5172	733	27.03	0.34	0.66	0.037567	2.1	0.26375	3.6	0.05093	2.9	0.58	238	5.0	238	7.70	237	5.0	100
A-15	1097	445	6.41	0.65	0.85	0.014801	2.3	0.08946	6.6	0.04385	6.2	0.34	95	2.1					
A-16	386	152	2.43	0.42	0.75	0.016286	2.5	0.10783	8.9	0.04804	8.6	0.28	104	2.6	104	8.82	100	2.6	104
A-17	1369	525	7.83	0.39	0.50	0.015227	2.5	0.10101	5.3	0.04812	4.6	0.48	97	2.4	98	4.90	105	2.4	93
A-18	2835	1179	16.27	0.50	0.64	0.014102	2.1	0.09280	4.3	0.04774	3.7	0.50	90	1.9	90	3.69	86	1.9	105
A-19	353	142	2.05	0.55	1.09	0.014888	2.7	0.08622	11.7	0.04201	11.4	0.23	95	2.5					
A-20	588	34	2.45	0.28	0.33	0.074134	2.9	0.57523	7.5	0.05629	6.9	0.39	461	13.0	461	27.87	463	13.0	100
A-21	2183	869	11.00	0.36	0.71	0.012939	2.5	0.08545	4.8	0.04791	4.1	0.52	83	2.1	83	3.86	94	2.1	88

spot number	$^{207}Pb^a$ (cps)	U^b (ppm)	Pb^b (ppm)	$\underline{Th^b}$ U	$\underline{^{206}Pb^c}$ ^{204}Pb	$\underline{^{206}Pb^c}$ ^{238}U	2 s %	$\underline{^{207}Pb^c}$ ^{235}U	2 s %	$\underline{^{207}Pb^c}$ ^{206}Pb	2 s %	rho^d	$\underline{^{206}Pb}$ ^{238}U	2 s (Ma)	$\underline{^{207}Pb}$ ^{235}U	2 s (Ma)	$\underline{^{207}Pb}$ ^{206}Pb	2 s (Ma)	conc %^{e}
A-23	880	320	4.96	0.47	0.38	0.015859	2.4	0.10520	6.1	0.04812	5.6	0.40	101	2.4	102	5.94	105	2.4	97
A-24	6106	844	31.56	0.40	1.19	0.038209	2.1	0.25222	3.6	0.04789	2.9	0.59	242	5.1	228	7.44	93	5.1	259
A-29	4341	614	22.38	0.47	0.39	0.037141	2.1	0.26208	3.4	0.05119	2.6	0.63	235	4.9	236	7.06	249	4.9	94
A-30	1583	694	9.31	0.16	0.42	0.013713	2.5	0.09015	4.8	0.04770	4.1	0.52	88	2.2	88	4.06	84	2.2	105
A-31	578	186	2.97	0.40	1.42	0.016413	2.6	0.09071	9.0	0.04010	8.7	0.28	105	2.7					
A-32	987	377	5.31	0.31	0.79	0.014412	2.2	0.09517	5.0	0.04791	4.4	0.45	92	2.0	92	4.38	94	2.0	98
A-33	3010	1262	18.93	0.52	0.31	0.015325	2.3	0.10155	4.0	0.04807	3.3	0.57	98	2.2	98	3.77	102	2.2	96
A-34	780	315	4.56	0.56	1.31	0.014927	2.5	0.07852	8.6	0.03816	8.2	0.29	96	2.3					
A-35	-8	139	0.01	0.56	-55.38	0.000084	42.1	0.00090											
A-36	738	220	3.69	0.48	2.13	0.017177	2.9	0.11548	8.4	0.04877	7.9	0.34	110	3.1	111	8.80	136	3.1	81
A-37	1098	7	0.11	0.58	0.78	0.015959	2.2	0.10556	5.6	0.04798	5.1	0.40	102	2.2	102	5.39	98	2.2	104
A-38	5069	2056	30.48	0.27	0.22	0.015152	2.1	0.10031	3.2	0.04803	2.4	0.66	97	2.1	97	2.98	100	2.1	97
A-39	1126	476	6.57	0.43	0.49	0.014120	2.3	0.09292	5.8	0.04774	5.3	0.40	90	2.1	90	5.04	86	2.1	105
A-40	55	173	0.30	0.38	2.20	0.001767	6.2	0.00965	36.6	0.03963	36.0	0.17	11	0.7					
A-41	104	295	0.66	0.34	0.46	0.002302	3.7	0.01470	16.4	0.04633	16.0	0.22	15	0.5	15	2.41			
A-42	221	337	0.80	0.35	3.83	0.002420	3.4	0.01534	15.5	0.04599	15.1	0.22	16	0.5					
A-43	2147	776	10.81	0.86	0.75	0.014242	2.3	0.09401	6.6	0.04789	6.2	0.35	91	2.1	91	5.73	93	2.1	98
A-44	2452	923	13.24	0.56	0.79	0.014655	2.2	0.09664	4.6	0.04784	4.1	0.47	94	2.0	94	4.13	91	2.0	103
A-45	2301	1049	12.45	0.45	0.68	0.012166	2.2	0.07355	7.3	0.04386	7.0	0.30	78	1.7					
A-46	2669	1103	15.98	0.47	0.55	0.014806	2.1	0.09790	3.6	0.04797	2.9	0.59	95	2.0	95	3.25	97	2.0	98
A-47	1283	480	7.29	0.69	0.70	0.015515	2.2	0.10293	5.7	0.04813	5.2	0.39	99	2.2	99	5.41	105	2.2	95
A-48	3443	1319	8.69	0.13	4.67	0.006771	5.3	0.03896	11.6	0.04175	10.3	0.46	44	2.3					
A-53	190	144	0.32	0.71	14.08	0.002278	6.1	0.01209	72.7	0.03849	72.4	0.08	15	0.9					
A-54	583	751	3.23	0.38	0.86	0.004405	2.5	0.02714	9.5	0.04469	9.1	0.27	28	0.7					
A-55	624	264	3.65	0.48	0.70	0.014126	2.7	0.09321	8.8	0.04787	8.4	0.31	90	2.4	90	7.62	92	2.4	98
A-56	1793	259	10.35	0.86	0.43	0.040651	2.2	0.28738	4.5	0.05129	3.9	0.48	257	5.5	256	10.17	253	5.5	101
A-57	706	249	4.41	0.36	0.48	0.018080	2.2	0.12044	6.4	0.04833	6.0	0.35	116	2.6	115	7.01	115	2.6	101

spot number	$^{207}Pb^a$ (cps)	U^b (ppm)	Pb^b (ppm)	Th^b / U	$\frac{^{206}Pb^c}{^{204}Pb}$	$\frac{^{206}Pb^c}{^{238}U}$	2 s %	$\frac{^{207}Pb^c}{^{235}U}$	2 s %	$\frac{^{207}Pb^c}{^{206}Pb}$	2 s %	rho^d	$\frac{^{206}Pb}{^{238}U}$	2 s (Ma)	$\frac{^{207}Pb}{^{235}U}$	2 s (Ma)	$\frac{^{207}Pb}{^{206}Pb}$	2 s (Ma)	conc %^{e}
A-58	953	406	5.62	0.44	0.47	0.014140	2.4	0.09338	5.7	0.04791	5.2	0.42	91	2.2	91	4.98	94	2.2	96
A-59	1409	513	7.78	0.56	0.59	0.015502	2.3	0.10240	4.9	0.04792	4.3	0.47	99	2.3	99	4.61	95	2.3	105
A-60	132	343	0.77	0.33	0.94	0.002304	3.6	0.01326	14.1	0.04175	13.6	0.26	15	0.5					
A-61	1155	429	7.12	0.38	0.49	0.016965	2.2	0.11275	5.0	0.04822	4.5	0.43	108	2.3	108	5.14	109	2.3	99
A-62	457	93	0.25	0.70	26.53	0.002801	7.7	0.00678	256.7	0.01756	256.6	0.03	18	1.4					
A-63	1042	177	5.78	2.36	2.12	0.033688	2.5	0.17277	14.6	0.03721	14.3	0.17	214	5.2					
A-64	1576	337	8.78	0.41	0.31	0.026582	2.2	0.18165	4.3	0.04958	3.7	0.52	169	3.7	169	6.70	174	3.7	97
A-65	1319	566	7.93	0.20	0.25	0.014313	2.3	0.09436	5.2	0.04783	4.7	0.44	92	2.1	92	4.59	90	2.1	102
A-66	112	176	0.03	0.53	33.99	0.000150	21.7	0.00682	88.3	0.32884	85.6	0.25	1	0.2					
A-67	816	283	4.48	0.40	0.59	0.016132	2.5	0.10731	6.9	0.04826	6.4	0.36	103	2.5	104	6.78	111	2.5	93
A-68	470	192	2.75	0.37	0.57	0.014687	2.5	0.08991	8.9	0.04441	8.6	0.28	94	2.3					
A-69	747	305	4.31	0.42	0.53	0.014418	2.6	0.09497	6.4	0.04779	5.8	0.42	92	2.4	92	5.59	88	2.4	105
A-70	954	6	0.10	0.60	0.62	0.016655	2.3	0.11104	6.9	0.04837	6.5	0.33	106	2.4	107	7.01	117	2.4	91
A-71	4546	754	25.19	0.26	0.43	0.034073	2.2	0.23670	3.9	0.05040	3.2	0.57	216	4.7	216	7.61	213	4.7	102
A-72	5065	738	26.96	0.33	0.30	0.037229	2.2	0.26186	3.5	0.05103	2.8	0.61	236	5.0	236	7.45	241	5.0	98
A-77	1870	731	11.32	0.55	0.64	0.015876	2.1	0.09473	6.3	0.04329	6.0	0.34	102	2.2					
A-78	878	321	5.20	0.28	0.31	0.016536	2.3	0.10953	6.2	0.04805	5.8	0.37	106	2.4	106	6.26	101	2.4	104
A-79	3070	1308	17.98	0.16	0.34	0.014044	2.5	0.09264	4.6	0.04786	3.9	0.55	90	2.2	90	3.96	91	2.2	98
A-80	1221	487	7.12	0.39	0.41	0.014941	2.3	0.09878	5.9	0.04796	5.5	0.40	96	2.2	96	5.42	97	2.2	99
A-81	2600	1256	16.00	0.47	0.40	0.013019	2.2	0.08563	3.3	0.04772	2.5	0.66	83	1.8	83	2.64	85	1.8	98
A-82	2274	768	10.86	0.36	0.61	0.014460	2.8	0.09568	5.8	0.04801	5.2	0.47	93	2.5	93	5.18	99	2.5	94
A-83	1780	628	8.08	0.14	0.77	0.013149	2.2	0.08660	3.9	0.04778	3.2	0.57	84	1.9	84	3.17	88	1.9	96
A-84	778	305	4.45	0.42	0.45	0.014905	2.6	0.09858	6.0	0.04798	5.5	0.43	95	2.4	95	5.51	98	2.4	98
A-85	939	359	5.76	0.42	0.93	0.016394	2.5	0.10912	6.4	0.04829	5.9	0.40	105	2.6	105	6.37	113	2.6	93
A-86	146	330	0.82	0.72	0.27	0.002549	3.6	0.01795	13.6	0.05109	13.1	0.27	16	0.6	18	2.43			
A-87	596	2382	3.36	0.94	0.84	0.001444	2.6	0.00921	8.6	0.04625	8.2	0.30	9	0.2	9	0.79			

spot number	$^{207}Pb^a$ (cps)	U^b (ppm)	Pb^b (ppm)	$\underline{Th^b}$ U	$\underline{^{206}Pb^c}$ ^{204}Pb	$\underline{^{206}Pb^c}$ ^{238}U	2 s %	$\underline{^{207}Pb^c}$ ^{235}U	2 s %	$\underline{^{207}Pb^c}$ ^{206}Pb	2 s %	rho^d	$\underline{^{206}Pb}$ ^{238}U	2 s (Ma)	$\underline{^{207}Pb}$ ^{235}U	2 s (Ma)	$\underline{^{207}Pb}$ ^{206}Pb	2 s (Ma)	conc %^{e}
A-89	432	162	2.69	0.87	0.86	0.016973	2.5	0.11309	6.8	0.04834	6.3	0.37	108	2.7	109	7.01	115	2.7	94
A-90	312	296	1.36	0.84	2.58	0.004678	3.1	0.03008	12.9	0.04664	12.5	0.24	30	0.9	30	3.82			
A-91	2565	1015	15.38	0.87	0.72	0.015478	2.2	0.10249	4.6	0.04804	4.1	0.47	99	2.2	99	4.37	100	2.2	99
A-92	627	270	3.90	0.50	0.53	0.014803	2.5	0.08994	6.7	0.04408	6.2	0.37	95	2.3					
A-93	1045	385	5.42	0.29	1.60	0.014393	2.3	0.09477	6.8	0.04777	6.4	0.34	92	2.1	92	5.97	87	2.1	106
A-94	261	99	1.61	0.41	1.14	0.016770	2.9	0.10012	11.6	0.04331	11.2	0.25	107	3.1					
A-95	420	161	2.41	0.40	0.77	0.015266	3.1	0.10092	9.2	0.04796	8.6	0.34	98	3.0	98	8.53	96	3.0	101
A-96	7591	1183	44.91	0.55	0.66	0.038876	2.1	0.24716	3.5	0.04612	2.8	0.60	246	5.1	224	7.10	3	5.1	7220
A-101	785	305	4.30	0.32	1.06	0.014460	2.4	0.08802	7.5	0.04416	7.1	0.32	93	2.2					
A-102	886	347	5.15	0.71	0.64	0.015194	2.5	0.10053	6.5	0.04800	6.0	0.38	97	2.4	97	6.02	99	2.4	99
A-103	5545	2461	32.92	0.18	0.35	0.013673	2.2	0.08979	3.1	0.04764	2.2	0.71	88	1.9	87	2.56	81	1.9	108
A-104	2393	1029	13.63	0.48	0.44	0.013540	2.3	0.08902	4.6	0.04769	4.1	0.49	87	1.9	87	3.86	83	1.9	104
A-105	104	187	0.44	0.81	3.96	0.002402	4.5	0.01193	40.2	0.03603	39.9	0.11	15	0.7					
A-106	524	3	0.05	0.42	0.91	0.015895	2.9	0.10538	9.6	0.04809	9.2	0.30	102	2.9	102	9.30	103	2.9	98
A-107	4674	1745	28.93	0.63	0.37	0.016939	2.2	0.11269	3.7	0.04826	3.0	0.59	108	2.3	108	3.84	112	2.3	97
A-108	202	313	0.77	0.86	3.76	0.002513	4.6	0.01672	25.3	0.04826	24.8	0.18	16	0.7	17	4.22			
A-109	547	213	3.10	0.33	0.32	0.014865	2.6	0.09804	7.4	0.04785	6.9	0.35	95	2.4	95	6.66	91	2.4	105
A-110	4749	661	23.99	0.28	0.29	0.036983	2.2	0.26191	4.7	0.05138	4.2	0.47	234	5.1	236	10.00	257	5.1	91
A-111	2074	590	8.08	0.26	3.02	0.013990	2.5	0.09231	6.6	0.04787	6.1	0.38	90	2.2	90	5.63	92	2.2	97
A-112	855	335	5.10	0.39	0.53	0.015611	2.7	0.09668	8.1	0.04493	7.6	0.33	100	2.6					
A-113	2399	1054	14.82	0.39	0.47	0.014375	2.2	0.09489	3.8	0.04789	3.1	0.57	92	2.0	92	3.32	93	2.0	99
A-114	10	289	0.04	0.46	-51.86	0.000120	18.9	0.00317	83.9	0.19170	81.7	0.22	1	0.1					
A-115	1279	541	7.30	0.40	0.42	0.013778	2.3	0.09080	5.7	0.04781	5.2	0.41	88	2.0	88	4.81	89	2.0	99
A-116	13234	227	46.47	0.27	0.40	0.202909	2.2	2.25214	2.8	0.08052	1.7	0.78	1191	23.7	1197	19.57	1209	23.7	98
A-117	4224	1816	24.98	0.35	0.25	0.014058	2.2	0.09308	3.7	0.04804	2.9	0.59	90	1.9	90	3.16	100	1.9	90
A-118	5078	58	0.34	0.61	88.75														
A-119	213	81	1.35	0.40	0.29	0.017012	3.4	0.11314	13.5	0.04825	13.1	0.25	109	3.6	109	13.93	111	3.6	98

spot number	$^{207}Pb^a$ (cps)	U^b (ppm)	Pb^b (ppm)	Th^b/U	$\frac{^{206}Pb^c}{^{204}Pb}$	$\frac{^{206}Pb^c}{^{238}U}$	2 s %	$\frac{^{207}Pb^c}{^{235}U}$	2 s %	$\frac{^{207}Pb^c}{^{206}Pb}$	2 s %	rho^d	$\frac{^{206}Pb}{^{238}U}$ (Ma)	2 s (Ma)	$\frac{^{207}Pb}{^{235}U}$ (Ma)	2 s (Ma)	$\frac{^{207}Pb}{^{206}Pb}$ (Ma)	2 s (Ma)	conc %^{e}
A-120	4192	1652	20.13	0.20	0.69	0.012393	2.3	0.09066	4.9	0.05308	4.3	0.47	79	1.8	88	4.13			
A-125	918	244	0.94	0.36	25.63	0.003923	4.5	0.02660	24.9	0.04920	24.5	0.18	25	1.1	27	6.56			
A-126	35910	0.59		8.04	25.00	-88.970436	26.4												
A-127	2123	908	12.73	0.25	0.46	0.014339	2.2	0.09456	4.8	0.04784	4.3	0.45	92	2.0	92	4.23	91	2.0	101
A-128	2317	773	8.11	0.71	0.71	0.010702	2.7	0.07289	5.3	0.04941	4.6	0.51	69	1.9	71	3.68			
A-129	3460	1280	20.15	0.74	0.67	0.016086	2.1	0.10673	4.7	0.04813	4.2	0.45	103	2.2	103	4.59	105	2.2	98
A-130	1925	889	11.42	0.23	0.46	0.013119	2.2	0.08637	4.2	0.04776	3.6	0.52	84	1.8	84	3.37	87	1.8	97
A-131	1557	528	7.56	0.28	0.99	0.014558	2.5	0.10723	6.4	0.05344	5.9	0.39	93	2.3	103	6.29	347	2.3	27
A-132	1640	1380	3.65	0.75	1.78	0.002723	6.6	0.01503	22.2	0.04005	21.2	0.30	18	1.2					
A-133	5292	795	29.43	0.46	0.65	0.037754	2.1	0.25853	3.9	0.04968	3.3	0.53	239	4.9	233	8.23	179	4.9	133
A-134	812	328	4.52	0.31	0.89	0.014071	2.5	0.09285	6.5	0.04787	6.0	0.39	90	2.2	90	5.57	92	2.2	98
A-135	704	261	3.64	0.37	0.62	0.014279	2.8	0.09408	10.1	0.04780	9.8	0.27	91	2.5	91	8.85	89	2.5	103
A-136	3569	1524	20.98	0.28	0.33	0.014071	2.2	0.09272	3.8	0.04780	3.1	0.58	90	2.0	90	3.28	89	2.0	101
A-137	420	994	2.12	0.38	1.75	0.002187	3.1	0.01398	13.2	0.04635	12.9	0.23	14	0.4	14	1.85			
A-138	4070	1669	23.29	0.28	0.32	0.014264	2.1	0.09406	3.6	0.04784	2.9	0.59	91	1.9	91	3.16	91	1.9	101
A-139	2559	815	16.50	0.14	1.32	0.020664	2.3	0.13990	4.2	0.04912	3.6	0.53	132	2.9	133	5.28	153	2.9	86
A-140	718	267	4.28	0.48	0.54	0.016377	2.4	0.10881	7.6	0.04820	7.2	0.32	105	2.5	105	7.57	109	2.5	96
A-141	731	255	3.76	0.39	2.24	0.015044	2.4	0.09990	11.6	0.04818	11.4	0.21	96	2.3	97	10.74	107	2.3	90
A-142	922	362	5.24	0.43	0.73	0.014791	2.4	0.09771	7.4	0.04792	7.0	0.33	95	2.3	95	6.70	95	2.3	100
A-143	163	334	0.09	0.54	49.94	0.000212	11.4	0.01003	42.8	0.34245	41.2	0.27	1	0.2					
A-144	9250	1386	51.24	0.47	0.63	0.037676	2.2	0.26462	3.1	0.05095	2.2	0.70	238	5.1	238	6.57	238	5.1	100
A-149	1830	643	8.04	0.14	1.79	0.012788	2.2	0.08412	5.2	0.04772	4.7	0.43	82	1.8	82	4.09	85	1.8	96
A-150	2768	22	0.27	0.42	0.64	0.012896	2.3	0.08475	4.5	0.04768	3.9	0.50	83	1.9	83	3.58	83	1.9	100
A-151	7527	594	35.95	1.22	1.21	0.061863	2.2	0.41327	8.9	0.04846	8.7	0.25	387	8.4	351	26.52	121	8.4	319
A-152	524	203	3.01	0.37	0.75	0.015151	2.5	0.10007	8.3	0.04791	7.9	0.30	97	2.4	97	7.69	94	2.4	103
A-153	8	367	0.03	0.49	5.50	0.000082	18.0	0.00152	123.3	0.13450	122.0	0.15	1	0.1					

spot	^{207}Pb[a]	U[b]	Pb[b]	Th[b]	$\frac{^{206}Pb[c]}{^{204}Pb}$	$\frac{^{206}Pb[c]}{^{238}U}$	2 s	$\frac{^{207}Pb[c]}{^{235}U}$	2 s	$\frac{^{207}Pb[c]}{^{206}Pb}$	2 s		$\frac{^{206}Pb}{^{238}U}$	2 s	$\frac{^{207}Pb}{^{235}U}$	2 s	$\frac{^{207}Pb}{^{206}Pb}$	2 s	
number	(cps)	(ppm)	(ppm)	U			%		%		%	rho[d]		(Ma)		(Ma)		(Ma)	conc %[e]
A-155	458	797	1.88	0.54	4.98	0.002411	3.1	0.01543	14.6	0.04643	14.2	0.21	16	0.5	16	2.25			
A-156	777	238	3.98	0.42	1.39	0.017043	2.6	0.12318	8.7	0.05244	8.3	0.30	109	2.8	118	9.73	304	2.8	36
A-157	1012	378	5.56	0.39	0.89	0.015050	2.4	0.09962	7.3	0.04802	6.9	0.33	96	2.3	96	6.71	100	2.3	97
A-158	4872	722	26.61	0.43	0.79	0.037550	2.2	0.26327	4.6	0.05086	4.0	0.47	238	5.0	237	9.67	234	5.0	102
A-159	1239	429	6.79	0.52	0.78	0.016175	2.3	0.10727	6.2	0.04811	5.7	0.37	103	2.4	103	6.09	104	2.4	99
A-160	2954	1291	17.05	0.53	0.57	0.013498	2.7	0.08902	5.5	0.04785	4.8	0.49	86	2.3	87	4.53	91	2.3	95
A-161	3549	1278	16.02	0.41	1.38	0.012813	2.5	0.08422	5.5	0.04769	4.9	0.46	82	2.1	82	4.35	83	2.1	99
A-162	3595	607	13.29	0.24	2.96	0.022339	2.9	0.15220	5.2	0.04943	4.4	0.55	142	4.1	144	7.03	168	4.1	85
A-163	612	178	2.86	0.89	2.97	0.016568	2.8	0.08266	21.6	0.03620	21.4	0.13	106	3.0					
A-164	15307	6432	85.35	0.35	0.39	0.013565	2.1	0.08918	3.8	0.04770	3.2	0.56	87	1.8	87	3.17	84	1.8	104
A-165	877	303	4.82	0.86	1.12	0.016236	2.4	0.10806	6.6	0.04828	6.2	0.36	104	2.4	104	6.55	112	2.4	92
A-166	287	226	1.01	0.62	3.74	0.004513	3.8	0.03550	19.8	0.05707	19.5	0.19	29	1.1	35	6.91			
A-167	676	279	0.44	0.48	26.93	0.001588	6.6	0.01425	42.4	0.06513	41.8	0.16	10	0.7	14	6.04			
A-168	117	284	0.66	0.28	1.27	0.002374	3.8	0.01516	18.2	0.04635	17.8	0.21	15	0.6	15	2.76			
A-173	659	196	2.83	0.39	2.25	0.014769	2.7	0.09943	8.6	0.04884	8.1	0.32	95	2.6	96	7.86	140	2.6	68
A-174	3434	1429	20.64	0.09	0.44	0.014762	2.3	0.09746	4.2	0.04790	3.5	0.55	94	2.1	94	3.76	94	2.1	101
A-175	2039	563	7.37	0.40	2.95	0.013382	2.4	0.08866	7.3	0.04807	6.9	0.33	86	2.1	86	6.05	102	2.1	84
A-176	1286	144	2.61	0.50	13.73	0.018430	5.1	0.14061	26.0	0.05535	25.5	0.20	118	5.9	134	32.54	426	5.9	28
A-177	81	258	0.02	0.66	41.18	0.000076	37.7	-0.00121	47.0	-0.11566	28.1	0.80	0	0.2					
A-178	950	411	5.94	0.31	0.39	0.014756	2.4	0.09767	4.5	0.04802	3.8	0.54	94	2.3	95	4.06	100	2.3	95
A-179	813	490	0.87	0.87	17.95	0.001801	5.1	0.01496	37.0	0.06027	36.7	0.14	12	0.6	15	5.54			
A-180	4749	1884	8.21	1.29	0.00	0.004127	4.1	0.07811	8.3	0.13732	7.2	0.50	27	1.1	76	6.13			
A-181	2630	981	15.35	0.56	0.73	0.015989	2.1	0.10632	5.7	0.04824	5.3	0.37	102	2.2	103	5.54	110	2.2	93
A-182	204	276	0.27	0.38	20.85	0.000995	5.9	0.00846	49.0	0.06173	48.7	0.12	6	0.4	9	4.18			
A-183	803	272	4.12	0.50	1.64	0.015518	2.6	0.10293	7.4	0.04812	6.9	0.35	99	2.5	99	7.01	104	2.5	95
A-184	2722	1082	16.08	0.36	1.02	0.015194	2.1	0.10050	5.1	0.04799	4.6	0.42	97	2.1	97	4.73	98	2.1	99
A-185	1463	556	8.73	0.59	0.38	0.016060	2.4	0.10634	5.4	0.04804	4.8	0.45	103	2.5	103	5.24	100	2.5	102

spot number	$^{207}Pb^a$ (cps)	U^b (ppm)	Pb^b (ppm)	Th^b/U	$^{206}Pb^c$/^{204}Pb	$^{206}Pb^c$/^{238}U	2 s %	$^{207}Pb^c$/^{235}U	2 s %	$^{207}Pb^c$/^{206}Pb	2 s %	rho^d	^{206}Pb/^{238}U (Ma)	2 s (Ma)	^{207}Pb/^{235}U (Ma)	2 s (Ma)	^{207}Pb/^{206}Pb (Ma)	2 s (Ma)	conc %^{e}
A-186	867	195	2.88	0.41	5.93	0.015057	2.9	0.09966	9.0	0.04802	8.5	0.32	96	2.8	96	8.30	99	2.8	97
A-187	745	211	3.10	0.46	2.14	0.015026	2.6	0.09952	11.9	0.04805	11.7	0.22	96	2.5	96	10.98	101	2.5	95
A-188	212	175	0.72	0.85	5.71	0.004174	3.7	0.02897	16.8	0.05034	16.4	0.22	27	1.0	29	4.80			
A-189	4497	1620	17.02	0.44	0.33	0.010500	3.2	0.10520	5.5	0.07269	4.4	0.58	67	2.1	102	5.28			
A-190	2798	109	0.40	0.35	69.26	0.003632	13.8	0.04319	147.2	0.08626	146.5	0.09	23	3.2	43	61.87			
A-191	2682	1241	17.05	0.47	0.52	0.014050	2.1	0.09251	4.3	0.04777	3.8	0.49	90	1.9	90	3.72	87	1.9	103
A-192	763	276	5.01	1.48	0.90	0.018652	2.5	0.11330	10.6	0.04407	10.3	0.23	119	2.9					
A-197	1350	566	8.40	0.28	0.40	0.015177	2.5	0.10039	4.6	0.04799	3.9	0.53	97	2.4	97	4.27	98	2.4	99
A-198	1010	7	0.11	0.32	0.37	0.016704	2.3	0.11088	5.6	0.04816	5.1	0.42	107	2.5	107	5.64	106	2.5	101
A-199	3316	1468	19.87	0.84	1.06	0.013919	2.2	0.07845	6.3	0.04089	5.9	0.36	89	2.0					
A-200	29	98	0.18	0.52	3.18	0.001868	6.2	0.02064	27.0	0.08016	26.3	0.23	12	0.7	21	5.55			
A-201	2561	1035	13.54	0.31	0.99	0.013362	2.4	0.08814	5.8	0.04786	5.3	0.41	86	2.0	86	4.80	91	2.0	94
A-202	137	284	0.67	0.26	1.83	0.002401	3.4	0.01535	19.8	0.04638	19.5	0.17	15	0.5	15	3.04			
A-203	2294	909	12.74	0.72	0.88	0.014329	2.2	0.09442	4.4	0.04780	3.8	0.51	92	2.0	92	3.86	89	2.0	103
A-204	858	317	4.56	0.45	1.14	0.014702	2.4	0.09718	6.6	0.04795	6.2	0.36	94	2.2	94	5.96	96	2.2	98
A-205	4939	2196	27.78	0.24	0.55	0.012931	2.2	0.08516	3.9	0.04778	3.3	0.56	83	1.8	83	3.13	88	1.8	94
A-206	1749	754	9.66	0.49	0.69	0.013104	2.3	0.08641	5.4	0.04784	4.9	0.42	84	1.9	84	4.38	91	1.9	93
A-207	866	329	4.60	0.57	1.83	0.014374	2.2	0.08215	9.6	0.04146	9.4	0.23	92	2.0					
A-208	1605	284	3.79	0.36	6.96	0.013514	2.7	0.10688	9.2	0.05738	8.8	0.29	87	2.3	103	9.00	505	2.3	17
A-209	26264	507		0.99	0.00	-0.018777	21.8	-0.47043	22.0	0.18175	3.2	0.99	-122	-26.9					
A-210	835	248	3.75	0.35	2.35	0.015444	2.6	0.10217	7.7	0.04799	7.2	0.34	99	2.5	99	7.22	98	2.5	101
A-211	1297	370	6.15	0.33	1.18	0.016959	2.5	0.11259	6.3	0.04817	5.8	0.40	108	2.7	108	6.49	107	2.7	102
A-212	1365	248	3.94	0.33	5.29	0.016213	2.6	0.11703	10.4	0.05236	10.1	0.25	104	2.6	112	11.10	301	2.6	34
A-213	5820	1091	30.48	0.53	1.13	0.028711	2.2	0.16635	5.0	0.04204	4.5	0.45	182	4.0					
A-214	584	248	0.50	0.67	23.89	0.002036	7.2	0.01663	45.4	0.05927	44.8	0.16	13	0.9	17	7.54			
A-215	1661	724	9.46	0.35	0.94	0.013359	2.2	0.08789	5.1	0.04773	4.6	0.44	86	1.9	86	4.16	85	1.9	100

spot number	$^{207}Pb^a$ (cps)	U^b (ppm)	Pb^b (ppm)	$\underline{Th^b}$ U	$\underline{^{206}Pb^c}$ ^{204}Pb	$\underline{^{206}Pb^c}$ ^{238}U	2 s %	$\underline{^{207}Pb^c}$ ^{235}U	2 s %	$\underline{^{207}Pb^c}$ ^{206}Pb	2 s %	rho^d	$\underline{^{206}Pb}$ ^{238}U	2 s (Ma)	$\underline{^{207}Pb}$ ^{235}U	2 s (Ma)	$\underline{^{207}Pb}$ ^{206}Pb	2 s (Ma)	conc %^{e}
A-221	4078	1291	19.84	0.18	1.32	0.015696	2.7	0.10439	4.2	0.04825	3.2	0.65	100	2.7	101	4.00	111	2.7	91
A-222	24136	3794	65.29	1.47	0.70	0.016543	3.2	0.27094	6.6	0.11882	5.8	0.48	106	3.3	243	14.35	1938	3.3	5
A-223	2044	123	0.27	0.56	63.84	0.002298	14.0	0.01211	245.7	0.03825	245.3	0.06	15	2.1					
A-224	1437	581	7.93	0.43	0.84	0.013936	2.5	0.09210	6.2	0.04794	5.7	0.40	89	2.2	89	5.32	96	2.2	93
A-225	966	390	5.60	0.46	0.43	0.014684	2.3	0.09673	5.3	0.04779	4.8	0.44	94	2.2	94	4.75	88	2.2	106
A-226	1399	3730	7.78	0.36	0.87	0.002134	2.2	0.01364	5.1	0.04637	4.6	0.44	14	0.3	14	0.69			
A-227	828	291	5.13	0.45	1.00	0.018007	2.4	0.11949	5.0	0.04814	4.5	0.47	115	2.7	115	5.47	106	2.7	109
A-228	5857	1569	20.69	0.20	4.34	0.013462	2.3	0.09154	5.1	0.04933	4.6	0.44	86	1.9	89	4.38	163	1.9	53
A-229	967	374	5.57	0.40	0.52	0.015231	2.3	0.10089	6.0	0.04805	5.5	0.38	97	2.2	98	5.55	101	2.2	96
A-230	442	599	2.72	0.40	0.81	0.004674	2.4	0.02646	10.3	0.04106	10.0	0.24	30	0.7					
A-231	859	327	5.87	0.38	0.37	0.018330	3.1	0.12230	6.4	0.04840	5.6	0.48	117	3.6	117	7.08	118	3.6	99
A-232	745	783	2.15	0.41	8.42	0.002807	2.9	0.01846	23.8	0.04771	23.6	0.12	18	0.5	19	4.38			
A-233	334	106	1.59	0.29	1.62	0.015360	2.9	0.10178	13.9	0.04807	13.6	0.21	98	2.9	98	13.01	102	2.9	96
A-234	764	267	4.49	0.70	0.57	0.017160	2.5	0.11427	7.2	0.04831	6.7	0.35	110	2.7	110	7.46	114	2.7	96
A-235	907	232	3.80	0.46	2.18	0.016683	2.7	0.11695	9.1	0.05086	8.7	0.29	107	2.8	112	9.68	234	2.8	46
A-236	580	213	3.51	0.68	0.73	0.016824	2.5	0.11152	6.6	0.04809	6.1	0.38	108	2.7	107	6.68	103	2.7	104
A-237	181	418	0.84	0.76	1.99	0.002061	3.8	0.01203	18.2	0.04235	17.9	0.21	13	0.5					
A-238	689	398	0.96	1.02	19.89	0.002527	4.1	0.00643	108.3	0.01847	108.2	0.04	16	0.7					
A-239	2182	907	12.64	0.38	0.73	0.014285	2.3	0.08715	4.9	0.04426	4.3	0.47	91	2.1					
A-240	1931	586	5.32	1.58	0.00	0.008962	3.5	0.10838	6.5	0.08773	5.4	0.54	58	2.0	104	6.41			
A-245	3848	1466	22.89	0.71	0.70	0.015961	2.1	0.10578	4.8	0.04808	4.3	0.44	102	2.2	102	4.70	103	2.2	99
A-246	1341	453	6.61	0.43	1.69	0.014912	2.3	0.09887	7.9	0.04810	7.5	0.29	95	2.2	96	7.20	104	2.2	92
A-247	1504	428	5.64	0.62	0.22	0.013223	2.6	0.12809	7.2	0.07027	6.7	0.36	85	2.2	122	8.32	936	2.2	9
A-248	2906	687	1.44	1.19	38.71	0.002185	5.1	0.00852	97.1	0.02827	97.0	0.05	14	0.7					
A-249	10721	1552	58.47	0.39	0.35	0.038394	2.1	0.26986	2.9	0.05099	2.0	0.73	243	5.1	243	6.25	240	5.1	101
A-250	403	160	0.37	0.51	25.76	0.002427	6.4	0.00689	216.4	0.02059	216.3	0.03	16	1.0					
A-251	95	191	0.44	0.33	3.14	0.002379	4.9	0.01365	25.7	0.04162	25.2	0.19	15	0.8					

spot number	$^{207}Pb^a$ (cps)	U^b (ppm)	Pb^b (ppm)	$\underline{Th^b}$ U	$\dfrac{^{206}Pb^c}{^{204}Pb}$	$\dfrac{^{206}Pb^c}{^{238}U}$	2 s %	$\dfrac{^{207}Pb^c}{^{235}U}$	2 s %	$\dfrac{^{207}Pb^c}{^{206}Pb}$	2 s %	rho^d	$\dfrac{^{206}Pb}{^{238}U}$	2 s (Ma)	$\dfrac{^{207}Pb}{^{235}U}$	2 s (Ma)	$\dfrac{^{207}Pb}{^{206}Pb}$	2 s (Ma)	conc %^{e}
A-252	979	188	2.90	0.31	5.66	0.015654	2.6	0.11279	12.7	0.05227	12.4	0.20	100	2.6	109	13.07	297	2.6	34
A-253	1999	835	12.13	0.46	0.49	0.014846	2.2	0.09780	6.4	0.04779	5.9	0.35	95	2.1	95	5.74	88	2.1	108
A-254	3821	96		1.00	94.92														
A-255	352	67	0.84	1.89	0.09	0.012352	4.4	0.17092	17.3	0.10039	16.7	0.25	79	3.5	160	25.66			
A-256	36215	1		3.06	100.00														
A-257	883	376	5.21	0.24	0.51	0.014191	2.4	0.08857	6.5	0.04528	6.1	0.37	91	2.2					
A-258	1154	453	6.46	0.41	0.83	0.014556	2.3	0.09609	5.2	0.04789	4.7	0.44	93	2.1	93	4.65	93	2.1	100
A-259	719	230	4.48	0.34	0.56	0.019915	2.5	0.13282	9.0	0.04839	8.7	0.27	127	3.1	127	10.75	117	3.1	108
A-260	1043	227	3.54	0.37	2.84	0.015930	2.5	0.10592	9.5	0.04824	9.2	0.26	102	2.5	102	9.27	110	2.5	92
A-261	3093	418	13.97	0.45	1.19	0.034158	2.9	0.22553	7.6	0.04790	7.0	0.39	217	6.3	207	14.26	94	6.3	231
A-262	1327	249	4.01	0.74	5.52	0.016290	3.1	0.13405	18.8	0.05970	18.6	0.16	104	3.2	128	22.62	592	3.2	18
A-263	3423	504	17.62	0.39	0.47	0.035619	2.2	0.24854	4.4	0.05062	3.8	0.50	226	4.8	225	8.87	223	4.8	101
A-264	2358	847	12.21	0.23	0.94	0.014684	2.7	0.10469	4.9	0.05173	4.1	0.55	94	2.5	101	4.67	273	2.5	34
A-269	593	154	2.34	0.38	3.46	0.015526	2.7	0.10277	12.2	0.04802	11.9	0.22	99	2.6	99	11.55	100	2.6	100
A-270	916	212	3.31	0.35	4.07	0.015949	2.6	0.10576	7.5	0.04811	7.0	0.34	102	2.6	102	7.24	104	2.6	98
A-271	2845	1323	16.98	0.28	0.51	0.013124	2.1	0.08623	3.8	0.04767	3.2	0.55	84	1.8	84	3.10	82	1.8	103
A-272	601	219	3.22	0.41	1.57	0.015100	2.3	0.09158	9.6	0.04400	9.3	0.24	97	2.2					
A-273	6822	566	20.83	0.31	5.85	0.037498	2.4	0.26443	5.4	0.05116	4.8	0.44	237	5.5	238	11.36	247	5.5	96
A-274	1173	262	7.08	0.68	0.36	0.027543	2.3	0.18762	4.7	0.04942	4.1	0.49	175	4.0	175	7.49	167	4.0	105
A-275	1025	407	5.90	0.41	0.51	0.014823	2.3	0.09768	5.5	0.04781	5.0	0.41	95	2.1	95	5.00	89	2.1	107
A-276	5805	261	26.52	0.85	0.09	0.102459	2.2	0.88888	2.8	0.06294	1.7	0.80	629	13.3	646	13.30	706	13.3	89
A-277	1482	591	8.79	0.82	0.63	0.015200	2.2	0.10035	6.7	0.04789	6.3	0.33	97	2.2	97	6.22	93	2.2	104
A-278	5346	2229	30.20	0.26	0.25	0.013845	2.2	0.09153	3.4	0.04796	2.6	0.63	89	1.9	89	2.90	97	1.9	92
A-279	676	270	4.05	0.28	0.51	0.015323	2.4	0.10142	6.3	0.04802	5.9	0.38	98	2.3	98	5.92	100	2.3	98
A-280	790	297	4.62	0.55	0.65	0.015916	2.6	0.10534	6.7	0.04802	6.1	0.39	102	2.6	102	6.46	100	2.6	102
A-281	2342	596	9.34	0.47	3.06	0.016041	2.3	0.10028	6.5	0.04535	6.1	0.35	103	2.3					

spot number	$^{207}Pb^a$ (cps)	U^b (ppm)	Pb^b (ppm)	$\underline{Th^b}$ U	$\underline{^{206}Pb^c}$ ^{204}Pb	$\underline{^{206}Pb^c}$ ^{238}U	2 s %	$\underline{^{207}Pb^c}$ ^{235}U	2 s %	$\underline{^{207}Pb^c}$ ^{206}Pb	2 s %	rho^d	$\underline{^{206}Pb}$ ^{238}U	2 s (Ma)	$\underline{^{207}Pb}$ ^{235}U	2 s (Ma)	$\underline{^{207}Pb}$ ^{206}Pb	2 s (Ma)	conc %^{e}
A-283	995	376	5.96	1.63	1.04	0.016274	2.3	0.09692	10.7	0.04321	10.4	0.22	104	2.4					
A-284	4458	287	0.82	0.40	65.60	0.002830	10.6	0.03025	76.1	0.07753	75.4	0.14	18	1.9	30	22.69			
A-285	1133	241	4.63	0.24	6.31	0.019692	2.7	0.12103	17.4	0.04459	17.2	0.15	126	3.3					
A-286	2105	287	9.05	0.57	2.14	0.032117	2.4	0.22351	4.8	0.05049	4.2	0.49	204	4.7	205	8.91	217	4.7	94
A-287	367	4	0.02	0.41	8.29	0.003585	3.6	0.02659	18.3	0.05381	17.9	0.20	23	0.8	27	4.81			
A-288	2201	890	12.82	0.33	0.77	0.014727	2.3	0.09727	4.5	0.04792	3.9	0.51	94	2.1	94	4.04	95	2.1	100
A-293	22	225	0.01	0.56	7.79	0.000086	35.3	-0.00267	104.9	-0.22422	98.7	0.34	1	0.2					
A-294	80	143		0.46	41.41	-0.000039	96.2	-0.00271	134.7	0.50647	94.3	0.71	0	-0.2					
A-295	335	469	1.62	0.41	1.54	0.003564	3.3	0.01936	14.8	0.03941	14.5	0.22	23	0.8					
A-296	1639	496	7.35	0.45	3.22	0.015040	2.4	0.11647	7.1	0.05618	6.6	0.34	96	2.3	112	7.49	459	2.3	21
A-297	3171	1376	19.35	0.49	0.47	0.014412	2.3	0.08995	4.4	0.04528	3.7	0.52	92	2.1					
A-298	1920	1914	4.14	0.58	9.55	0.002192	2.4	0.01704	10.9	0.05639	10.6	0.22	14	0.3	17	1.85			
A-299	709	276	4.19	0.47	0.68	0.015566	2.6	0.09676	9.4	0.04510	9.1	0.28	100	2.6					
A-300	442	171	2.76	0.33	0.40	0.016488	2.6	0.10956	7.9	0.04821	7.5	0.33	105	2.7	106	7.93	109	2.7	97
A-301	373	642	1.48	0.33	3.48	0.002360	3.4	0.01508	16.5	0.04636	16.1	0.21	15	0.5	15	2.49			
A-302	2614	706	10.76	0.21	2.91	0.015571	2.2	0.10319	5.5	0.04808	5.0	0.40	100	2.2	100	5.19	103	2.2	97
A-303	726	253	3.67	0.44	1.49	0.014848	2.4	0.09806	8.3	0.04791	8.0	0.28	95	2.2	95	7.53	94	2.2	101
A-304	6995	258	5.27	1.14	23.94	0.019666	2.7	0.32407	10.2	0.11955	9.8	0.26	126	3.3	285	25.33	1949	3.3	6
A-305	1614	515	8.86	0.40	0.82	0.017573	2.6	0.11721	5.7	0.04839	5.1	0.45	112	2.9	113	6.08	118	2.9	95
A-306	95	113	0.25	0.54	4.33	0.002199	5.8	0.02028	53.3	0.06689	53.0	0.11	14	0.8	20	10.76			
A-307	1307	504	7.29	0.37	1.48	0.014783	2.3	0.09765	6.8	0.04792	6.4	0.34	95	2.2	95	6.17	95	2.2	100
A-308	495	1191	2.55	0.91	2.20	0.002213	2.9	0.01126	16.1	0.03693	15.9	0.18	14	0.4					
A-309	11602	642	22.19	0.08	0.50	0.034875	7.8	0.29648	8.4	0.06167	2.9	0.94	221	17.0	264	19.44	662	17.0	33
A-310	1106	382	6.09	0.46	0.69	0.016264	2.5	0.10801	5.3	0.04818	4.6	0.47	104	2.5	104	5.21	107	2.5	97
A-311	3175	693	9.02	0.50	6.96	0.013417	2.6	0.07045	12.0	0.03810	11.7	0.22	86	2.2					
A-312	453	167	2.30	0.33	1.40	0.014142	3.1	0.08151	17.1	0.04182	16.8	0.18	91	2.8					
A-317	1462	539	8.61	0.47	1.10	0.016392	2.4	0.09732	11.1	0.04307	10.8	0.22	105	2.5					

spot number	$^{207}Pb^a$ (cps)	U^b (ppm)	Pb^b (ppm)	$\underline{Th^b}$ U	$\underline{^{206}Pb^c}$ ^{204}Pb	$\underline{^{206}Pb^c}$ ^{238}U	2 s %	$\underline{^{207}Pb^c}$ ^{235}U	2 s %	$\underline{^{207}Pb^c}$ ^{206}Pb	2 s %	rho^d	$\underline{^{206}Pb}$ ^{238}U	2 s (Ma)	$\underline{^{207}Pb}$ ^{235}U	2 s (Ma)	$\underline{^{207}Pb}$ ^{206}Pb	2 s (Ma)	conc %^{e}
A-318	1592	68	6.80	0.55	0.85	0.101647	2.7	0.85482	6.9	0.06101	6.4	0.40	624	16.3	627	32.47	639	16.3	98
A-319	3309	382	14.91	0.34	1.34	0.039804	2.4	0.28231	5.0	0.05145	4.4	0.47	252	5.8	252	11.26	261	5.8	97
A-320	9351	4194	56.98	0.20	0.23	0.013885	2.2	0.09155	3.5	0.04783	2.7	0.63	89	1.9	89	2.97	90	1.9	98
A-321	294	108	1.62	0.35	1.16	0.015336	3.6	0.09360	9.4	0.04428	8.7	0.38	98	3.5					
A-322	3882	518	20.63	0.39	1.73	0.040682	2.2	0.27209	5.0	0.04852	4.5	0.44	257	5.6	244	10.95	124	5.6	207
A-323	726	290	4.84	0.50	0.85	0.017035	3.4	0.11456	13.8	0.04879	13.4	0.25	109	3.7	110	14.41	137	3.7	80
A-324	1180	347	5.08	0.38	2.50	0.015014	3.0	0.08632	9.5	0.04171	9.0	0.31	96	2.8					
A-325	1093	422	6.59	0.48	0.36	0.015965	2.5	0.10589	6.2	0.04812	5.7	0.40	102	2.5	102	6.06	104	2.5	98
A-326	9960	4436	60.52	0.22	0.51	0.013942	2.4	0.09180	3.3	0.04777	2.3	0.72	89	2.1	89	2.84	87	2.1	102
A-327	2211	611	9.64	0.34	3.55	0.016083	2.4	0.11029	5.8	0.04975	5.3	0.41	103	2.4	106	5.90	183	2.4	56
A-328	587	215	3.13	0.28	1.15	0.014879	2.4	0.09829	9.6	0.04793	9.3	0.25	95	2.3	95	8.76	95	2.3	100
A-329	924	382	5.60	0.29	0.80	0.015000	2.4	0.09923	5.0	0.04799	4.4	0.48	96	2.3	96	4.56	98	2.3	98
A-330	740	154	2.38	0.41	5.75	0.015821	2.9	0.09449	18.8	0.04333	18.6	0.15	101	2.9					
A-331	1258	69	0.44	0.42	46.16	0.006483	8.3	0.05656	74.0	0.06329	73.6	0.11	42	3.4	56	40.24			
A-332	17781	54	0.88	0.20	6.15	0.016659	2.3	0.10242	5.1	0.04460	4.6	0.45	107	2.4					
A-333	33	127	0.17	0.32	3.09	0.001387	5.8	0.01007	27.1	0.05268	26.5	0.21	9	0.5	10	2.74			
A-334	1252	517	7.39	0.33	0.48	0.014608	2.3	0.09627	4.9	0.04781	4.3	0.48	93	2.1	93	4.33	89	2.1	105
A-335	1083	452	6.63	0.36	0.60	0.014982	2.3	0.09936	4.6	0.04811	4.0	0.50	96	2.2	96	4.26	104	2.2	92
A-336	1230	501	6.52	0.10	0.86	0.013282	2.4	0.08713	5.7	0.04759	5.2	0.43	85	2.1	85	4.65	78	2.1	109
A-341	981	359	5.64	0.59	0.95	0.016105	2.3	0.09881	7.9	0.04451	7.6	0.29	103	2.4					
A-342	1312	517	7.47	0.38	0.86	0.014752	2.3	0.09765	6.6	0.04802	6.2	0.35	94	2.2	95	6.00	100	2.2	95
A-343	2021	810	11.28	0.39	0.44	0.014237	2.2	0.09399	4.7	0.04790	4.1	0.48	91	2.0	91	4.07	93	2.0	97
A-344	2746	1272	16.61	0.35	0.40	0.013349	2.2	0.08767	3.4	0.04764	2.6	0.64	85	1.9	85	2.82	81	1.9	106
A-345	765	780	1.81	0.40	9.87	0.002386	2.9	0.01380	18.4	0.04197	18.1	0.16	15	0.4					
A-346	1258	447	7.48	0.20	0.63	0.017110	2.3	0.11378	4.6	0.04824	4.0	0.50	109	2.5	109	4.77	110	2.5	99
A-347	2743	799	11.59	0.30	2.80	0.014814	2.2	0.09768	7.1	0.04784	6.7	0.31	95	2.0	95	6.40	90	2.0	105

spot number	$^{207}Pb^{a}$ (cps)	U^{b} (ppm)	Pb^{b} (ppm)	Th^{b}/U	$^{206}Pb^{c}$/^{204}Pb	$^{206}Pb^{c}$/^{238}U	2 s %	$^{207}Pb^{c}$/^{235}U	2 s %	$^{207}Pb^{c}$/^{206}Pb	2 s %	rho^{d}	^{206}Pb/^{238}U (Ma)	2 s (Ma)	^{207}Pb/^{235}U (Ma)	2 s (Ma)	^{207}Pb/^{206}Pb (Ma)	2 s (Ma)	conc %^{e}
A-349	1043	423	6.21	0.30	0.42	0.015002	2.3	0.09899	6.0	0.04787	5.6	0.37	96	2.2	96	5.50	92	2.2	104
A-350	3266	1096	14.92	0.56	1.67	0.013919	2.2	0.09176	5.2	0.04783	4.7	0.43	89	2.0	89	4.46	90	2.0	99
A-351	183	175	0.31	0.56	15.00	0.001778	5.3	0.01340	27.9	0.05467	27.4	0.19	11	0.6	14	3.75			
A-352	414	172	2.51	0.33	0.76	0.014963	2.5	0.08503	11.1	0.04123	10.8	0.23	96	2.4					
A-353	1243	328	4.70	0.36	3.58	0.014645	2.5	0.09693	6.4	0.04802	5.9	0.38	94	2.3	94	5.77	99	2.3	94
A-354	4195	1672	25.03	0.24	0.58	0.015298	2.2	0.10090	4.3	0.04785	3.6	0.52	98	2.2	98	3.97	91	2.2	107
A-355	103	70	0.14	0.62	16.78	0.002121	7.6	0.00403	267.2	0.01380	267.1	0.03	14	1.0					
A-356	715	241	3.75	0.34	0.87	0.015905	2.5	0.10573	6.5	0.04823	5.9	0.39	102	2.6	102	6.26	110	2.6	93
A-357	610	243	3.63	0.73	0.77	0.015261	2.6	0.10119	6.6	0.04810	6.1	0.39	98	2.5	98	6.17	104	2.5	94
A-358	1490	635	8.90	0.39	0.32	0.014335	2.3	0.09458	5.9	0.04787	5.4	0.39	92	2.1	92	5.16	92	2.1	100
A-359	237	85	1.24	0.45	2.00	0.015126	3.2	0.08539	15.3	0.04096	14.9	0.21	97	3.0					
A-360	513	177	2.60	0.47	1.65	0.015031	2.5	0.09173	9.6	0.04428	9.3	0.26	96	2.4					

6.3.2 SI 3 Grainsize diagrams of samples of the present study (except for PL)

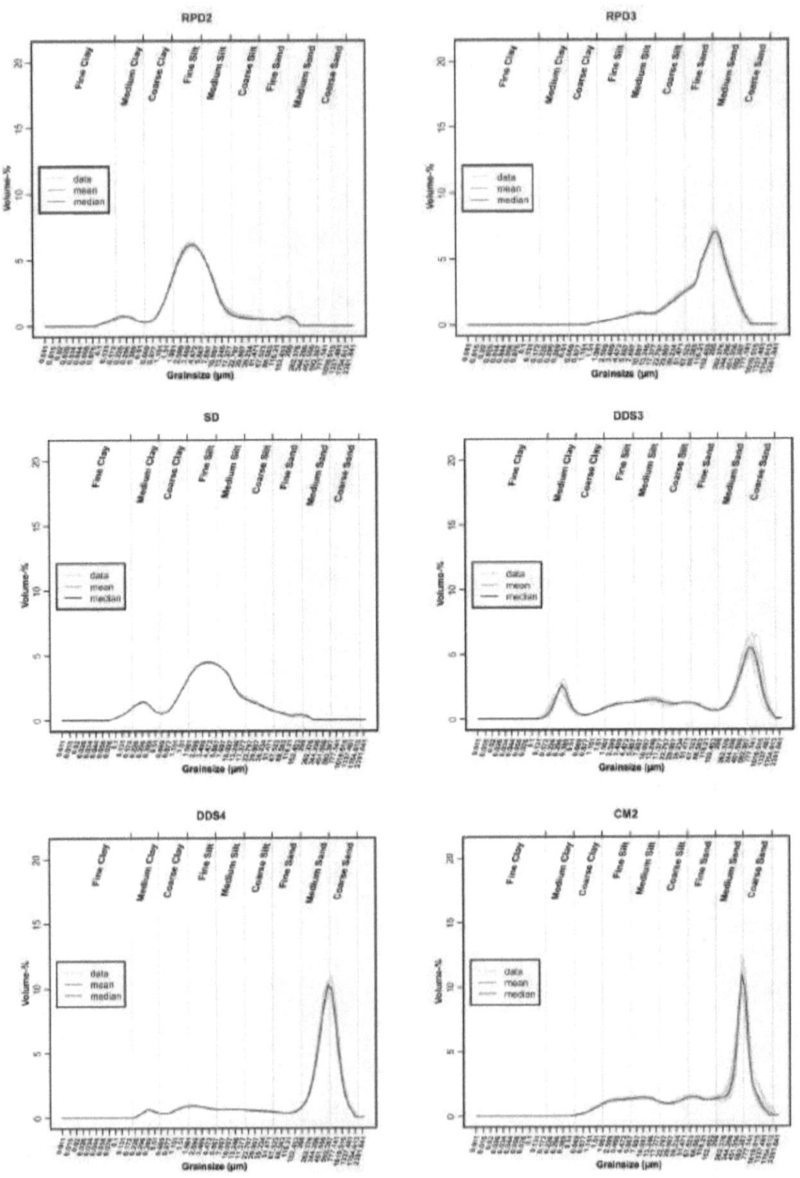

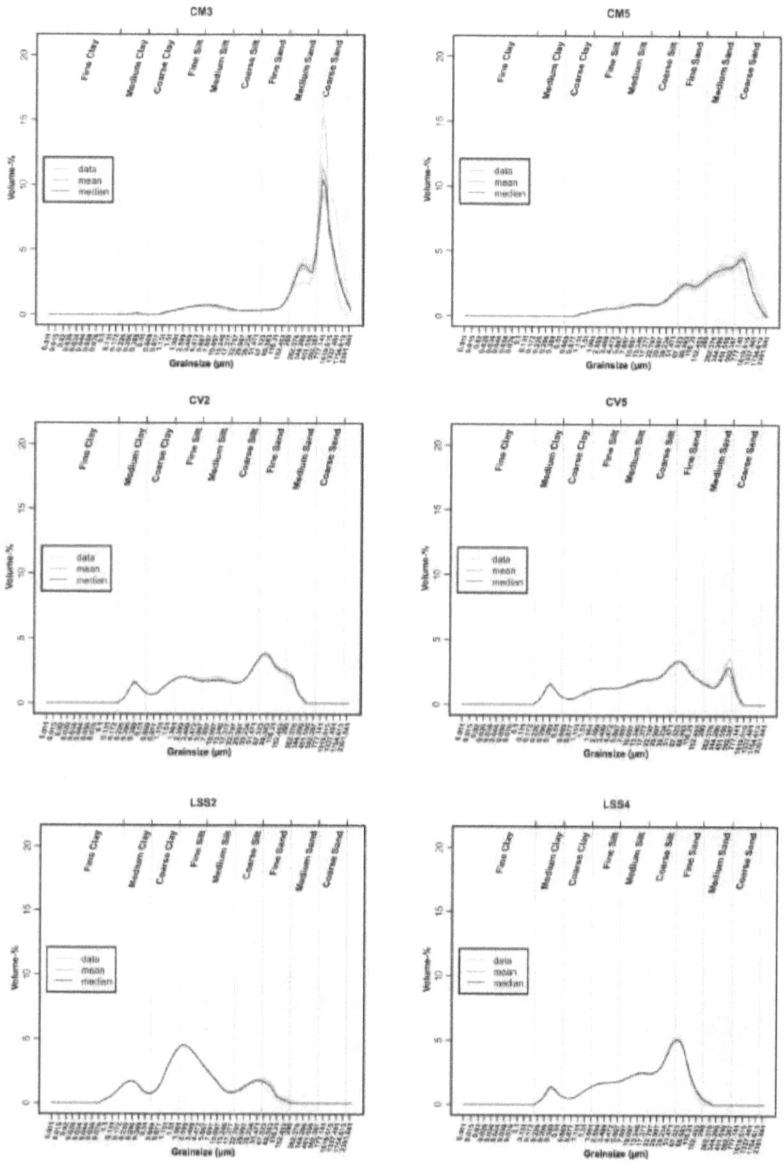

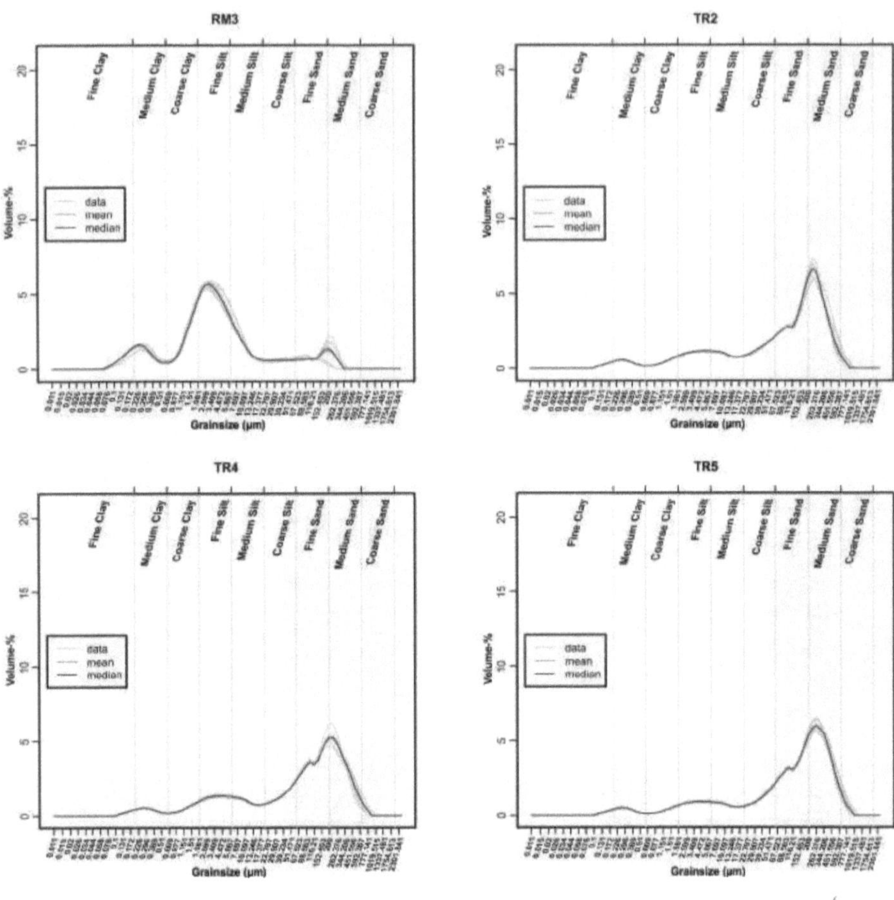

6.3.3 SI 4 Zircon morphology data

6.3.3.1 Great Basin

profile layer	zirconnumber	length	width	roundness	surface	age (Ma)
DDS3	A-89	265	75	7	2	34
	A-90	159	43	4	1	31
	A-91	76	47	8	4	1442
	A-92	75	42	9	3	457
	A-93	126	43	2	2	43
	A-94	91	39	2	1	33
	A-95	134	43	5	2	33
	A-96	130	61	3	2	34
	A-101	71	38	2	1	33
	A-102	90	40	4	2	30
	A-103	108	46	3	2	38
	A-104	212	46	3	1	29
	A-105	96	50	4	1	29
	A-106	56	42	4	3	33
	A-107	81	48	4	2	37
	A-108	78	57	9	3	1348
	A-109	115	53	4	2	34
	A-110	170	61	2	1	30
	A-111	84	44	3	1	33
	A-112	72	53	8	2	882
	A-113	81	35	3	2	39
	A-114	102	20	3	1	96
	A-115	94	35	2	2	33
	A-116	82	34	2	2	28
	A-117	54	37	7	2	392
	A-118	104	21	3	1	32
	A-119	73	25	4	1	31
	A-120	97	46	8	3	959
	A-125	112	43	4	1	29
	A-126	103	38	3	2	29
	A-127	110	48	4	1	34
	A-128	72	40	3	2	31
	A-129	92	61	4	2	34
	A-130	157	57	7	2	35
	A-131	63	42	6	1	1250
	A-132	82	35	6	1	29
	A-133	79	33	5	1	34

269

A-134	74	42	3	2	26
A-135	124	49	4	2	33
A-136	136	51	3	1	39
A-137	167	62	3	2	36
A-138	99	38	4	2	36
A-139	73	41	2	2	32
A-140	112	45	5	1	35
A-141	123	88	8	3	33
A-142	185	41	2	1	36
A-143	145	82	1	2	32
A-144	123	38	3	2	31
A-149	102	43	5	1	35
A-150	103	56	6	2	34
A-151	51	48	4	1	32
A-152	87	57	9	4	428
A-153	70	40	3	1	37
A-154	188	31	2	1	30
A-155	147	26	5	1	32
A-156	196	56	4	2	27
A-157	207	81	4	3	34
A-158	94	49	3	2	37
A-159	107	55	3	1	32
A-160	82	37	4	2	12
A-161	70	54	3	1	32
A-162	199	47	6	2	29
A-163	75	34	4	1	32
A-164	120	72	5	1	31
A-165	210	53	7	2	30
A-166	77	29	7	1	430
A-167	190	72	4	3	32
A-168	127	33	4	2	34
A-173	193	43	3	1	33
A-174	131	36	5	2	37
A-175	89	79	1	2	102
A-176	74	71	3	1	30
A-177	77	41	9	1	30
A-178	110	52	5	1	45
A-179	111	46	4	2	31
A-180	166	34	2	3	36
A-181	132	62	3	1	33
A-182	140	53	4	1	29
A-183	67	51	3	2	28

270

	A-184	95	42	8	2	29
	A-185	82	58	7	4	426
	A-186	137	59	4	1	39
	A-187	135	45	5	1	39
	A-188	75	59	9	4	1746
DDS4	A-5	247	122	2	1	28
	A-6	349	114	5	3	36
	A-7	213	87	3	2	32
	A-8	222	90	3	2	33
	A-9	114	43	3	2	35
	A-10	160	73	4	2	35
	A-11	157	62	5	3	34
	A-12	148	53	3	2	35
	A-13	208	82	3	1	35
	A-14	248	92	2	2	31
	A-15	99	42	4	1	31
	A-16	243	44	4	2	35
	A-17	108	44	3	1	32
	A-18	130	45	2	1	31
	A-19	98	33	3	2	35
	A-20	149	74	4	2	29
	A-21	48	46	7	4	1500
	A-22	166	75	7	3	36
	A-23	190	95	2	1	33
	A-24	90	61	2	1	31
	A-29	150	38	5	3	38
	A-30	218	97	2	2	36
	A-31	184	90	4	2	34
	A-32	177	69	6	2	33
	A-33	455	120	3	2	33
	A-34	204	173	1	1	33
	A-35	93	50	6	1	33
	A-36	192	88	5	2	32
	A-37	82	34	1	1	1210
	A-38	98	40	3	1	36
	A-39	174	70	2	2	36
	A-40	100	50	5	3	36
	A-41	219	96	3	2	35
	A-42	150	81	4	2	35
	A-43	127	59	5	3	33
	A-44	104	52	2	1	8
	A-45	90	55	4	3	43

271

	A-46	286	103	3	2	32
	A-47	220	103	5	2	32
	A-48	47	37	8	4	661
	A-53	207	64	4	1	37
	A-54	193	74	3	1	38
	A-55	203	107	2	2	35
	A-56	146	110	5	2	33
	A-57	153	80	6	2	36
	A-58	84	46	4	2	108
	A-59	80	49	8	3	652
	A-60	253	94	4	2	34
	A-61	229	86	3	1	33
	A-62	59	40	4	1	33
	A-63	145	76	5	3	35
	A-64	144	77	4	1	34
	A-65	175	79	4	1	31
	A-66	110	36	4	2	32
	A-67	123	92	1	2	36
	A-68	70	50	6	3	37
	A-69	344	125	3	2	40
	A-70	60	29	5	2	180
	A-71	152	73	4	1	40
	A-72	164	54	5	1	39
	A-77	166	91	2	2	34
	A-78	71	53	9	4	1568
	A-79	154	72	3	2	34
	A-80	109	51	3	1	37
	A-81	177	55	5	2	42
	A-82	217	124	2	2	36
	A-83	265	104	3	1	39
	A-84	132	107	4	3	33
	A-85	182	78	3	2	33
	A-86	120	70	2	2	34
	A-87	152	40	3	1	34
	A-88	89	48	5	2	31
CM2	a57/73/A-1	79	40	1	1	25
	a58/74/A-2	85	39	2	2	30
	a59/75/A-3	58	32	5	2	199
	a60/76/A-4	103	27	10	1	725
	a61/80/A-5	95	38	9	2	1113
	a62/81/A-6	54	25	6	4	659
	a63/82/A-7	45	27	7	1	107

272

	a64/83/A-8	81	42	1	1	16
	a65/84/A-9	61	40	4	2	11
	a66/85/A-10	57	37	9	4	1192
	a67/86/A-11	44	42	3	1	32
	a68/87/A-12	89	64	9	4	265
	a69/88/A-13	71	34	5	4	217
	a70/89/A-14	120	35	5	1	17
	a71/92/A-15	76	58	10	4	1534
	a72/93/A-16	81	43	2	2	1423
	a73/94/A-17	60	35	3	1	21
	a74/95/A-18	70	38	8	4	168
	a75/96/A-19	165	35	1	1	35
	a76/97/A-20	87	46	10	0	403
	a77/98/A-21	43	50	5	2	100
	a78/99/A-22	70	36	1	1	79
	a79/100/A-23	112	50	10	4	614
	a80/101/A-24	56	40	10	0	362
	a81/105/A-25	49	34	7	2	9
	a82/106/A-26	45	42	10	2	11
	a83/107/A-27	68	50	1	1	17
	a84/108/A-28	73	36	7	1	600
	a85/109/A-29	52	31	0	2	146
	a86/110/A-30	78	63	10	4	1066
	a87/111/A-31	68	34	3	2	266
	a88/112/A-32	74	40	5	4	1361
	a89/113/A-33	77	28	3	2	95
	a90/114/A-34	76	51	0	3	1362
	a91/117/A-35	73	28	1	1	34
	a92/118/A-36	110	46	2	2	84
	a93/119/A-37	60	41	2	3	86
	a94/120/A-38	61	48	6	4	102
	a95/121/A-39	62	40	5	3	473
	a96/122/A-40	135	28	1	1	6
	a97/123/A-41	59	33	8	2	10
	a98/124/A-42	145	80	10	4	238
	a99/125/A-43	112	46	6	4	456
	a100/126/A-44	55	74	5	4	430
CM2, 2. Mount	A-144	93	61	10	4	2211
	A-149	61	61	8	3	1287
	A-150	72	49	2	1	95
	A-151	55	36	4	2	215
	A-152	99	28	4	1	220

273

A-153	84	24	4	4	36
A-154	109	36	2	2	111
A-155	77	53	8	4	1218
A-156	76	29	0	0	77
A-157	79	36	8	3	384
A-158	82	51	5	3	111
A-159	175	31	1	1	11
A-160	90	39	3	1	97
A-161	95	55	5	4	30
A-162	89	38	1	2	1112
A-163	142	84	1	2	104
A-164	130	78	3	2	113
A-165	96	25	1	2	89
A-166	98	31	0	4	475
A-167	156	51	3	4	113
A-168	132	44	1	2	77
A-173	122	21	0	1	33
A-174	67	42	0	3	1181
A-175	81	37	3	3	330
A-176	68	48	3	1	31
A-177	60	47	1	2	1428
A-178	74	31	1	3	105
A-179	74	38	5	4	31
A-180	72	32	6	1	117
A-181	80	39	4	3	114
A-182	62	38	1	1	103
A-183	73	38	2	1	71
A-184	46	44	7	4	1314
A-185	76	44	1	3	28
A-186	75	47	0	1	31
A-187	49	50	7	2	630
A-188	99	49	6	1	98
A-189	119	37	1	2	34
A-190	87	45	10	4	1970
A-191	89	38	0	0	100
A-192	54	50	2	3	187
A-197	92	41	10	4	1294
A-198	105	30	4	1	131
A-199	72	53	4	2	100
A-200	34	53	2	4	88
A-201	88	49	3	1	24
A-202	87	50	2	2	25

274

	A-203	68	35	0	2	28
	A-204	148	29	3	4	4
	A-205	69	49	7	4	2947
	A-206	87	36	5	4	370
	A-207	54	39	0	0	25
	A-208	93	48	2	2	101
	A-209	72	38	0	1	91
	A-210	83	35	1	1	103
	A-211	75	48	2	2	104
	A-212	70	34	9	4	1669
	A-213	85	39	1	1	341
	A-214	136	32	2	1	103
	A-215	71	48	2	4	103
	A-216	96	53	3	3	86
	A-221	104	38	2	2	84
	A-222	59	42	3	2	1740
	A-223	62	21	0	3	5
	A-224	141	27	1	2	85
	A-225	65	26	0	0	108
	A-226	101	27	9	2	457
CM3	a47	79	48	0	0	26
	a48	76	42	10	4	891
	a49	116	36	1	1	417
CM3, 2. Mount	A-5	100	31	0	0	123
	A-6	125	35	1	2	449
	A-7	93	44	4	2	117
	A-8	83	56	3	1	108
	A-9	122	37	4	1	910
	A-10	82	46	7	2	1534
	A-11	104	50	1	1	451
	A-12	79	48	4	2	20
	A-13	79	47	3	1	510
	A-14	63	51	0	4	1769
	A-15	75	50	10	4	1285
	A-16	90	41	6	3	432
	A-17	56	33	0	0	869
	A-18	106	54	10	4	1271
	A-19	98	58	10	4	1232
	A-20	62	57	3	2	466
	A-21	44	40	10	4	1050
	A-22	67	35	5	3	2811
	A-23	156	86	1	2	-12

A-24	34	44	10	4	1023
A-29	62	36	0	2	1715
A-30	123	76	4	3	15
A-31	94	85	10	4	1274
A-32	53	37	10	4	391
A-33	261	1	1	1	4
A-34	83	39	4	1	221
A-35	89	36	7	1	1160
A-36	55	39	9	1	1211
A-37	96	41	2	2	524
A-38	90	40	9	4	528
A-39	67	37	0	4	2683
A-40	62	38	0	0	1739
A-41	79	45	5	0	1042
A-42	51	33	0	0	394
A-43	85	38	3	2	97
A-44	53	38	7	0	431
A-45	89	48	0	0	397
A-46	131	47	7	4	516
A-47	39	36	0	0	113
A-48	160	59	7	2	4
A-53	77	47	6	4	435
A-54	75	34	7	1	411
A-55	68	44	8	2	830
A-56	64	44	7	1	104
A-57	93	82	1	2	126
A-58	100	44	4	1	940
A-59	41	34	0	4	297
A-60	57	39	4	2	1663
A-61	69	37	6	3	836
A-62	61	47	0	2	1137
A-63	55	55	6	2	220
A-64	90	44	10	4	367
A-65	46	86	3	1	550
A-66	83	37	10	4	1256
A-67	55	43	10	3	1403
A-68	62	39	10	4	934
A-69	73	40	10	3	1550
A-70	98	51	6	1	4
A-71	80	38	1	1	31
A-72	84	39	8	2	1124
A-77	65	36	5	4	1163

276

A-78	86	43	7	1	397
A-79	68	41	6	0	109
A-80	65	43	9	0	1076
A-81	115	68	6	1	337
A-82	77	41	6	3	1159
A-83	171	53	5	1	118
A-84	71	41	0	0	97
A-85	112	62	10	2	409
A-86	62	48	8	1	1044
A-87	99	68	5	4	89
A-88	80	45	10	4	536
A-89	71	51	3	4	87
A-90	85	45	5	1	303
A-91	134	49	3	2	528
A-92	58	36	9	4	1171
A-93	72	46	3	1	558
A-94	60	33	3	1	223
A-95	100	40	5	2	88
A-96	56	48	8	1	320
A-101	56	42	3	0	106
A-102	91	52	3	4	411
A-103	80	55	9	4	478
A-104	58	51	10	4	2323
A-105	52	46	0	4	1522
A-106	49	37	0	4	1990
A-107	81	47	1	1	210
A-108	136	33	6	3	4
A-109	75	49	3	2	4
A-110	63	43	5	4	566
A-111	87	44	10	4	719
A-112	81	39	9	4	272
A-113	69	54	6	4	1026
A-114	65	42	8	3	16
A-115	64	33	1	1	25
A-116	98	56	1	1	23
A-117	54	40	9	2	1197
A-118	59	52	2	1	663
A-119	96	59	8	3	1720
A-120	56	36	0	0	100
A-125	65	41	0	0	604
A-126	90	47	0	0	1399
A-127	53	25	0	0	136

277

	A-128	71	44	0	0	652
	A-129	60	66	0	0	628
	A-130	60	58	0	0	195
	A-131	58	26	0	0	233
	A-132	67	36	0	0	1427
	A-133	122	53	0	0	4
	A-134	29	53	0	0	1532
	A-135	36	85	0	0	540
	A-136	88	99	0	0	1098
	A-137	71	53	0	0	2101
	A-138	59	37	0	0	43
	A-139	58	45	0	0	437
	A-140	101	40	0	0	236
	A-141	73	35	0	0	461
	A-142	39	76	2	1	22
	A-143	66	32	9	4	1227
CM5	A-4	113	29	1	1	105
	A-5	156	37	1	1	33
	A-6	145	64	10	3	1048
	A-7	82	47	10	4	430
	A-8	80	37	4	3	601
	A-9	148	41	5	1	603
	A-10	52	21	0	0	388
	A-11	45	35	0	1	591
	A-13	75	42	3	1	24
	A-14	81	38	8	2	1321
	A-15	66	48	11	c	30
	A-16	103	48	7	2	285
	A-17	68	60	10	4	2558
	A-18	160	66	1	1	250
	A-19	74	58	2	2	1256
	A-20	79	47	8	4	113
	A-21	63	88	10	4	1547
	A-22	81	41	9	2	1619
	A-23	114	61	10	4	1006
	A-27	36	51	10	3	1516
	A-28	50	53	0	4	93
	A-29	114	38	7	2	788
	A-30	69	77	4	1	719
	A-31	147	61	1	2	559
	A-32	73	41	9	1	3
	A-33	72	78	0	0	945

278

A-34	95	71	3	1	1035
A-35	182	44	2	1	24
A-36	126	51	8	2	1092
A-37	144	63	6	2	258
A-38	81	81	10	2	862
A-39	108	52	1	1	4
A-40	144	32	4	1	10
A-41	71	77	10	4	637
A-42	101	70	10	4	1409
A-43	57	74	10	4	960
A-44	81	41	10	4	1728
A-45	44	53	1	1	111
A-46	74	40	7	2	970
A-50	101	41	3	1	433
A-51	91	58	7	4	345
A-52	76	44	1	1	210
A-53	74	87	10	4	2140
A-54	65	45	8	4	1288
A-55	110	76	4	2	211
A-56	71	37	7	1	1664
A-57	74	58	1	3	356
A-58	73	37	1	1	216
A-59	106	32	1	1	405
A-60	103	36	1	1	230
A-61	131	48	6	4	989
A-62	61	35	0	3	1701
A-63	102	43	3	1	195
A-64	55	33	2	2	98
A-65	66	53	1	1	5
A-66	72	59	0	4	492
A-67	103	49	2	1	126
A-68	105	52	0	1	21
A-69	70	50	10	4	300
A-73	86	37	7	1	357
A-74	87	30	7	3	406
A-75	120	48	7	2	450
A-76	106	48	0	4	962
A-77	80	46	3	1	936
A-78	125	47	0	1	5
A-79	82	65	9	4	1044
A-80	60	39	3	1	28
A-81	58	49	1	1	216

A-82	81	52	10	1	1657
A-83	78	37	0	1	469
A-84	72	54	0	0	1347
A-85	82	47	1	1	92
A-86	66	40	4	3	1008
A-87	91	46	1	1	78
A-88	116	32	0	1	13
A-89	103	50	2	3	466
A-90	134	57	8	4	954
A-91	102	53	6	1	1637
A-92	87	44	7	3	1515
A-96	70	52	1	1	85
A-97	85	56	10	4	999
A-98	67	65	10	4	592
A-99	93	54	0	1	23
A-100	60	43	1	1	88
A-101	143	37	1	1	14
A-102	67	41	9	1	683
A-103	121	67	1	2	414
A-104	68	38	7	3	1086
A-105	86	30	4	2	1610
A-106	99	36	1	1	1279
A-107	70	39	7	2	1241
A-108	77	45	4	2	504
A-109	144	58	1	1	101
A-110	99	57	2	1	3
A-111	62	44	0	0	984
A-112	94	54	9	1	621
A-113	97	50	6	4	1064
A-114	77	43	8	1	639
A-115	71	43	6	2	24
A-119	120	39	0	0	7
A-120	81	48	1	1	198
A-121	45	53	8	2	1096
A-122	53	55	0	0	26
A-123	77	48	2	1	87
A-124	75	41	10	4	203
A-125	43	53	8	3	1203
A-126	77	42	0	4	873
A-127	66	39	0	1	103
A-129	103	46	2	1	111
A-130	66	53	6	1	1627

280

	A-131	95	49	1	1	1042
	A-132	67	43	0	2	25
	A-133	87	35	1	1	59
	A-134	94	41	1	1	111
	A-135	187	48	1	1	28
	A-136	83	44	1	1	4
	A-137	63	43	1	2	1036
	A-138	106	45	1	1	6
	A-142	68	44	9	4	442
	A-143	68	54	1	1	5
	A-144	59	40	3	2	4
	A-145	73	35	7	1	960
	A-146	76	76	9	3	1201
	A-147	77	43	4	1	66
	A-148	97	34	2	2	700
	A-149	70	43	0	3	1670
	A-150	84	48	1	1	371
	A-151	83	41	7	2	388
	A-152	102	82	0	0	1209
	A-153	84	43	3	1	103
PL2	A-96	99	72	1	1	21
	A-101	57	25		4	28
	A-102	211	57	7	2	110
	A-103	111	40	1	4	28
	A-104	58	40	1	3	26
	A-105	61	38	1	1	31
	A-106	78	49	9	4	12
	A-107	89	50	1	1	28
	A-108	125	57	1	2	30
	A-109	90	58	3	3	29
	A-110	51	44	5	1	32
	A-111	75	49	6	2	35
	A-112	40	32			34
	A-113	51	28			32
	A-114	64	40		1	28
	A-115	53	29	8	4	32
	A-116	58	58			30
	A-117	78	40			32
	A-118	88	39		4	29
	A-119	345	87	2	1	24
	A-120	76	47	4	2	27
	A-125	90	50		4	32

281

	A-126	43	45			27
	A-127	31	40	4	1	27
	A-128	33	37	4	1	30
	A-129	88	44	1	1	31
	A-130	45	50	3	3	90
	A-131	60	32	5	4	25
	A-132	75	68	1	1	28
	A-133	64	35		3	24
	A-134	115	60	3	4	24
	A-135	104	59	2	2	14
	A-136	57	39	1	2	28
PL3	A-5	58	41			32
	A-6	103	75			30
	A-7	123	35	5	1	11
	A-8	125	28	7	1	25
	A-9	108	73	10	2	22
	A-10	140	62	7	4	218
	A-11	75	36	8	3	26
	A-12	86	33	3	1	26
	A-13	62	40			24
	A-14	198	71	1	1	25
	A-15	99	24			80
	A-16	108	43	7	2	25
	A-17	144	79	5	1	11
	A-18	101	42	1	1	25
	A-19	85	50	3	3	16
	A-20	106	47	3	1	26
	A-21	46	26	5	2	153
	A-22	127	67		2	16
	A-23	138	45	1	1	12
	A-24	99	24	1	3	30
	A-29	94	44	4		23
	A-30	117	59	5	1	22
	A-31	45	30	10		13
	A-32	89	33	6	1	26
	A-33	84	27	2	3	21
	A-34	86	39			25
	A-35	76	57	5		23
	A-36	117	54	4	2	24
	A-37	126	58		1	27
	A-38	51	48	6	1	30
	A-39	132	57	1	1	24

282

A-40	120	17			26
A-41	102	41	1	1	74
A-42	78	40	3	1	13
A-43	75	51	7	3	32
A-44	175	78	8	3	24
A-45	64	24	5	2	22
A-46	90	48	6	2	18
A-47	64	48	1	1	24
A-48	67	24	7		109
A-53	84	40	10	4	28
A-54	85	40	3	2	21
A-55	82	49	7	3	28
A-56	102	38			22
A-57	46	23			26
A-58	139	102	1	2	21
A-59	85	61	4	1	30
A-60	76	33	9	2	15
A-61	88	31			29
A-62	125	25	4	1	30
A-63	134	75	10	4	22
A-64	110	49	1	1	32
A-65	73	46	6	1	16
A-66	64	40	3	1	25
A-67	82	47	2	1	25
A-68	65	38	7	3	30
A-69	48	26			25
A-70	57	34	2	2	25
A-71	99	29	10		11
A-72	58	32	8		23
A-77	95	52	4	4	15
A-78	106	62	1	4	23
A-79	93	50	6	4	23
A-80	78	48	3	2	35
A-81	214	42	1	1	18
A-82	63	45			17
A-83	105	76	1	1	15
A-84	89	38	10	4	130
A-85	65	37	6		15
A-86	76	34	5	1	85
A-87	93	39	1	1	28
A-88	61	29	3		23
A-89	114	36	8	1	25

283

A-90	63	33	4	1	25
A-91	109	46	3	1	22
A-92	65	46	10	1	32
A-93	67	31			29
A-94	121	48	7	1	32
A-95	153	45	5	2	27
A-96	83	39	7		13
A-101	71	34	3	1	35
A-102	108	44	3	1	26
A-103	120	65	2	3	24
A-104	36	31			32
A-105	66	35	2	1	25
A-106	79	38	7	3	23
A-107	82	53	4	3	29
A-109	60	43			32
A-110	67	59			30
A-111	47	29			12
A-112	67	30	10	2	15
A-113	68	32	1	1	26
A-114	61	26	9		32
A-115	59	33			24
A-116	57	36	10		25
A-117	67	36	5	2	30
A-118	51	26	8	4	24
A-119	60	41	9	4	32
A-120	79	60	10	4	16
A-125	51	30	10	3	24
A-126	127	52	8	4	21
A-127	58	33	4	2	25
A-128	40	28			1470
A-129	56	29	9	4	33
A-130	91	72	6	2	28
A-131	143	58	5	4	26
A-132	169	92	4	2	24
A-133	93	48	5		23
A-134	86	45			24
A-135	66	32			25
A-136	79	32	2		26
A-137	94	48	7	4	26
A-138	57	41	6	4	26
A-139	78	43	4	1	25
A-140	126	64	4	1	23

284

A-141	160	67	1	1	27
A-142	99	52	3	1	14
A-143	208	70	6	3	14
A-144	141	83	3	1	15
A-149	133	89	4	2	23
A-150	82	41	2	3	31
A-151	129	56	3		23
A-152	50	30	4	1	26
A-153	156	60	1	1	21
A-154	60	40			31
A-155	100	53	7	1	22
A-156	131	36	6	1	22
A-157	125	53	2	1	25
A-158	57	31			27
A-159	64	30			27
A-160	120	77	4	1	25
A-161	125	68	6	3	24
A-162	75	36	3	1	26
A-163	72	40	7	1	30
A-164	52	24	6	2	14
A-165	80	57		4	100
A-166	57	54	9	4	13
A-167	82	32	1	4	26
A-168	57	32	2	1	25
A-173	54	40	8	4	177
A-174	88	48	10	3	210
A-175	107	54	5	4	29
A-176	75	50	5	1	25
A-177	51	32	9	4	31
A-178	30	28	9	4	28
A-179	67	38	5	4	27
A-180	87	41	7	4	23
A-181	69	55			30
A-182	66	30	1	1	24
A-183	49	39	10	4	30
A-184	53	30	3	2	24
A-185	61	32	3	1	30
A-186	70	28	5	4	27
A-187	94	60	10	4	30
A-188	62	32	7	3	31
A-189	69	38	5		28
A-190	81	32	8	3	13

285

A-191	113	33	8	2	27
A-192	66	38	3	1	28
A-197	69	51	7	4	30
A-198	61	40	5	2	31
A-199	71	46	4	2	31
A-200	66	36	3	4	22
A-201	84	46	5	3	23
A-202	124	39	7	3	23
A-203	59	29	5	2	39
A-204	164	77	1	1	25
A-205	173	90	1	1	23
A-206	137	71	4	2	32
A-207	215	93	1	1	26
A-208	252	137	4	3	17
A-209	185	134	1	1	14
A-210	165	104	1	1	22
A-211	186	74	1	1	28
A-212	82	64	3	1	7
A-213	113	67	1	1	27
A-214	204	121	2	1	23
A-215	163	75	1	1	22
A-216	191	82	3	2	25
A-221	141	68	2	1	13
A-222	149	91	3	1	21
A-224	235	146	2	1	23
A-225	310	155	1	1	24
A-226	93	88	3	1	25
A-227	144	62	2	2	25
A-228	139	100	2	1	30
A-229	113	51	3	a	32
A-230	94	47	6	1	23
A-231	123	39	6	4	20
A-232	84	51	6	3	22
A-233	120	59	8	1	25
A-234	109	55		2	97
A-235	92	49	5	4	26
A-236	93	58	8	3	29
A-237	69	48	7	3	14
A-238	118	81	3	1	23
A-239	72	83	9	4	28
A-240	109	69	7	1	89
A-245	156	79	7	3	44

286

	A-246	122	65	4	2	24
	A-247	151	82	3	3	23
	A-248	201	60		4	14
	A-249	109	66	9	4	23
	A-250	65	75	3	4	24
	A-251	55	40	7	3	103
	A-252	44	44	6	3	14
	A-253	74	45	7	4	13
	A-254	77	43	4	4	25
	A-255	133	38	7	4	17
	A-256	63	32	8	2	31
	A-257	94	40	5	1	30
	A-258	75	46	6	2	25
RM3	a41	246	86	2	1	87
	a42	64	38	3	2	38
	a43	108	40	5	2	233
	a44	113	32			36
	a45	71	45	8	3	16
	a46	175	69	3	2	1004
RM3, 2. Mount	A-05	91	48	3	4	117
	A-06	108	81	8	2	69
	A-07	37	54	2	2	27
	A-08	184	117	3	3	528
	A-09	58	41	8	4	1092
	A-10	69	43	3	2	202
	A-11	135	46			146
	A-12	27	47			1101
	A-15	95	45	6	3	251
	A-16	45	33	10	3	983
	A-17	145	38	2	1	33
	A-18	125	63	3	3	1262
	A-21	70	32			0
	A-22	138	54	2	4	87
	A-23	107	58	5	3	83
	A-24	198	98	1	3	763
	A-29	110	47	4	2	68
	A-30	80	41	7	2	752
	A-31	66	35	8	2	280
	A-32	105	80	10	4	1050
	A-33	97	65	8	1	67
	A-34	96	35	1	1	35
	A-35	147	54	1	1	37

287

A-36	100	50	8	4	86
A-37	92	56	6	2	408
A-38	234	105	3	2	92
A-39	101	35	9	3	82
A-40	168	57	8	4	1026
A-41	91	45	2	3	38
A-42	174	874	4	2	753
A-43	125	104	4	1	419
A-45	70	42	9	2	67
A-46	87	40	4	3	82
A-47	224	94	7	1	1589
A-48	124	75	8	2	1757
A-53	84	47	6	3	37
A-54	61	35	6	2	77
A-55	54	59	10	4	789
A-56	106	53	3	3	78
A-57	123	53	7	4	1190
A-58	183	70	5	2	374
A-59	132	112	4	3	1033
A-60	161	96	5	2	1164
A-61	132	66	7	1	964
A-62	124	47	2	1	48
A-63	102	35	3	2	484
A-67	102	43			448
A-68	42	39			83
A-69	72	39	5	2	39
A-70	74	24	4	3	143
A-71	90	45			62
A-72	98	44	6	1	158
A-77	161	85	3	1	270
A-78	215	92	1	2	1086
A-79	182	74	1	2	67
A-80	84	39		2	1282
A-81	110	82	8	3	681
A-82	171	79	3	1	1388
A-83	43	18			39
A-84	57	40	5	2	17
A-85	139	69	2	1	41
A-86	80	41	2	1	43
A-88	61	37	7	1	35
A-89	69	51	1	2	79
A-90	145	85	5	2	737

288

	A-91	70	37	3	1	35
	A-92	45	29			95
	A-93	77	37			1220
	A-94	90	40	4	2	40
	A-95	46	38			24
	A-96	73	35	4	2	143
	A-101	238	97	2	1	38
	A-102	78	30	8	3	40
	A-103	62	54	10	2	1051
	A-104	237	96	1	1	124
	A-105	274	90	4	3	70
	A-107	51	23	4	1	188
	A-108	64	45			242
	A-109	136	37	7	2	78
	A-111	180	45	9	3	68
	A-112	45	41		4	1437
	A-113	72	36	9	4	234
	A-114	57	49	2	4	78
	A-115	60	27	7	3	185
	A-116	58	21	4	4	39
	A-117	60	35	6	2	38
	A-118	92	36	3	1	30
	A-119	61	23			17
	A-120	251	80	6	2	79
	A-125	56	29			431
	A-126	75	39	2	2	1342
	A-127	122	32	2	4	23
	A-128	87	39	5	2	39
	A-129	119	66	9	1	58
	A-130	104	84	4		713
	A-131	115	57	4		386
	A-132	90	50	5		1442

6.3.3.2 Colorado Plateau

profile layer	zirconnumber	length	width	roundness	surface	age (Ma)
CV2	a1	74	35	1	1	168
	a2	110	37	0	0	(-1360)
	a3	84	47	3	4	197
	a4	96	42	7	3	993
	a5	64	31	8	2	427
	a6	77	82	10	2	1204
	a7	92	36	5	1	842

289

	a8	28	34	0	2	471
	a9	83	32	2	1	1756
	a10	76	63	8	2	1097
	a11	84	48	6	1	994
	a12	57	37	0	2	212
	a13	55	36	0	0	1292
	a14	58	33	0	4	964
	a15	53	46	8	3	660
	a16	81	66	6	3	637
	a17	79	32	10	4	1035
	a18	92	50	7	2	333
	a19	86	33	0	1	199
	a20	81	33	3	1	1342
	a21	82	41	7	2	1713
	a22	87	33	10	4	1155
	a23	140	46	6	4	446
	a24	59	36	3	4	2167
	a25	154	43	9	2	376
	a26	80	41	7	2	1026
	a27	98	45	3	2	572
	a28	77	31	10	4	631
	a29	53	40	6	1	1465
	a30	65	60	10	4	1513
	a31	102	49	6	4	1414
	a32	91	33	8	1	1297
	a33	79	42	4	1	1695
	a34	91	70	7	4	1005
	a35	95	43	5	4	264
	a36	28	53	8	1	1658
	a37	69	36	0	1	1082
	a38	56	57	4	1	364
	a39	64	58	9	4	393
	a40	117	35	7	1	440
CV5	A-133	56	44	5	2	361
	A-134	71	51			963
	A-135	59	48	6	3	431
	A-136	77	43	3	1	1481
	A-137	73	57	8	1	947
	A-139	95	45	2	1	1000
	A-140	89	54	1	1	93
	A-141	72	49	5	4	1326
	A-142	159	61	1	2	174

290

A-143	109	47	1	1	28
A-144	56	61			939
A-150	101	53	1	1	27
A-151	84	49		3	1685
A-152	82	45	1	2	1239
A-153	69	48	4	3	656
A-154	58	52	3	2	475
A-155	69	39	7	4	1498
A-156	85	41	4	1	1206
A-157	64	48	5	1	1999
A-158	99	49	8	2	178
A-159	65	45	10	4	1437
A-160	70	47	9	2	2938
A-161	69	48	2	1	1418
A-162	95	47	10	4	1294
A-163	47	33			423
A-164	62	38			459
A-165	122	68	10	4	1908
A-166	73	49	1	1	401
A-167	91	41	1	1	369
A-168	121	67	8	4	1577
A-173	116	50	1	1	106
A-174	90	75	9	3	708
A-175	155	52	0	1	1625
A-176	83	96	7	4	1305
A-177	162	111	9	3	440
A-178	106	58	5	1	146
A-179	80	51	2	4	452
A-180	55	55	0	1	1139
A-182	55	68	0	4	575
A-183	136	68	1	1	36
A-185	45	65	0	2	1465
A-186	85	63	8	1	1150
A-187	52	64	9	1	1929
A-189	81	37	4	2	418
A-190	68	52	3	3	340
A-191	144	31	1	1	452
A-192	78	54	6	1	1508
A-197	87	44	6	4	1586
A-198	62	48	5	4	1105
A-199	84	41	5	4	244
A-200	94	63	3	2	1657

291

	A-201	153	38	7	3	1068
	A-202	148	52	7	1	1338
	A-203	50	44	3	4	1541
	A-204	194	87	4	1	1385
	A-205	128	47	9	3	628
	A-206	130	63	8	4	1102
	A-207	96	53	6	1	577
	A-208	132	53	10	4	502
LSS2	a1	158	41	7	4	403
	a2	85	44	9	3	502
	a3	70	37	2	2	1804
	a4	59	35	2	2	204
	a5	70	47	3	2	217
	a6	83	31	6	2	456
	a7	75	38	1	1	1209
	a8	88	63	7	3	1301
	a9	68	45	9	3	1114
	a10	82	68	6	4	1861
	a11	61	36	2	4	1290
	a12	86	44	1	4	615
	a13	55	47	7	2	1498
	a14	53	41	5	4	2072
	a15	65	67	10	4	1106
	a16	67	58	8	4	561
	a17	27	87	5	3	384
	a18	130	65	1	2	257
	a19	53	56	7	4	1095
	a20	48	52	10	3	162
	a21	52	48	1	1	624
	a22	92	38	4	1	1620
	a23	152	23	1	2	28
	a24	71	58	4	2	1215
	a25	49	62	10	4	1203
	a26	69	50	2	2	419
	a27	43	35	2	3	1687
	a28	69	36	2	4	1095
	a29	99	39	5	2	1837
	a30	63	47	7	3	1540
	a31	54	41	9	4	1777
	a32	50	40	10	4	291
	a33	74	56	5	2	580
	a34	58	32	5	3	378

292

a35	61	42	7	4	582
a36	89	48	9	4	1221
a37	99	58	5	1	1369
a38	96	47	7	4	1749
a39	71	43	3	3	756
a40	65	49	3	1	438
a41	87	59	9	4	1008
a42	96	93	10	4	1941
a43	78	54	5	3	974
a44	108	48	4	1	296
a45	36	58	2	4	299
a46	38	53	5	3	1143
a47	80	46	8	2	1437
a48	59	44	1	1	176
a49	75	42	6	4	456
a50	60	65	9	3	1668
a51	70	58	8	4	455
a52	58	47	2	4	625
a53	101	58	10	4	1252
a54	65	37	3	2	1732
a55	85	39	3	2	581
a56	76	40	6	4	568
a57	63	45	2	1	436
a58	78	49	10	4	700
a59	72	39	1	1	968
a60	70	52	10	4	1254
a61	48	68	6	4	968
a62	87	66	9	4	545
a63	68	49	9	1	2479
a64	66	42	8	2	232
a65	79	42	2	1	234
a66	95	40	1	1	289
a67	19	42	1	2	1712
a68	149	36	1	2	1652
a69	102	24	3	1	403
a70	56	35	1	1	1438
a71	92	46	5	3	1548
a72	78	51	10	4	1037
a73	70	43	1	1	1820
a74	72	45	3	2	99
a75	47	37	2	1	137
a76	72	42	3	4	1516

293

	a77	90	42	7	2	670
	a78	59	48	5	3	1077
	a79	76	35	4	2	1238
	a80	93	40	1	1	492
	a81	69	27	2	2	1674
	a82	67	53	9	2	759
	a83	103	51	4	1	1253
	a84	105	58	8	2	1108
	a85	84	69	10	4	420
	a86	127	45	1	2	950
	a87	94	69	2	2	171
	a88	91	47	1	1	107
	a89	65	55	6	2	1618
	a90					1344
	a91	90	32	1	1	33
	a92	68	45	1	1	1784
	a93	54	55	7	10	1627
	a94	83	42	8	2	1375
	a95	81	52	4	1	1492
	a96	81	62	6	3	1183
	a97	73	43	1	1	189
	a98	88	45	8	3	1409
	a99	31	72			1294
	a100	44	73	7	2	1302
	a101	82	47	4	1	404
	a102	57	73	8	3	1062
	a103	79	57	9	3	673
	a104	50	36	7	2	985
	a105	77	53	8	3	1091
	a106	80	49	4	2	1724
	a107	39	43	4	2	162
LSS4	A-189	63	47	7	3	540
	A-190	67	40	6	4	605
	A-191	92	58	10	4	1309
	A-192	80	38	3	3	605
	A-197	65	38	1	1	289
	A-198	130	48	0	2	263
	A-199	58	49	6	2	1811
	A-200	70	49	2	1	1684
	A-201	90	28	2	1	614
	A-202	43	54	5	4	575
	A-203	103	36	1	1	1568

294

	A-204	60	33	8	2	630
	A-205	152	49	10	3	309
	A-206	48	38	10	4	1559
	A-207	85	49	5	2	584
	A-208	70	81	0	1	419
	A-209	61	51	10	4	521
	A-210	63	47	3	3	798
	A-211	109	24	1	1	31
	A-212	123	55	1	1	1671
	A-213	95	44	10	4	1104
	A-214	61	33	10	4	515
	A-215	50	55	6	3	1364
	A-216	60	40	3	1	607
	A-221	68	47	4	2	502
	A-222	142	37	1	1	-101
	A-223	125	33	3	2	221
	A-224	45	66	10	4	178
	A-225	88	24	7	3	914
	A-226	42	64	0	0	1395
	A-227	134	48	10	4	433
	A-228	49	41	3	1	195
	A-229	46	41	2	2	1266
	A-230	44	59	0	1	2125
	A-231	88	40	5	4	516
	A-232	60	51	0	2	1105
	A-233	104	34	6	1	1106
	A-234	82	35	8	4	366
	A-235	65	47	10	4	1747
	A-236	73	39	0	3	1436
	A-237	57	31	8	4	1326
	A-238	197	53	10	4	1585
TR2	a1/5	74	45	7	2	949
	a2/6	57	39	5	2	672
	a3/7	106	72	5	2	1206
	a4/8	150	35	2	1	219
	a5/9	84	50	7	4	576
	a6/10	109	72	7	3	1071
	a7/11	65	44	3	1	1570
	a8/12	87	32	6	2	300
	a9/13	94	56	7	3	1666
	a10/14	69	44	2	1	209
	a11/17	122	78	7	3	677

295

a12/18	104	51	7	2	549
a13/19	93	62	2	1	1723
a14/20	87	63	4	3	428
a15/21	128	46	9	3	1260
a16/22	58	57	7	1	484
a17/23	80	63	8	4	1274
a18/24	121	48	8	3	428
a19/25	61	43	5	1	444
a20/26	91	54	2	2	182
a21/30	56	40	5	1	578
a22/31	82	47	7	3	640
a23/32	131	52	9	1	1143
a24/33	98	88	7	3	1095
a25/34	122	65	6	2	953
a26/35	158	72	4	2	461
a27/36	107	45	6	2	332
a28/37	82	53	3	1	2664
a29/38	80	35	4	1	189
a30/39	90	60	1	1	1615
a31/42	46	43	2	1	596
a32/43	68	63	6	4	1123
a33/44	83	50	7	1	1455
a34/45	78	49	8	3	1277
a35/46	56	51	1	2	553
a36/47	66	50	6	4	374
a37/48	85	58	5	2	1269
a38/49	86	75	4	1	432
a39/50	81	69	4	3	1085
a40/51	134	56	4	1	1179
a41/55	79	56	2	1	194
a42/56	91	45	6	2	1011
a43/57	84	42	3	1	357
a44/58	62	43	4	1	910
a45/59	90	57	9	2	1376
a46/60	77	52	6	3	989
a47/61	95	59	4	2	2811
a48/62	70	49	4	1	437
a49/63	83	40	2	1	550
a50/64	80	63	7	2	1165
a51/67	69	68	7	4	328
a52/68	102	46	1	1	257
a53/69	90	56	6	3	1184

296

a54/70	79	51	3	3	1066
a55/71	107	43	3	2	576
a56/72	91	51	4	2	413
a57/73	68	52	7	2	1434
a58/74	85	57	5	4	1082
a59/75	149	74	10	2	394
a60/76	95	36	2	1	73
a61/80	107	72	7	2	1182
a62/81	95	32	1	1	144
a63/82	81	43	3	1	601
a64/83	64	55	4	2	1706
a65/84	69	45	8	3	607
a66/85	72	65	5	2	380
a67/86	89	61	9	3	472
a68/87	63	41	2	3	1666
a69/88	94	46	8	2	296
a70/89	79	43	2	1	396
a71/92	99	66	9	2	1153
a72/93	78	60	5	4	1338
a73/94	147	41	9	3	597
a74/95	86	60	1	2	89
a75/96	84	57	3	2	1461
a76/97	68	36	2	1	632
a77/98	56	55	2	1	1246
a78/99	91	45	1	1	1550
a79/100	99	42	2	1	384
a80/101	73	66	10	3	1291
a81/105	85	50	5	2	198
a82/106	122	72	8	2	1530
a83/107	78	67	2	2	333
a84/108	85	45	1	1	239
a85/109	91	58	2	1	1826
a86/110	75	40	3	1	682
a87/111	77	48	7	2	1019
a88/112	263	68	1	4	153
a89/113	96	54	6	2	281
a90/114	91	33	2	1	97
a91/117	98	55	7	3	1093
a92/118	55	52	4	3	306
a93/119	225	50	2	1	1536
a94/120	80	71	8	3	1354
a95/121	62	61	5	2	294

297

a96/122	92	42	9	4	1529
a97/123	120	52	4	1	399
a98/124	93	53	4	4	1043
a99/125	91	48	2	1	170
a100/126	78	40	3	2	391
a101/130	84	44	8	3	1016
a102/131	98	73	2	1	26
a103/132	62	45	1	2	265
a1/5	99	43	4	4	185
a2/6	105	41	3	1	1869
a3/7	116	55	5	3	1066
a4/8	117	46	5	2	383
a5/9	114	63	6	2	2712
a6/10	93	73	2	1	1701
a7/11	119	41	8	2	257
a8/12	116	64	6	3	227
a9/13	147	49	6	2	351
a10/14	76	36	2	2	
a11/15	99	63	5	4	541
a12/16	122	38	7	1	398
a13/17	88	48	3	1	1655
a14/18	142	55	2	2	189
a15/19	116	69	6	3	241
a16/20	75	57	7	4	1335
a17/21	73	55	5	1	1426
a18/22	114	56	7	2	697
a19/23	158	46	4	1	1767
a20/24	116	47	4	1	328
a21/29	166	71	6	3	1031
a22/30	90	38	3	1	477
a23/31	84	50	8	4	540
a24/32	140	49	9	3	1215
a25/33	114	43	7	2	1134
a26/34	69	40	7	3	1073
a27/35	122	48	2	1	1399
a28/36	196	47	1	1	127
a29/37	207	36	5	3	415
a30/38	87	53	6	3	1447
a31/39	127	54	5	2	479
a32/40	99	51	7	3	595
a33/41	121	58	7	2	786
a34/42	72	41	8	2	1652

298

	a35/43	100	38	3	1	232
	a36/44	146	39	5	3	472
	a37/45	63	46	1	1	139
	a38/46	163	41	2	1	311
	a39/47	77	51	3	1	432
	a40/48	88	51	4	1	1364
	a41/53	142	66	1	1	153
	a42/54	80	49	5	3	549
	a43/55	138	40	1	1	170
	a44/56	89	63	2	1	146
	a45/57	87	43	4	3	304
	a46/58	115	42	4	2	576
	a47/59	125	44	2	1	66
	a48/60	88	58	5	2	1219
	a49/61	132	51	8	4	978
	a50/62	83	49	2	1	260
	a51/63	91	54	2	1	1076
	a52/64	109	53	7	3	634
	a53/65	82	30	1	1	182
	a54/66	158	50	4	2	480
	a55/67	199	53	7	3	640
	a56/68	125	51	9	4	382
	a57/69	94	27	2	1	570
	a58/70	114	60	8	4	604
	a59/71	53	45	4	2	1933
	a60/72	144	60	8	3	1842
	a61/77	70	39	7	2	1439
	a62/78	104	63	7	4	1193
TR4	a1	74	44	4	2	376
	a2	111	76	3	2	855
	a3	76	50	6	3	884
	a4	107	52	8	3	404
	a5	114	55	8	4	368
	a6	90	62	4	2	421
	a7	57	60	3	1	704
	a8	114	74	1	1	26
	a9	53	31	3	1	870
	a10	96	33	2	1	336
	a11	75	48	3	2	423
	a12	98	74	3	2	1469
	a13	136	83	8	3	996
	a14	44	42	3	1	1737

299

a15	156	57	9	3	1230
a16	59	45	2	1	649
a17	80	57	3	2	456
a18	66	41	4	2	639
a19	95	43	4	3	1798
a20	71	42	3	1	1106
a21	89	47	6	3	1056
a22	86	69	5	3	465
a23	66	64	5	2	1122
a24	108	67	4	2	435
a25	75	41	3	2	614
a26	64	49	4	2	1349
a27	71	44	2	1	112
a28	62	41	3	4	1738
a29	70	39	3	1	1342
a30	56	37	8	4	1748
a31	92	51	4	2	441
a32	81	48	3	1	1230
a33	68	59	7	4	569
a34	60	50	2	1	226
a35	100	32	4	1	1342
a36	64	39	5	2	552
a37	84	62	8	3	551
a38	92	50	1	1	89
a39	135	71	8	3	1307
a40	60	49	7	3	424
a41	72	42	8	4	1703
a42	100	36	7	1	321
a43	81	59	2	1	165
a44	128	60	8	3	302
a45	38	33	4	4	1357
a46	60	32	7	3	506
a47	44	38	2	1	538
a48	63	39	5	1	450
a49	63	50	6	2	282
a50	55	45	4	2	1271
a51	89	50	5	2	418
a52	80	54	7	3	2248
a53	82	42	5	2	489
a54	61	52	6	3	1834
a55	71	50	7	2	1791
a56	101	24	3	1	312

300

a57	68	49	7	4	887
a58	87	49	6	4	583
a59	74	57	6	3	1094
a60	65	51	5	4	547
a61	79	47	5	4	693
a62	54	47	5	3	411
a63	105	68	8	3	873
a64	98	83	3	2	164
a65	89	62	6	4	1566
a66	83	48	5	3	681
a67	34	24	4	1	854
a68	72	43	6	3	451
a69	107	48	5	2	641
a70	77	37	5	2	480
a71	65	59	4	2	561
a72	75	38	6	2	2076
a73	69	41	7	4	1144
a74	93	56	7	2	1536
a75	62	50	8	4	1169
a76	86	52	9	3	68
a77	103	50	8	4	1522
a78	90	49	4	1	165
a79	121	49	6	3	412
a80	73	44	5	2	412
a81	72	40	6	2	346
a82	69	31	4	2	1735
a83	98	42	8	3	1012
a84	75	41	4	1	1138
a85	79	51	6	3	957
a86	94	86	7	3	398
a87	67	38	4	2	408
a88	96	61	7	3	569
a89	89	46	7	2	580
a90	113	66	5	1	168
a91	88	45	3	1	336
a92	92	52	6	3	452
a93	107	54	3	1	1705
a1	88	46	7	4	634.24
a2	88	44	8	4	1942.25
a3	86	45	6	4	1342.22
a4	52	36	4	2	1165.26
a5	71	44	5	2	1754.46

301

a6	129	35	1	1	150.08
a7	109	62	4	2	1568.66
a8	113	44	8	4	682.72
a9	129	51	4	3	1006.23
a10	95	58	6	4	1022.17
a11	53	40	8	2	644.60
a12	58	35	2	1	1733.28
a13	76	35	2	1	252.12
a14	86	34	5	1	761.98
a15	86	47	2	1	1124.76
a16	60	46	5	4	937.18
a17	80	31	4	4	1053.22
a18	97	50	7	4	392.84
a19	72	43	2	2	1701.52
a20	76	50	1	2	158
a21	61	45	1	1	161
a22	76	41	2	2	365
a23	86	32	3	1	358
a24	60	42	3	1	1041
a25	80	53	7	4	1014
a26	67	54	7	3	793
a27	94	35	4	1	898
a28	98	59	7	3	409
a29	52	36	2	1	388
a30	95	33	3	1	221
a31	102	44	3	3	635
a32	85	46	2	2	1640
a33	62	30	4	1	629
a34	127	33	6	1	448
a35	43	37	7	2	1094
a36	56	40	2	1	399
a37	121	63	2	3	209
a38	84	44	7	3	664
a39	63	48	2	2	91
a40	119	53	5	1	1716
a41	68	44	4	1	1542
a42	54	51	4	2	235
a43	76	42	4	2	1361
a44	62	37	6	3	622
a45	108	39	5	4	1162
a46	73	41	4	1	172
a47	60	39	7	1	1712

302

	a48	67	55	5	4	722
	a49	104	46	6	3	578
	a50	49	47	6	4	656
	a51	68	46	6	3	1279
	a52	88	37	4	1	604
	a53	48	31	5	1	292
	a54	96	55	6	2	1072
	a55	78	53	7	3	2588
	a56	84	42	7	1	1925
	a57	91	48	5	1	417
	a58	109	48	2	1	205
	a59	134	55	3	1	1092
	a60	110	59	5	3	587
	a61	82	41	6	4	609
	a62	83	44	4	2	347
	a63	112	48	3	1	200
	a64	81	42	8	4	2671
	a65	135	48	7	4	401
	a66	74	45	5	4	1258
	a67	89	64	5	2	1760
	a68	72	39	6	2	983
	a69	64	54	8	3	1054
	a70	123	44	1	1	300
	a71	109	57	7	4	627
	a72	100	34	6	2	896
	a73	88	63	8	4	346
	a74	78	43	5	2	912
	a75	74	40	3	1	201
	a76	76	63	5	4	1085
	a77	89	43	2	1	251
	a78	111	38	6	2	194
	a79	62	29	1	1	245
TR5	a1	437	206	2	1	25
	a2	118	81	9	3	1301
	a3	99	43	5	1	408
	a4	76	74	6	2	1281
	a5	68	63	4	2	815
	a6	100	68	5	2	1064
	a7	134	76	6	2	1844
	a8	106	86	2	1	163
	a9	65	42	3	1	1641
	a10	81	53	7	2	439

303

a11	59	56	2	3	247
a12	67	48	7	3	1628
a13	67	44	6	3	-318
a14	109	47	7	2	413
a15	128	61	8	4	2837
a16	111	73	8	3	616
a17	86	32	5	1	1926
a18	63	58	5	1	1059
a19	139	74	8	3	1940
a20	80	53	8	3	1831
a21	97	72	6	4	681
a22	108	73	6	4	389
a23	101	65	6	2	629
a24	93	67	7	2	417
a25	103	54	7	3	581
a26	106	66	6	3	889
a27	78	75	7	4	1178
a28	128	43	3	2	1170
a29	89	52	3	2	1521
a30	117	74	7	3	1346
a31	70	62	7	3	636
a32	55	32	2	2	1493
a33	118	64	8	3	1171
a34	91	44	5	1	1402
a35	109	45	7	4	378
a36	147	57	7	4	1013
a37	73	41	2	1	195
a38	148	64	9	3	1755
a39	139	91	8	4	2487
a40	108	62	7	2	1925
a41	92	40	6	3	315
a42	102	34	4	3	313
a43	103	63	4	4	340
a44	120	24	2	1	1082
a45	75	51	1	2	1772
a46	89	41	1	1	1994
a47	63	45	7	2	1641
a48	206	31	1	1	98
a49	76	40	1	1	164
a50	85	45	5	3	585
a51	156	68	9	4	415
a52	148	70	10	4	1215

304

a53	116	47	1	1	221
a54	90	55	2	1	1388
a55	88	40	2	1	392
a56	99	46	5	2	625
a57	106	41	4	3	624
a58	117	91	8	4	1226
a59	89	44	5	2	1434
a60	144	59	8	4	1301
a61	104	40	7	3	444
a62	205	97	9	3	966
a63	166	54	2	1	117
a64	46	37	2	1	1241
a65	132	48	6	2	413
a66	129	55	6	2	412
a67	62	50	2	2	379
a68	66	40	7	3	624
a69	89	46	1	1	166
a70	85	53	9	3	216
a71	64	54	3	1	1054
a72	66	46	1	2	1308
a73	88	53	6	3	1044
a74	100	62	7	4	540
a75	48	46	2	3	1144
a76	156	74	7	4	451
a77	96	43	3	2	1025
a78	228	59	1	1	26
a79	61	40	5	1	1489
a80	74	52	6	4	682
a81	124	65	7	2	400
a82	90	41	1	1	147
a83	92	64	8	4	1091
a84	73	53	5	4	478
a85	111	54	1	1	151
a86	76	34	3	1	106
a87	97	42	5	2	405
a88	94	52	3	1	351
a89	131	82	8	3	1180
a90	119	39	4	2	393
a91	132	49	6	3	378
a92	81	37	3	1	605
a93	78	48	2	1	308
a94	143	68	9	4	629

305

a95	103	57	3	2	996
a96	116	59	6	3	413
a97	57	45	2	1	1627
a98	66	64	1	1	234
a99	108	60	5	2	281
a100	58	53	7	4	334
a101	48	30	2	1	314
a102	145	53	5	2	648
a103	122	80	8	3	931
a104	115	62	7	3	1263
a105	74	41	2	1	1459
a106	72	56	2	1	230
a107	208	51	6	3	408
a108	138	72	6	2	1457
a109	179	59	9	3	601
a110	60	45	4	1	1686
a111	98	62	7	4	482
a112	106	49	1	1	173
a113	84	44	6	3	429
a114	58	50	7	2	2190
a115	134	41	4	1	408
a116	63	43	3	1	412
a117	61	47	6	2	509
a118	58	48	2	1	1017
a119	83	55	8	4	1059
a120	64	42	2	2	162
a121	84	56	6	2	428
a122	106	49	3	1	369
a123	88	54	3	1	1616
a124	166	48	1	1	171
a125	56	47	2	1	458
a126	75	64	6	4	1791
a127	70	68	5	2	1325
a128	99	53	6	3	387
a129	108	47	4	2	567
a130	111	57	8	4	1121
a131	82	34	2	1	575
a132	89	51	2	1	189
a133	107	33	2	1	1558
a134	58	53	5	2	1104
a135	133	58	1	1	1729
a136	110	43	6	3	497

306

a137	94	70	8	4	363
a138	85	53	5	2	2040
a139	105	54	5	2	1239
a140	67	32	3	1	339
a141	54	47	7	2	785
a142	72	39	3	1	1030
a143	129	72	8	4	988
a144	73	67	5	2	1074
a145	87	56	3	1	310
a146	64	54	8	4	611
a147	113	52	3	2	1761
a148	95	46	3	1	457
a149	135	49	4	2	143
a150	141	66	7	4	411
a151	125	61	2	1	175
a152	130	55	3	2	1155
a153	76	55	3	2	1440
a154	162	83	8	2	1069
a155	112	78	7	4	535
a156	109	44	3	1	1322
a157	105	51	8	3	1628
a158	134	48	1	1	167
a159	97	67	7	3	1304
a160	127	43	6	3	584
a161	72	54	4	2	1186
a162	138	52	7	3	441
a163	80	54	6	3	445
a164	61	45	1	1	454
a165	93	41	5	2	344
a166	94	47	2	1	429
a167	104	50	7	2	249
a168	82	39	1	1	199
a169	70	58	3	1	1091
a170	95	55	1	1	174

307

7 References (excluding chapters 2, 3 and 4)

Bally, A. W., Palmer, Allison R. (Eds.) (1989): The Geology of North America. An Overview. The Geological Society of America. Boulder, Colorado: Geological Society of America Inc.

Blij, Harm J. de; Muller, P. O. (2006): Geography. Realms, regions, and concepts. 12 ed. Hoboken, NJ: Wiley.

Corfu, F.; Hanchar, J. M.; Hoskin, P. W.O.; Dinny, P. (2003): Atlas of Zircon Textures. In: Rev. Mineral. Geochem. 53 (1), 469–500. DOI: 10.2113/0530469.

Davis, D. W.; Williams, I. S.; Krogh T. E. (2003): Historical Development of Zircon Geochronology. In: Rev. Mineral. Geochem. 53 (1), 145–181. DOI: 10.2113/0530145.

DeCourten, F.; Biggar, N. (2017): Roadside geology of Nevada. Missoula, Mont.: Mountain Press Publishing Company (Roadside geology series).

Fagan, D. (2012): Canyon country wildflowers. A guide to common wildflowers, shrubs, and trees. 2nd. ed. Guilford: FalconGuides.

Gärtner, A. (2011): Morphologische, geochronologische und isotopengeochemische Untersuchungen an Zirkonen aus rezenten Sedimenten der Elbe. Diplomarbeit. Technische Universität, Dresden. Institut für Geographie.

Gärtner, A., Linnemann, U., Sagawe, A., Hofmann, M., Ullrich, B. & Kleber, A. (2013) Morphology of zircon crystal grains in sediments – characteristics, classifications, definitions // Morphologie von Zirkonen in Sedimenten – Merkmale, Klassifikationen, Definitionen. Geologica Saxonia, 59, 65–73

Hülle, D.; Hilgers, A.; Kühn, P.; Radtke, U. (2009): The potential of optically stimulated luminescence for dating periglacial slope deposits — A case study from the Taunus area, Germany. In: Geomorphol. 109 (1-2), 66–78. DOI: 10.1016/j.geomorph. 2008.08.021.

Jensen, M.; Kowallis, B.; Christiansen, E.; Webb, C.; Dorais, M.; Sprinkel, D.; Jicha, B. (2020): Fallout tuffs in the Eocene Duchesne River Formation, northeastern Utah. ages, compositions, and likely source. In: Geology of the Intermountain West 7, 1–27. DOI: 10.31711/giw.v7.pp1.

308

Kleber, A. (1997): Cover-beds as soil parent materials in midlatitude regions. In: *Catena* 30 (2-3), 197–213. DOI: 10.1016/S0341-8162(97)00018-0.

Kleber, A. (2013): Heavy-mineral analysis as a tool in tephrochronology, with an example from the La Sal Mountains, Utah, U.S.A. In: Geologos 19 (6), 87–94.

Kleber, A.; Terhorst, B. (Eds.) (2013a): Mid-latitude slope deposits (cover beds). Amsterdam: Elsevier (Developm. Sedimentol., 66).

Kleber, A.; Terhorst, B. (2013b): Introduction. In: Arno Kleber and Brigit Terhorst (Eds.): Mid-latitude slope deposits (cover beds). Amsterdam: Elsevier (Developm. Sedimentol., 66), 1–8.

Krautz, J. (2015): Tephrochronologie in den La Sal Mountains, Utah, USA. Staatsexamensarbeit. TU Dresden, Fakultät Umweltwissenschaften, Institut für Geographie. Dresden.

Krippner, A.; Bahlburg, H. (2013): Provenance of Pleistocene Rhine River Middle Terrace sands between the Swiss–German border and Cologne based on U–Pb detrital zircon ages. In: Int. J. Earth Sci. (Geol. Rundsch.) 102 (3), 917–932. DOI: 10.1007/s00531-012-0842-8.

Lerch, M. (2017): Lipid biomarker characterisation in aeolian sediments under Desert Pavements - potential and first results from Black Rock Desert, Utah, USA and Fuerteventura, Spain. Masterarbeit. Technische Universität, Dresden. Institut für Geographie.

Lerch, M.; Bliedtner, M.; Roettig, C.-B.; Schmidt, J.-U.; Szidat, S.; Salazar, G.; Zech, R.; Glaser, B.; Kleber, A.; Zech, M. (2018): Lipid biomarkers in aeolian sediments under desert pavements – potential and first results from the Black Rock Desert, Utah, USA, and Fuerteventura, Canary Islands, Spain. In: E&G Quaternary Sci. J., 66, 103–108. DOI: 10.5194/egqsj-66-103-2018

Lorz, C.; Frühauf, M.; Mailänder, R.; Phillips, J. D.; Kleber, A. (2013): Influence of cover beds on soils. In: Arno Kleber and Birgit Terhorst (Eds.): Mid-latitude slope deposits (cover beds). Amsterdam: Elsevier (Developm. Sedimentol., 66), 95–126.

Machette, M.N. (1985): Calcic soils of the southwestern United States. In: Geol. Soc. Am. Spec. Pap 203, 1e22. DOI: 10.1130/SPE203-p1.

Neil, B. J.C.; Gibson, H. D.; Pehrsson, S. J.; Martel, E.; Thiessen, E. J.; Crowley, J. L. (2021): Provenance, stratigraphic and precise depositional age constraints for an outlier of the 1.9 to 1.8 Ga Nonacho Group, Rae craton, Northwest Territories, Canada. In: Precambrian Res. 352, S. 105999. DOI: 10.1016/j.precamres.2020.105999.

Rønnevik, C., Ksienzyk, A.K., Fossen, H., Jacobs, J. (2017): Thermal evolution and exhumation history of the Uncompahgre Plateau (northeastern Colorado Plateau), based on apatite fission track and (U-Th)-He thermochronology and zircon U-Pb dating. In: Geosphere 13, 518-537. DOI:10.1130/GES01415.1.

Schoene, B. (2014): U-Th-Pb Geochronology. In: Heinrich D. Holland and Karl K. Turekian (Eds.): Treatise on Geochemistry (Second Edition). Vol. 4. Amsterdam: Elsevier, 341-378.

Sigloch, K.; Mihalynuk, M. G. (2013): Intra-oceanic subduction shaped the assembly of Cordilleran North America. In: Nature 496 (7443), S. 50–56. DOI: 10.1038/nature12019.

Snyder, K. A.; Evers, L.; Chambers, J. C.; Dunham, J.; Bradford, J. B.; Loik, M. E. (2019): Effects of Changing Climate on the Hydrological Cycle in Cold Desert Ecosystems of the Great Basin and Columbia Plateau. In: Rangeland Ecology & Management 72 (1), 1–12. DOI: 10.1016/j.rama.2018.07.007.

Stacey, J.S.; Kramers, J.D. (1975): Approximation of terrestrial lead isotope evolution by a two-stage model. In: Earth Planetary Sci. Lett., 26, 207-221.

USGS (2002): Precipitation History of the Colorado Plateau Region, 1900–2000. Online verfügbar unter https://pubs.usgs.gov/fs/2002/fs119-02, zuletzt geprüft am 28.12.2020.

Waroszewski, J.; Sprafke, T.; Kabala, C.; Musztyfaga, E.; Kot, A.; Tsukamoto, S.; Frechen, M. (2020): Chronostratigraphy of silt-dominated Pleistocene periglacial slope deposits on Mt. Ślęża (SW, Poland): Palaeoenvironmental and pedogenic significance. In: *Catena* 190, S. 104549. DOI: 10.1016/j.catena.2020.104549.

Wetherill, G. W. (1956): Discordant uranium–lead ages. Transactions of the American Geophysical Union 37: 320–326.

Whitmeyer, S. J.; Karlstrom, K. E. (2007): Tectonic model for the Proterozoic growth of North America. In: Geosphere 3 (4), S. 220. DOI: 10.1130/GES00055.1.

Williams, F.; Chronic, L. M.; Chronic, H. (2014): Roadside geology of Utah. Missoula, Montana: Mountain Press Publishing Company.